International Workshop on Quantum Chromodynamics - Theory and Experiment (QCD@Work 2022)

EPJ Web of Conferences Volume 270 (2022)

Leece, Italy
27 – 30 June 2022

Editors:

G.E. Bruno
G. Chiodini
P. Colangelo
C. Corianò
D.M. Creanza
F. De Fazio
E. Nappi

ISBN: 978-1-7138-6618-3

Printed from e-media with permission by:

Curran Associates, Inc.
57 Morehouse Lane
Red Hook, NY 12571

Some format issues inherent in the e-media version may also appear in this print version.

For additional information, please contact EDP Sciences – Web of Conferences
at the address below.

EDP Sciences – Web of Conferences
17, Avenue du Hoggar
Parc d'Activité de Courtabœuf
BP 112
F-91944 Les Ulis Cedex A
France

Phone: +33 (0) 1 69 18 75 75
Fax: +33 (0) 1 69 28 84 91

contact-edps@webofconferences.org

Additional copies of this publication are available from:

Curran Associates, Inc.
57 Morehouse Lane
Red Hook, NY 12571 USA
Phone: 845-758-0400
Fax: 845-758-2633
Email: curran@proceedings.com
Web: www.proceedings.com

TABLE OF CONTENTS

Editors

Giuseppe E. Bruno
Dipartimento Interateneo di Fisica "M. Merlin", Università e Politecnico di Bari
Via Orabona 4, 70126 Bari, Italy
e.mail: giuseppe.bruno@ba.infn.it

Gabriele Chiodini
Istituto Nazionale di Fisica Nucleare – Sezione di Lecce
Via Arnesano, 73100 Lecce, Italy
e.mail: gabriele.chiodini@le.infn.it

Pietro Colangelo
Istituto Nazionale di Fisica Nucleare – Sezione di Bari
Via Orabona 4, 70126 Bari, Italy
e.mail: pietro.colangelo@ba.infn.it

Claudio Corianò
Dipartimento di Matematica e Fisica "E. De Giorgi", Università del Salento
Via Arnesano, 73100 Lecce, Italy
e.mail: claudio.coriano@le.infn.it

Donato M. Creanza
Dipartimento Interateneo di Fisica "M. Merlin", Università e Politecnico di Bari
Via Orabona 4, 70126 Bari, Italy
e.mail: donato.creanza@ba.infn.it

Fulvia De Fazio
Istituto Nazionale di Fisica Nucleare – Sezione di Bari
Via Orabona 4, 70126 Bari, Italy
e.mail: fulvia.defazio@ba.infn.it

Eugenio Nappi
Istituto Nazionale di Fisica Nucleare – Sezione di Bari
Via Orabona 4, 70126 Bari, Italy
e.mail: eugenio.nappi@ba.infn.it

Scientific Advisory Committee

Elke Caroline Aschenauer (Brookhaven National Laboratory)
Roberto Casalbuoni (Firenze University)
Albert de Roeck (CERN Geneva)
Fernando Ferroni (GSSI – L'Aquila and INFN)
Michelangelo Mangano (CERN Geneva)
Mikhail A. Shifman (Minnesota University)
George Sterman (SUNY Stony Brook)

Local Organizing Committee

Giuseppe E. Bruno (Bari Polytechnic)
Gabriele Chiodini (INFN-Lecce)
Pietro Colangelo (INFN-Bari)
Claudio Corianò (Università del Salento)
Donato M. Creanza (Bari Polytechnic)
Fulvia De Fazio (INFN-Bari)
Eugenio Nappi (INFN-Bari)

Participants

ALFORD, Mark	Washington University, St Louis, USA
BARSBAY, Birten	Dogus University, Istanbul, TURKEY
BOCHICCHIO, Marco	INFN - Sezione di Roma 1, Roma, ITALY
BOGLIONE, Mariaelena	Università di Torino and INFN, Torino, ITALY
BRANCHINA, Carlo	LPTHE Sorbonne University, Paris, FRANCE
BRANCHINA, Vincenzo	Università di Catania and INFN, Catania, ITALY
BRUNO, Giuseppe Eugenio	Politecnico di Bari and INFN, Bari, ITALY
CAFAGNA, Francesco Saverio	INFN - Sezione di Bari, Bari, ITALY
CALIVÁ, Alberto	Università di Salerno and INFN, Salerno, ITALY
CAPPIELLO, Luigi	Università di Napoli and INFN, Napoli, ITALY
CHIODINI, Gabriele	INFN - Sezione di Lecce, Lecce, ITALY
CHIRILLI, Giovanni Antonio	University of Regensburg, Regensburg, GERMANY
COLANGELO, Pietro	INFN - Sezione di Bari, Bari, ITALY
CORIANÓ, Claudio	Università del Salento and INFN, Lecce, ITALY
CORVINO, Michele	CERN, Geneva, SWITZERLAND
CREANZA, Donato	Politecnico di Bari and INFN, Bari, ITALY
CRETÍ, Mario	Università del Salento and INFN, Lecce, ITALY
DEBERNARDIS, Francesco	Università di Bari and INFN, Bari, ITALY
DE FAZIO, Fulvia	INFN - Sezione di Bari, Bari, ITALY
DE LA INCERA, Vivian	University of Texas Rio Grande Valley, Edinburg, USA
DELLE ROSE, Luigi	Universitá della Calabria and INFN, Cosenza, ITALY
DE PAOLIS, Luca	INFN - LNF, Frascati, ITALY
ELIA, Domenico	INFN - Sezione di Bari, Bari, ITALY
EVANS, Geraint	University of Southampton, Southampton, USA
FERRER, Efrain J.	University of Texas Rio Grande Valley, Edinburg, USA
FINI, Rosa Anna	INFN - Sezione di Bari, Bari, ITALY
FRAMPTON, Paul	Università del Salento, Lecce, ITALY
GENSINI, Paolo	Università del Salento, Lecce, ITALY
GIACALONE, Marco	Università di Bologna and INFN, Bologna, ITALY
GIANNUZZI, Floriana	INFN - Sezione di Bari, Bari, ITALY
GRAZIANI, Giacomo	INFN - Sezione di Firenze, Firenze, ITALY
GRAZZINI, Massimiliano	University of Zurich, Zurich, SWITZERLAND
GUZZI, Marco	Kennesaw State University, Kennesaw, USA
HARRIS, Ynir	Oxford University, Oxford, UNITED KINGDOM
HAYS, Chris	Oxford University, Oxford, UNITED KINGDOM
KUMAR, Shyam	INFN - Sezione di Bari, Bari, ITALY
LANDSTEINER, Carl	Universidad Autonoma de Madrid, Madrid, SPAIN
LEAL GÓMEZ, Sergio	Universität Wien, Wien, AUSTRIA
LIONETTI, Stefano	Università del Salento and INFN, Lecce, ITALY

LLORENTE MERINO, Javier	Simon Fraser University, Burnaby, CANADA
LOMBARDO, Maria Paola	INFN - Sezione di Firenze, Firenze, ITALY
LOPARCO, Francesco	INFN - Sezione di Bari, Bari, ITALY
LOSACCO, Nicola	INFN - Sezione di Bari, Bari, ITALY
LUCHA, Wolfgang	Austrian Academy of Sciences, Wien, AUSTRIA
MAGLIO, Matteo Maria	Heidelberg University, Heidelberg, GERMANY
MAIO, Lorenzo	Università di Pisa and INFN, Pisa, ITALY
MARGETIS, Spyridon	Kent State University, Kent, USA
MASTROSERIO, Annalisa	Università di Foggia and INFN, Foggia, ITALY
MEGIAS, Eugenio	University of Granada, Granada, SPAIN
MELIKHOV, Dmitri	Heidelberg University, Heidelberg, GERMANY
MELLE, Dario	Università del Salento and INFN, Lecce, ITALY
MY, Salvatore	Università di Bari and INFN, Bari, ITALY
NAPPI, Eugenio	INFN - Sezione di Bari, Bari, ITALY
NAVIGLIO, Manuel	Università di Pisa and INFN, Pisa, ITALY
NICOTRI, Stefano	INFN - Sezione di Bari, Bari, ITALY
NOCERA, Emanuele Roberto	University of Edinburgh, Edinburgh, UNITED KINGDOM
NOVOA-BRUNET, Martin	INFN - Sezione di Bari, Bari, ITALY
PAPA, Alessandro	Universitá della Calabria and INFN, Cosenza, ITALY
PASTORE, Alessandra	INFN - Sezione di Bari, Bari, ITALY
PITSINIGKOS, Savas	University of Southampton, Southampton, UNITED KINGDOM
POOLE, Aaron	University of Southampton, Southampton, UNITED KINGDOM
RIDOLFI, Giovanni	Università di Genova and INFN, Genova, ITALY
ROSELL, Ignasi	Universidad CEU Cardenal Herrera, Valencia, SPAIN
RUGGIERI, Marco	Lanzhou University, Lanzhou, CHINA
SANDACZ, Andrzej	National Centre for Nuclear Research, Warsaw, POLAND
SANTOPINTO, Elena	INFN - Sezione di Genova, Genova, ITALY
SADJIAN, Hagop	Université Paris-Saclay, IJCLab, Paris, FRANCE
SCODEGGIO, Marco	INFN - Sezione di Ferrara, Ferrara, ITALY
SIMULA, Silvano	INFN - Sezione di Roma 3, Roma, ITALY
TARONNA, Massimo	Università di Napoli and INFN, Napoli, ITALY
TETLAMATZI-XOLOCOTZI, Gilberto	Siegen University, Siegen, GERMANY
TOMMASI, Riccardo	Università del Salento and INFN, Lecce, ITALY
TORRES RAMOS, Arianna Grisel	INFN - Sezione di Bari, Bari, ITALY
VOSSEN, Anselm	Duke University, Durham, USA
WANG, Qun	DESY, Hamburg, GERMANY

Preface

The 2022 edition of **QCD@Work - International Workshop on Quantum Chromodynamics - Theory and Experiment** was held in the auditorium of Museo Provinciale "Sigismondo Castromediano" in Lecce, Italy, from 27 to 30 June 2022. It was the 10th edition of a series started in 2001 in Martina Franca and continued in Conversano (2003 and 2005), Martina Franca (2007, 2010 and 2016), Lecce (2012), Giovinazzo (2014) and Matera (2018). The 2022 edition was one of the first in-person scientific events of the year, after the difficult period of the Covid-19 pandemic which obliged the cancellation of the 2020 edition.

This edition was dedicated to Prof. Sheldon Stone, who served in the International Scientific Advisory Committee in all the editions of the workshop and sadly passed away on October 2021.

The aim of this workshop is to bring together senior and young researchers, theorists and experimentalists putting QCD to work in different frameworks, for discussions, dissemination of the results, promotion of new collaborations, education of the young scientists. The main topics discussed in this edition were:

- low energy and non-perturbative aspects of QCD
- progress in perturbative QCD
- physics of heavy quarks
- QCD in extreme conditions of baryon density and temperature
- new holographic methods for strong interactions.

About 80 researchers (with a large fraction of young scientists) actively participated with presentations and discussions. There was a lively poster session: the prize for the best poster was awarded to Dr. Savvas Pitsinigkos (Southampton).

In this volume we collect the outcome of the meeting.

We acknowledge the institutions and individuals who have contributed to the successful realization of the event: Istituto Nazionale di Fisica Nucleare (INFN), Commissione Scientifica Nazionale Teorica INFN, Bari INFN Theory Group for financial support, and Prof. Daniele Martello and Dr. Vito Manzari, Directors of the INFN units of Lecce and Bari. We thank Dr. Luigi De Luca, Director of Museo Provinciale "Sigismondo Castromediano" and of Polo Biblio-Museale di Lecce, Provincia di Lecce and Regione Puglia for granting the use of the auditorium. We are grateful to Ing. Andrea Colitta and Dr. Marcella Nuzzo for kind assistance at the museum. We thank the staff of Motus Animi in Lecce, in particular Drs. Federica Di Benedetto, Federica Golia, Valentina Rainò and Ulrico Viganotti, for the on-site organization. We are grateful to Drs. Francesca Assisi and Enrico Fasanelli, of Bari and Lecce INFN units, for administrative and technical assistance.

Giuseppe E. Bruno, Gabriele Chiodini, Pietro Colangelo, Claudio Corianò,
Donato M. Creanza, Fulvia De Fazio, Eugenio Nappi

Statement of Peer review

In submitting conference proceedings to *Web of Conferences*, the editors of the proceedings certify to the Publisher that

1. They adhere to its **Policy on Publishing Integrity** in order to safeguard good scientific practice in publishing.
2. All articles have been subjected to peer review administered by the proceedings editors.
3. Reviews have been conducted by expert referees, who have been requested to provide unbiased and constructive comments aimed, whenever possible, at improving the work.
4. Proceedings editors have taken all reasonable steps to ensure the quality of the materials they publish and their decision to accept or reject a paper for publication has been based only on the merits of the work and the relevance to the journal.

Title, date and place of the conference

QCD@Work 2022 - International workshop on Quantum Chromodynamics
- Theory and Experiment
27-30 June 2022, Lecce (Italy)

Proceedings editor(s):

G. E. Bruno, G. Chiodini, P. Colangelo, C. Coriano', D. M. Creanza, F. De Fazio, E. Nappi

Date and editor's signature 16 October 2022

The high-energy limit of perturbative QCD: Theory and phenomenology

Francesco Giovanni Celiberto[1,2,3,*], *Michael* Fucilla[4,5,**], and *Alessandro* Papa[4,5,***]

[1]ECT*, I-38123 Villazzano, Trento, Italy
[2]Fondazione Bruno Kessler (FBK), I-38123 Povo, Trento, Italy
[3]INFN-TIFPA Trento Institute of Fundamental Physics and Applications, I-38123 Povo, Trento, Italy
[4]Università della Calabria, I-87036 Rende, Cosenza, Italy
[5]Istituto Nazionale di Fisica Nucleare, Gruppo collegato di Cosenza, I-87036 Rende, Cosenza, Italy

Abstract. After a brief introduction of formal and phenomenological progresses in the study of the high-energy limit of perturbative QCD, we present arguments supporting the statement that the inclusive emission of Higgs bosons or heavy-flavored hadrons acts as fair stabilizer of high-energy resummed differential distributions. We come out with the message that the hybrid high-energy and collinear factorization, built in term of the next-to-leading logarithmic resummation *à la* BFKL and supplemented by collinear parton distributions and fragmentation functions, is a valid and powerful tool to gauge the feasibility of precision analyses of QCD in its high-energy limit.

1 Introductory remarks

The Balitsky-Fadin-Kuraev-Lipatov (BFKL) resummation [1–4] is the most powerful and adequate tool to account for large energy logarithms arising when QCD observables are studied in the Regge-Gribov or semi-hard regime [5] (see [6] for applications). The BFKL approach permits us to resum leading logarithms (LL) $(\alpha_s \ln s)^n$, with s is the center-of-mass energy squared and $\{Q\}$ one or a set of hard scales typical of the reaction, as well as next-to-leading ones (NLL), $\alpha_s^{n+1} \ln^n s$. BFKL cross sections take the form of a high-energy convolution between a process-independent Green's function, which encodes the logarithmic resummation, and two impact factors, describing the fragmentation of each incoming particle. The BFKL Green's function evolves according to an integral equation, whose kernel is known within the next-to-leading order (NLO) for any fixed momentum transfer t and for any possible two-gluon color exchange in the t-channel [7–11]. Impact factors are process dependent and often represent the most challenging pieces in the calculation of the cross section. They have been computed at the NLO for a very limited number of forward outgoing particles: (*i*) quarks and gluons, (*ii*) light jets, (*iii*) light hadrons, (*iv*) electroproduced vector mesons, (*v*) photons, and (*vi*) Higgs bosons in the infinite top-mass limit. To test the BFKL resummation and, more in general, the high-energy dynamics of QCD at LHC energies and kinematic configurations, a *hybrid* high-energy and collinear factorization was built [12–15] (see [16–18] for a similar formalism) to encode collinear parton distribution functions (PDFs) and fragmentation

*e-mail: fceliberto@ectstar.eu
**e-mail: michael.fucilla@unical.it
***e-mail: alessandro.papa@fis.unical.it (corresponding author)

functions (FFs) inside BFKL-factorized cross sections. Probe channels for the hybrid factorization are: Mueller-Navelet jets [19–27], two-hadron [28–32], hadron + jet [13, 33–37], Higgs + jet [14, 38–42], heavy-hadron + jet [43–55], heavy-light two-jet [15, 56], and multi-jet [57–61] inclusive tags. In this work we present phenomenological arguments supporting the statement that large transverse masses observed in Higgs-boson emissions, as well as heavy-hadron detections, allow for clear stabilization of high-energy resummed observables under higher-order corrections and energy-scale variations.

2 Natural stability

In this section we present phenomenological analyses to provide evidence that Higgs-boson emissions, as well as heavy-hadron detections, act as *natural stabilizers* of the high-energy series. For our numerical analyses we use JETHAD, a hybrid PYTHON3.0/FORTRAN2008 modular package [35, 62] aimed at the calculation, management, and processing of observables defined via distinct formalisms. Calculations are done in the $\overline{\text{MS}}$ renormalization scheme and sensitivity to scale variation of our predictions has been evaluated by varying renormalization and factorization scales from half to two times their natural values suggested by kinematics. The error connected to the multi-dimensional integrations has constantly been kept below 1%.

2.1 Higgs + jet

We investigate the inclusive emission of a Higgs boson in association with a light-flavored jet at high rapidity distance, ΔY. We investigate two different distributions: a) the total cross section differential in ΔY, and b) the Higgs transverse-momentum distribution at $\Delta Y = 5$. The jet transverse momentum is in both cases integrated between 20 and 60 GeV, whereas the Higgs one is in the range 20 GeV and $2M_t$ (with M_t the top mass) in the first distribution, while it is instead left to vary in the same range in the second one. The jet rapidity lies in the range $y_J < |4.7|$ and the Higgs one in the $y_H < |2.5|$ range. The hadronic center-of-mass energy, $\sqrt{s}$, is fixed at 14 TeV.

The total cross section is shown in the left panel of Fig. 1. Different curves are for: the LL-resummed prediction (blue), the partially NLL-resummed series (red), and the fixed-order NLO result, obtained via the POWHEG method [63–66] (bars). The cross section shows a decreasing trend as the rapidity interval increases. This behavior is the net result of two opposite effects in the hybrid factorization. On the one hand, the BFKL resummed partonic cross section grows as the squared partonic center-of-mass energy, $\hat{s}$, increases. Since s is fixed, this corresponds to saying that the partonic cross section grows with increasing ΔY, as expected. For the same reason, however, going to larger values of $\hat{s}$ means exploring kinematic regions associated with increasing value of the product $x_1 x_2$, where x_1 and x_2 are the Bjorken variables which characterize the two PDFs describing the incoming protons. The rapid falloff of PDFs in these kinematic regimes dominates and generates the downward trend. Notably, the NLL prediction has a narrower uncertainty band, which is always contained within the LL one. This represents a clear signal of the achieved stability at natural scales. Remarkably, fixed-order results are systematically lower than the resummed ones.

In the right panel of Fig. 1 the Higgs transverse-momentum distribution is shown. We distinguish three kinematic subregions:

- The low-p_T region, where $|\vec{p}_H| < 10$ GeV; since it is dominated by large transverse-momentum logarithms, neglected by our formalism, we excluded it from our analysis;

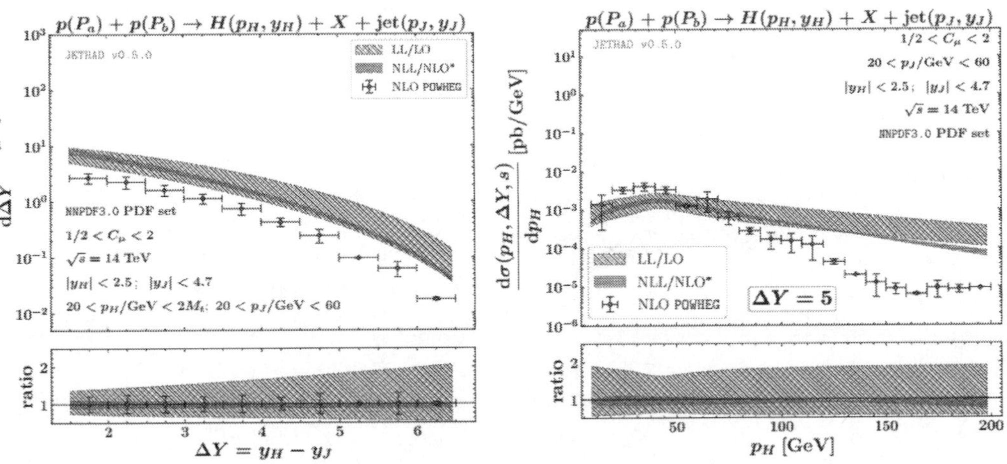

Figure 1. Rapidity (left) and transverse-momentum distribution for the inclusive Higgs + jet production at 14 TeV LHC. Proton PDFs are described via the NNPDF4.0 NLO set [67, 68].

- The intermediate-p_T region, namely when $|\vec{p}_H| \sim |\vec{p}_J|$; this is the region where our formalism is expected to be adequate and we observe an impressive stability of the perturbative series;
- The large-p_T region, which is essentially the last part of the tail; here DGLAP-type and threshold-type logarithms are relevant and our formalism is not the most appropriate one.

We conclude from this analysis that the presence of a large energy-scale (the mass of the Higgs boson) improves the stability of the BFKL series.

2.2 Heavy-flavored emissions

We analyze total cross sections of two processes involving heavy-flavored emissions. It is important to specify that the presented studies refer to the production of heavy-flavored hadrons at large transverse momentum, a kinematic condition which allows us to adopt the variable-flavor number-scheme (VFNS) [69, 70].

The first reaction is the inclusive hadroproduction of a charged pion, $\pi^\pm$, detected at the planned Forward Physics Facility (FPF) [71, 72], in association with a $D^{*\pm}$-meson tagged in the CMS barrel. Both particles feature large transverse momenta and are widely separated in rapidity. We choose as kinematic windows: 10 GeV $< |\vec{p}_{T_\pi}| <$ 20 GeV $< |\vec{p}_{T_D}| <$ 60 GeV, $5 < y_\pi < 7$ and $|y_D| < 2.4$. In the left panel of Fig. 2 we compare the NLL-resummed ΔY-differential cross section (blue) with corresponding fixed-order NLO taken in the high-energy limit (blue). Different curves are obtained through the *replica* method [73, 74]. We observe that the weight of the resummation is quite large and the discrepancy between BFKL and fixed-order predictions is amplified by making use of asymmetric kinematic windows genuinely offered by a FPF + ATLAS coincidence setup [54, 62].

Finally, in the right panel of Fig. 2 we present a study on the sensitivity on energy-scale variations of the ΔY-distribution for the inclusive hadroproduction of a Υ in association with a light-flavored jet. The jet is tagged in kinematic configurations typical of current studies at the CMS detector [75], namely 35 GeV $< |\vec{p}_J| <$ 60 GeV and $|y_J| < 4.7$. The quarkonium transverse momentum is chosen to be in the range 20 GeV $< |\vec{p}_Q| <$ 60 GeV, in the spirit of a VFNS treatment. The Υ is detected by the CMS barrel detector, thus having $|y_Q| < 2.4$.

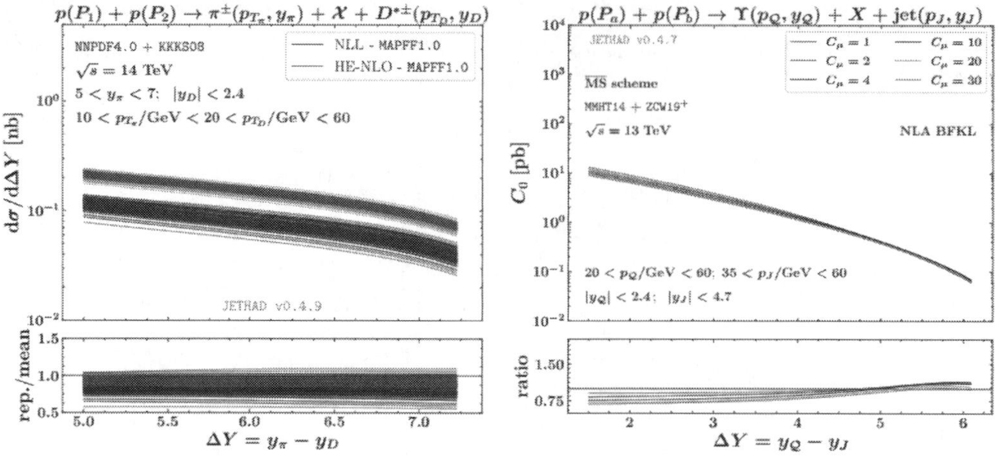

Figure 2. Left panel: rapidity distribution for the inclusive $\pi^{\pm} + D^{*\pm}$ at 14 TeV FPF + ATLAS. `NNPDF4.0` proton NLO PDFs [67, 68] are employed together with `MAPFF1.0` pion NLO FFs [76] and `KKKS08` D-meson NLO FFs [77]. Right panel: rapidity distribution for the inclusive Υ + jet channel at 13 TeV LHC. A study on progressive energy-scale variation in the range $1 < C_\mu < 30$ is made. Υ fragmentation is described in terms of `ZCW19`$^+$ NLO FFs [48, 55, 78, 79].

Predictions are stable even by varying scales by a factor 30 and stability improves as the ΔY increases.

We conclude this section by stressing the important difference between the two observed stability mechanisms. In the Higgs production case, the presence of a large energy scale, given by the Higgs transverse mass, provides us with the stabilization pattern. Conversely, in the heavy-flavor production at large transverse momentum, NLO impact factors are calculated for light partons; here, the smooth and non-decreasing with μ_F behavior of the gluon FFs has a stabilizing effect on physical distributions.

3 Future perspectives

We presented arguments supporting the statement that the inclusive emission of Higgs bosons or heavy-flavored bound states acts as fair stabilizer of high-energy resummed differential distributions. We came out with the important message that the hybrid high-energy and collinear factorization, built in term of the NLL resummation *à la* BFKL and supplemented by collinear PDFs and FF, is a valid and powerful tool to gauge the feasibility of precision analyses of QCD in its high-energy limit.

Future studies will extend this work to: (*i*) full NLL/NLO analyses of Higgs-boson emissions by making use of the recently calculated NLO forward-Higgs impact factor [42, 80], (*ii*) the development of a *multi-lateral* formalism in which distinct resummations are concurrently and consistently implemented, (*iii*) accessing the low-x proton structure at new-generation colliders [62, 71, 72, 81–89] via the connection of BFKL with unintegrated gluon densities [90–103], improved PDFs [104, 105] and polarized gluon transverse-momentum-dependent distributions [106–113].

References

[1] V.S. Fadin, E. Kuraev, L. Lipatov, Phys. Lett. B **60**, 50 (1975)
[2] E.A. Kuraev, L.N. Lipatov, V.S. Fadin, Sov. Phys. JETP **44**, 443 (1976)
[3] E. Kuraev, L. Lipatov, V.S. Fadin, Sov. Phys. JETP **45**, 199 (1977)
[4] I. Balitsky, L. Lipatov, Sov. J. Nucl. Phys. **28**, 822 (1978)
[5] L.V. Gribov, E.M. Levin, M.G. Ryskin, Phys. Rept. **100**, 1 (1983)
[6] F.G. Celiberto, Ph.D. thesis, Università della Calabria and INFN-Cosenza (2017), 1707.04315
[7] V.S. Fadin, L.N. Lipatov, Phys. Lett. B **429**, 127 (1998), hep-ph/9802290
[8] M. Ciafaloni, G. Camici, Phys. Lett. B **430**, 349 (1998), hep-ph/9803389
[9] V.S. Fadin, R. Fiore, A. Papa, Phys. Rev. D **60**, 074025 (1999), hep-ph/9812456
[10] V.S. Fadin, D.A. Gorbachev, JETP Lett. **71**, 222 (2000)
[11] V.S. Fadin, D.A. Gorbachev, Phys. Atom. Nucl. **63**, 2157 (2000)
[12] D. Colferai, F. Schwennsen, L. Szymanowski, S. Wallon, JHEP **12**, 026 (2010), 1002.1365
[13] A.D. Bolognino, F.G. Celiberto, D.Yu. Ivanov, M.M. Mohammed, A. Papa, Eur. Phys. J. C **78**, 772 (2018), 1808.05483
[14] F.G. Celiberto, D.Yu. Ivanov, M.M.A. Mohammed, A. Papa, Eur. Phys. J. C **81**, 293 (2021), 2008.00501
[15] A.D. Bolognino, F.G. Celiberto, M. Fucilla, D.Yu. Ivanov, A. Papa, Phys. Rev. D **103**, 094004 (2021), 2103.07396
[16] M. Deak, F. Hautmann, H. Jung, K. Kutak, JHEP **09**, 121 (2009), 0908.0538
[17] M. Deak, A. van Hameren, H. Jung, A. Kusina, K. Kutak, M. Serino, Phys. Rev. D **99**, 094011 (2019), 1809.03854
[18] A. van Hameren, L. Motyka, G. Ziarko (2022), 2205.09585
[19] B. Ducloué, L. Szymanowski, S. Wallon, JHEP **05**, 096 (2013), 1302.7012
[20] B. Ducloué, L. Szymanowski, S. Wallon, Phys. Rev. Lett. **112**, 082003 (2014), 1309.3229
[21] F. Caporale, D.Yu. Ivanov, B. Murdaca, A. Papa, Eur. Phys. J. C **74**, 3084 (2014), [Erratum: Eur.Phys.J.C 75, 535 (2015)], 1407.8431
[22] F.G. Celiberto, D.Yu. Ivanov, B. Murdaca, A. Papa, Eur. Phys. J. C **75**, 292 (2015), 1504.08233
[23] F.G. Celiberto, D.Yu. Ivanov, B. Murdaca, A. Papa, Acta Phys. Polon. Supp. **8**, 935 (2015), 1510.01626
[24] F.G. Celiberto, D.Yu. Ivanov, B. Murdaca, A. Papa, Eur. Phys. J. C **76**, 224 (2016), 1601.07847
[25] F.G. Celiberto, D.Yu. Ivanov, B. Murdaca, A. Papa, PoS **DIS2016**, 176 (2016), 1606.08892
[26] F. Caporale, F.G. Celiberto, G. Chachamis, D. Gordo Gómez, A. Sabio Vera, Nucl. Phys. B **935**, 412 (2018), 1806.06309
[27] F.G. Celiberto, A. Papa (2022), 2207.05015
[28] F.G. Celiberto, D.Yu. Ivanov, B. Murdaca, A. Papa, Phys. Rev. D **94**, 034013 (2016), 1604.08013
[29] F.G. Celiberto, D.Yu. Ivanov, B. Murdaca, A. Papa, AIP Conf. Proc. **1819**, 060005 (2017), 1611.04811
[30] F.G. Celiberto, D.Yu. Ivanov, B. Murdaca, A. Papa, Eur. Phys. J. C **77**, 382 (2017), 1701.05077

[31] F.G. Celiberto, D.Yu. Ivanov, B. Murdaca, A. Papa, *Inclusive charged light di-hadron production at 7 and 13 TeV LHC in the full NLA BFKL approach*, in *25th Low-x Meeting* (2017), 1709.01128

[32] F.G. Celiberto, D.Yu. Ivanov, B. Murdaca, A. Papa, *Inclusive dihadron production at the LHC in NLA BFKL*, in *17th conference on Elastic and Diffractive Scattering* (2017), 1709.04758

[33] A.D. Bolognino, F.G. Celiberto, D.Yu. Ivanov, M.M.A. Mohammed, A. Papa, PoS **DIS2019**, 049 (2019), 1906.11800

[34] A.D. Bolognino, F.G. Celiberto, D.Yu. Ivanov, M.M. Mohammed, A. Papa, Acta Phys. Polon. Supp. **12**, 773 (2019), 1902.04511

[35] F.G. Celiberto, Eur. Phys. J. C **81**, 691 (2021), 2008.07378

[36] F.G. Celiberto, D.Yu. Ivanov, A. Papa, Phys. Rev. D **102**, 094019 (2020), 2008.10513

[37] F.G. Celiberto (2022), 2208.14577

[38] F.G. Celiberto, D.Y. Ivanov, M.M.A. Mohammed, A. Papa, SciPost Phys. Proc. **8**, 039 (2022), 2107.13037

[39] F.G. Celiberto, M. Fucilla, A. Papa, D.Yu. Ivanov, M.M.A. Mohammed, PoS **EPS-HEP2021**, 589 (2022), 2110.09358

[40] F.G. Celiberto, M. Fucilla, D.Yu. Ivanov, M.M.A. Mohammed, A. Papa, PoS **PANIC2021**, 352 (2022), 2111.13090

[41] F.G. Celiberto, M. Fucilla, D.Yu. Ivanov, M.M.A. Mohammed, A. Papa (2021), 2110.12649

[42] F.G. Celiberto, M. Fucilla, D.Y. Ivanov, M.M.A. Mohammed, A. Papa, JHEP **08**, 092 (2022), 2205.02681

[43] R. Boussarie, B. Ducloué, L. Szymanowski, S. Wallon, Phys. Rev. D **97**, 014008 (2018), 1709.01380

[44] F.G. Celiberto, D.Yu. Ivanov, B. Murdaca, A. Papa, Phys. Lett. B **777**, 141 (2018), 1709.10032

[45] A.D. Bolognino, F.G. Celiberto, M. Fucilla, D.Yu. Ivanov, B. Murdaca, A. Papa, PoS **DIS2019**, 067 (2019), 1906.05940

[46] A.D. Bolognino, F.G. Celiberto, M. Fucilla, D.Yu. Ivanov, A. Papa, Eur. Phys. J. C **79**, 939 (2019), 1909.03068

[47] A.D. Bolognino, F.G. Celiberto, M. Fucilla, D.Yu. Ivanov, B. Murdaca, A. Papa, PoS **DIS2019**, 067 (2019), 1906.05940

[48] F.G. Celiberto, M. Fucilla, D.Yu. Ivanov, A. Papa, Eur. Phys. J. C **81**, 780 (2021), 2105.06432

[49] F.G. Celiberto, M. Fucilla, D.Yu. Ivanov, M.M.A. Mohammed, A. Papa, Phys. Rev. D **104**, 114007 (2021), 2109.11875

[50] A.D. Bolognino, F.G. Celiberto, M. Fucilla, D.Yu. Ivanov, A. Papa, PoS **EPS-HEP2021**, 389 (2022)

[51] A.D. Bolognino, F.G. Celiberto, M. Fucilla, D.Yu. Ivanov, A. Papa, PoS **EPS-HEP2021**, 389 (2022), 2110.12772

[52] F.G. Celiberto, M. Fucilla, under revision in Eur. Phys. J. C (2022), 2202.12227

[53] F.G. Celiberto (2022), 2206.09413

[54] F.G. Celiberto, M. Fucilla, M.M.A. Mohammed, A. Papa, Phys. Rev. D **105**, 114056 (2022), 2205.13429

[55] F.G. Celiberto, M. Fucilla, Zenodo, in press (2022), 2208.07206

[56] A.D. Bolognino, F.G. Celiberto, M. Fucilla, D.Yu. Ivanov, A. Papa, SciPost Phys. Proc. **8**, 068 (2022), 2103.07396
[57] F. Caporale, F.G. Celiberto, G. Chachamis, A. Sabio Vera, Eur. Phys. J. C **76**, 165 (2016), 1512.03364
[58] F. Caporale, F.G. Celiberto, G. Chachamis, D. Gordo Gómez, A. Sabio Vera, Nucl. Phys. B **910**, 374 (2016), 1603.07785
[59] F. Caporale, F.G. Celiberto, G. Chachamis, D. Gordo Gómez, A. Sabio Vera, Eur. Phys. J. C **77**, 5 (2017), 1606.00574
[60] F.G. Celiberto, Frascati Phys. Ser. **63**, 43 (2016), 1606.07327
[61] F. Caporale, F.G. Celiberto, G. Chachamis, D. Gordo Gómez, A. Sabio Vera, Phys. Rev. D **95**, 074007 (2017), 1612.05428
[62] F.G. Celiberto, Phys. Rev. D **105**, 114008 (2022), 2204.06497
[63] P. Nason, JHEP **11**, 040 (2004), hep-ph/0409146
[64] S. Frixione, P. Nason, C. Oleari, JHEP **11**, 070 (2007), 0709.2092
[65] S. Alioli, P. Nason, C. Oleari, E. Re, JHEP **06**, 043 (2010), 1002.2581
[66] J.M. Campbell, R.K. Ellis, R. Frederix, P. Nason, C. Oleari, C. Williams, JHEP **07**, 092 (2012), 1202.5475
[67] R.D. Ball et al. (NNPDF), Eur. Phys. J. C **81**, 958 (2021), 2109.02671
[68] R.D. Ball et al. (NNPDF), Eur. Phys. J. C **82**, 428 (2022), 2109.02653
[69] B. Mele, P. Nason, Nucl. Phys. B **361**, 626 (1991), [Erratum: Nucl.Phys.B 921, 841–842 (2017)]
[70] M. Cacciari, M. Greco, Nucl. Phys. B **421**, 530 (1994), hep-ph/9311260
[71] L.A. Anchordoqui et al., Phys. Rept. **968**, 1 (2022), 2109.10905
[72] J.L. Feng et al., *The Forward Physics Facility at the High-Luminosity LHC* (2022), 2203.05090
[73] S. Forte, L. Garrido, J.I. Latorre, A. Piccione, JHEP **05**, 062 (2002), hep-ph/0204232
[74] R.D. Ball, S. Forte, R. Stegeman, Eur. Phys. J. C **81**, 1046 (2021), 2110.08274
[75] V. Khachatryan et al. (CMS), JHEP **08**, 139 (2016), 1601.06713
[76] R.A. Khalek, V. Bertone, E.R. Nocera, Phys. Rev. D **104**, 034007 (2021), 2105.08725
[77] T. Kneesch, B.A. Kniehl, G. Kramer, I. Schienbein, Nucl. Phys. B **799**, 34 (2008), 0712.0481
[78] X.C. Zheng, C.H. Chang, X.G. Wu, Phys. Rev. D **100**, 014005 (2019), 1905.09171
[79] E. Braaten, T.C. Yuan, Phys. Rev. Lett. **71**, 1673 (1993), hep-ph/9303205
[80] M. Hentschinski, K. Kutak, A. van Hameren, Eur. Phys. J. C **81**, 112 (2021), [Erratum: Eur. Phys. J. C 81, 262 (2021)], 2011.03193
[81] E. Chapon et al., Prog. Part. Nucl. Phys. **122**, 103906 (2022), 2012.14161
[82] M. Hentschinski et al., *White Paper on Forward Physics, BFKL, Saturation Physics and Diffraction*, in *2022 Snowmass Summer Study* (2022), 2203.08129
[83] R. Abdul Khalek et al. (2021), 2103.05419
[84] R. Abdul Khalek et al., *Snowmass 2021 White Paper: Electron Ion Collider for High Energy Physics*, in *2022 Snowmass Summer Study* (2022), 2203.13199
[85] D. Acosta, E. Barberis, N. Hurley, W. Li, O.M. Colin, D. Wood, X. Zuo, *The Potential of a TeV-Scale Muon-Ion Collider*, in *2022 Snowmass Summer Study* (2022), 2203.06258
[86] I. Adachi et al. (ILC International Development Team and ILC Community) (2022), 2203.07622
[87] A. Arbuzov et al., Prog. Part. Nucl. Phys. **119**, 103858 (2021), 2011.15005

[88] S. Amoroso et al., *Snowmass 2021 whitepaper: Proton structure at the precision frontier*, in *2022 Snowmass Summer Study* (2022), 2203.13923

[89] F.G. Celiberto, M. Fucilla, D.Yu. Ivanov, M.M.A. Mohammed, A. Papa, *High-energy QCD at colliders: semi-hard reactions and unintegrated gluon densities: Letter of Interest for SnowMass 2021*, in *2022 Snowmass Summer Study* (2021)

[90] M. Hentschinski, A. Sabio Vera, C. Salas, Phys. Rev. Lett. **110**, 041601 (2013), 1209.1353

[91] A.D. Bolognino, F.G. Celiberto, D.Yu. Ivanov, A. Papa, Eur. Phys. J. **C78**, 1023 (2018), 1808.02395

[92] A.D. Bolognino, F.G. Celiberto, D.Yu. Ivanov, A. Papa, Frascati Phys. Ser. **67**, 76 (2018), 1808.02958

[93] A.D. Bolognino, F.G. Celiberto, D.Yu. Ivanov, A. Papa, Acta Phys. Polon. Supp. **12**, 891 (2019), 1902.04520

[94] A.D. Bolognino, A. Szczurek, W. Schaefer, Phys. Rev. D **101**, 054041 (2020), 1912.06507

[95] F.G. Celiberto, Nuovo Cim. **C42**, 220 (2019), 1912.11313

[96] A.D. Bolognino, F.G. Celiberto, D.Yu. Ivanov, A. Papa, W. Schäfer, A. Szczurek, Eur. Phys. J. C **81**, 846 (2021), 2107.13415

[97] A.D. Bolognino, F.G. Celiberto, D.Y. Ivanov, A. Papa, SciPost Phys. Proc. **8**, 089 (2022), 2107.12725

[98] A.D. Bolognino, F.G. Celiberto, M. Fucilla, D.Yu. Ivanov, A. Papa, W. Schäfer, A. Szczurek (2022), 2202.02513

[99] F.G. Celiberto, *Phenomenology of the hadronic structure at small-x* (2022), 2202.04207

[100] A.D. Bolognino, F.G. Celiberto, D.Yu. Ivanov, A. Papa, W. Schäfer, A. Szczurek, Zenodo, in press (2022), 2207.05726

[101] L. Motyka, M. Sadzikowski, T. Stebel, JHEP **05**, 087 (2015), 1412.4675

[102] F.G. Celiberto, D. Gordo Gómez, A. Sabio Vera, Phys. Lett. **B786**, 201 (2018), 1808.09511

[103] A. Arroyo Garcia, M. Hentschinski, K. Kutak, Phys. Lett. B **795**, 569 (2019), 1904.04394

[104] R.D. Ball, V. Bertone, M. Bonvini, S. Marzani, J. Rojo, L. Rottoli, Eur. Phys. J. **C78**, 321 (2018), 1710.05935

[105] M. Bonvini, F. Giuli, Eur. Phys. J. Plus **134**, 531 (2019), 1902.11125

[106] A. Bacchetta, F.G. Celiberto, M. Radici, P. Taels, Eur. Phys. J. C **80**, 733 (2020), 2005.02288

[107] F.G. Celiberto, Nuovo Cim. **C44**, 36 (2021), 2101.04630

[108] A. Bacchetta, F.G. Celiberto, M. Radici, P. Taels, SciPost Phys. Proc. **8**, 040 (2022), 2107.13446

[109] A. Bacchetta, F.G. Celiberto, M. Radici, PoS **EPS-HEP2021**, 376 (2022), 2111.01686

[110] A. Bacchetta, F.G. Celiberto, M. Radici, PoS **PANIC2021**, 378 (2022), 2111.03567

[111] A. Bacchetta, F.G. Celiberto, M. Radici (2022), 2201.10508

[112] A. Bacchetta, F.G. Celiberto, M. Radici (2022), 2206.07815

[113] A. Bacchetta, F.G. Celiberto, M. Radici, A. Signori, Zenodo, in press (2022), 2208.06252

High-p_T results and parton density functions from ATLAS

Javier Llorente Merino[1,*], on behalf of the ATLAS Collaboration

[1]Simon Fraser University

Abstract. The ATLAS detector at the LHC is an ideal experiment to rigorously test QCD in a large variety of final states. This talk will focus on recent results over a wide range of energies, which are particularly sensitive to parton density functions. We will review results with high-p_T jets, photons and vector bosons. We will then discuss fits to determine parton distribution functions (PDFs) using these diverse sets of measurements from the ATLAS experiment. These measurements are used in combination with deep-inelastic scattering data from HERA. Particular attention is paid to the correlation of systematic uncertainties within and between the various ATLAS data sets and to the impact of model, theoretical and parameterisation uncertainties. Finally, we will also present results on the determination of α_s.

1 Introduction

By virtue of the factorisation theorem [1], the differential QCD cross section for $2 \to 2$ processes in hadron-hadron collisions can be written as a convolution of the partonic cross section $\hat{\sigma}(\vec{x}, \mu_R^2)$, which is of perturbative nature and depends on the renormalisation scale μ_R, with functions parameterising the non-perturbative properties of the initial and final states, *i.e.*

$$d\sigma = \sum_{i,j,a,b} \int_{\Omega} d^2\vec{x} d^2\vec{z}\, f_i(x_1, \mu_F^2) f_j(x_2, \mu_F^2) \times d\hat{\sigma}_{ij\to ab}(\vec{x}, \mu_R^2) \times D_a^h(z_3, \mu_f^2) D_b^h(z_4, \mu_f^2). \quad (1)$$

The initial state is characterised by the functions $f_i(x_j, \mu_F^2)$, which parameterise the fraction of momentum x carried by partons i and j, as a function of the factorisation scale μ_F. On the other hand, the final state is characterised by the fragmentation functions, $D_a^h(z, \mu_f^2)$, parameterising the fraction of momentum z carried by a hadron h when fragmented out of a final-state parton a as a function of the fragmentation scale μ_f.

The ATLAS detector provides a unique environment for testing each of the components of Equation 1. This is achieved through the interpretation of different measurements, sensitive to each of the aspects mentioned above, by means of the comparison of the experimental data to different models.

*e-mail: javier.llorente.merino@cern.ch

2 The initial state: Parton distribution functions

A determination of the parton distribution functions has been recently published by the ATLAS Collaboration in Ref. [2]. This determination makes use of a combination of deep inelastic scattering data from HERA [3], together with various ATLAS measurements at different values of the centre-of-mass energy: $\sqrt{s}$ = 7, 8 and 13 TeV. This includes the production of W and Z bosons as well as photons, both inclusively and in association with jets. Measurements of inclusive jets and $t\bar{t}$ production are also considered. Table 1 summarises the input datasets used in the fit.

Table 1. The ATLAS data sets used for the PDF determination, together with the HERA DIS data [2].

Data set	$\sqrt{s}$ [TeV]	Luminosity [fb^{-1}]	Decay channel	Observables entering the fit
Inclusive $W, Z/\gamma^*$	7	4.6	e, μ combined	η_ℓ (W), y_Z (Z)
Inclusive Z/γ^*	8	20.2	e, μ combined	$\cos\theta^*$ in bins of $y_{\ell\ell}$, $m_{\ell\ell}$
Inclusive W	8	20.2	μ	η_μ
$W^\pm$ + jets	8	20.2	e	p_{T}^W
Z + jets	8	20.2	e	$p_{\mathrm{T}}^{\mathrm{jet}}$ in bins of $\lvert y^{\mathrm{jet}}\rvert$
$t\bar{t}$	8	20.2	lepton + jets, dilepton	$m_{t\bar{t}}$, p_{T}^t, $y_{t\bar{t}}$
$t\bar{t}$	13	36	lepton + jets	$m_{t\bar{t}}$, p_{T}^t, y_t, $y_{t\bar{t}}^{\mathrm{b}}$
Inclusive isolated γ	8, 13	20.2, 3.2	-	E_{T}^γ in bins of η^γ
Inclusive jets	7, 8, 13	4.5, 20.2, 3.2	-	$p_{\mathrm{T}}^{\mathrm{jet}}$ in bins of $\lvert y^{\mathrm{jet}}\rvert$

The measurements are compared to NNLO QCD + NLO EW predictions, which are then fitted to the data. The fit consists in the minimisation of a χ^2 function in a 21-dimensional parameter space, which fully takes into account the correlations between the measurements and their uncertainties. The results of the fit are compared to global PDF sets such as CT18 [4], NNPDF 3.1 [5] and MSHT20 [6].

Figure 1 shows the determined parton distribution functions for the valence d-quark as well as for gluons at a scale Q^2 = 1.9 GeV^2, together with the results from other global PDF fits, including CT18 and CT18A [4] as well as HERAPDF [3].

3 The hard scattering: Fixed-order matrix elements

ATLAS has recently published several measurements sensitive to the calculation of the fixed-order matrix element in Equation 1. The data are compared to theoretical predictions at different orders of the perturbative expansion, up to NNLO pQCD.

Reference [7] presents a measurement of the Z+jets cross section with high-p_T jets in different angular configurations between the Z boson and the leading jet. The results are compared to NNLO QCD + NLO EW theoretical predictions and include a comparison of the ratio of the p_T of the Z boson to the p_T of the closest jet in collinear and back-to-back topologies. The left part of Figure 2 shows the Z+jets cross section as a function of the angular distance $\Delta R = \sqrt{(\Delta y)^2 + (\Delta\varphi)^2}$, being y the rapidity and φ the azimuthal coordinate, between the Z boson and the closest jet. The measurements are compared to different versions of MadGraph5_aMC@NLO [8] and Sherpa [9] with different parton shower matching schemes. Sherpa 2.2.11 and MG5_aMC@NLO with NLO matrix elements and FxFx matching [10] provide the best description of the data.

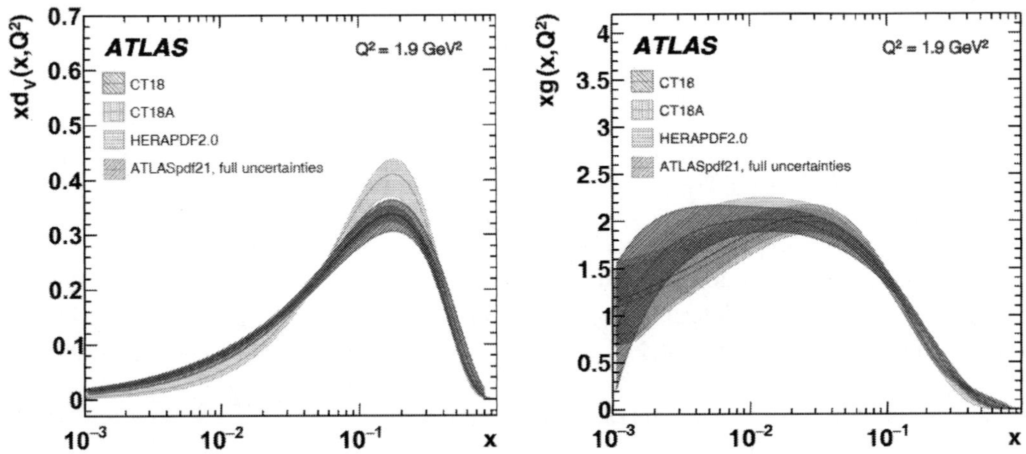

Figure 1. The valence d-quark parton density, $x \cdot d_v(x, Q^2)$ (left) and gluon parton density, $x \cdot g(x, Q^2)$ (right) at $Q^2 = 1.9$ GeV^2, as determined from the ATLAS analysis [2].

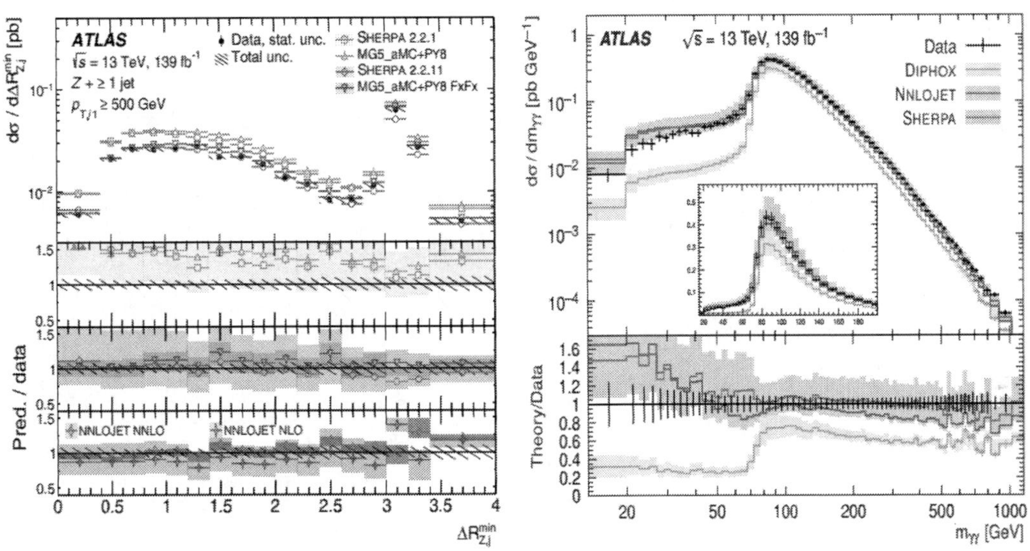

Figure 2. The Z+jets cross section as a function of the ΔR distance between the Z boson and the closest jet (left, [7]). The diphoton cross section as a function of the invariant mass of the two photons (right, [12]).

The diphoton cross section is presented by ATLAS in Ref. [12]. Measurements are shown as a function of different kinematic variables including, among others, the invariant mass and the transverse momentum of the diphoton system, as well as the angular separation between the two photons. The results are compared to theoretical predictions up to NNLO. The production cross section for pairs of photons as a function of the invariant mass of the diphoton pair is shown on the right-hand side of Figure 2. While the fixed-order NLO predictions by DiPhox [13] fail to model the shape of the distributions and underestimate the total cross section, the NNLO predictions show a large improvement with respect to the previous order. The resummed predictions from Sherpa [9] agree well with the data in all regions of the phase space.

A measurement of Transverse Energy-Energy Correlations, i.e. the energy weighted distributions of the differences in azimuth between pairs of jets, and their azimuthal asymmetry, is presented in Ref. [11]. The results are compared to theoretical predictions up to NLO, corrected for non-perturbative effects, and the value of the strong coupling constant at the Z mass pole, $\alpha_s(m_Z)$, is determined. A study of the running of $\alpha_s(Q)$, obtained from fits in different kinematic regions, is also presented. Figure 3 shows the values of $\alpha_s(Q)$ together with the NLO solution to the Renormalisation Group Equation. The value of the strong coupling constant at the Z mass pole, obtained from a global fit to the TEEC distributions in different kinematic regimes, is

$$\alpha_s(m_Z) = 0.1196 \pm 0.0001\ (\text{stat.}) \pm 0.0004\ (\text{sys.})^{+0.0071}_{-0.0104}\ (\text{scale}) \pm 0.0011\ (\text{PDF}) \pm 0.0002\ (\text{NP}),$$

where experimental uncertainties (statistical and systematic sources) and theoretical uncertainties (due to the choice of the QCD scales, PDFs and non-perturbative corrections) are included.

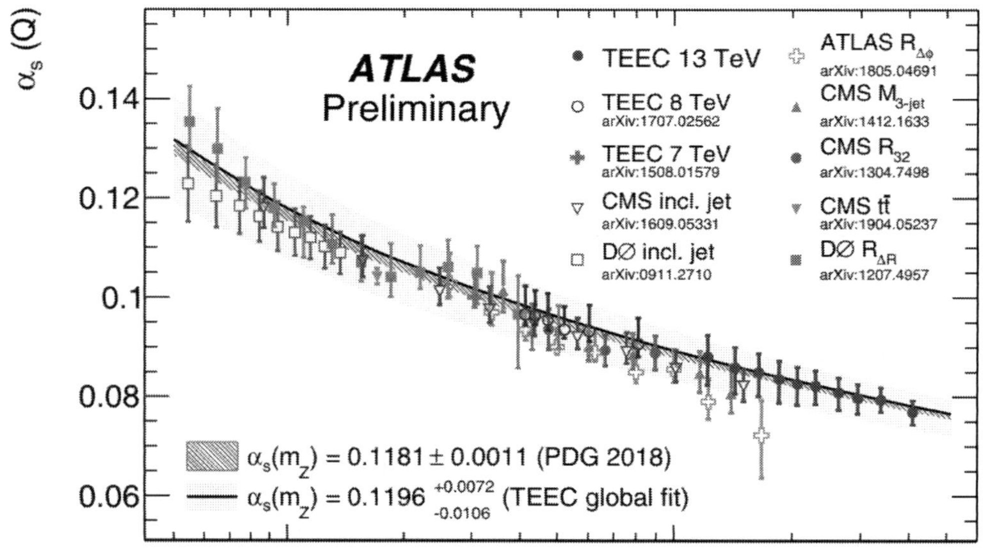

Figure 3. Comparison of the values of $\alpha_s(Q)$ determined from fits to the TEEC functions with the QCD prediction using the world average as input (blue band) and the value obtained from the global fit (yellow band). Results from previous analyses, both from ATLAS and from other experiments, are also included, showing an excellent agreement with the current measurements and with the world average.

4 The final state: Heavy quark fragmentation functions

The fragmentation of heavy quarks into hadrons is explored by the ATLAS Collaboration in Refs. [14, 15]. Reference [14] presents a measurement of the fragmentation properties of charged B mesons, decaying through $B^{\pm} \to J/\psi K^{\pm}$, inside jets. More specifically, the

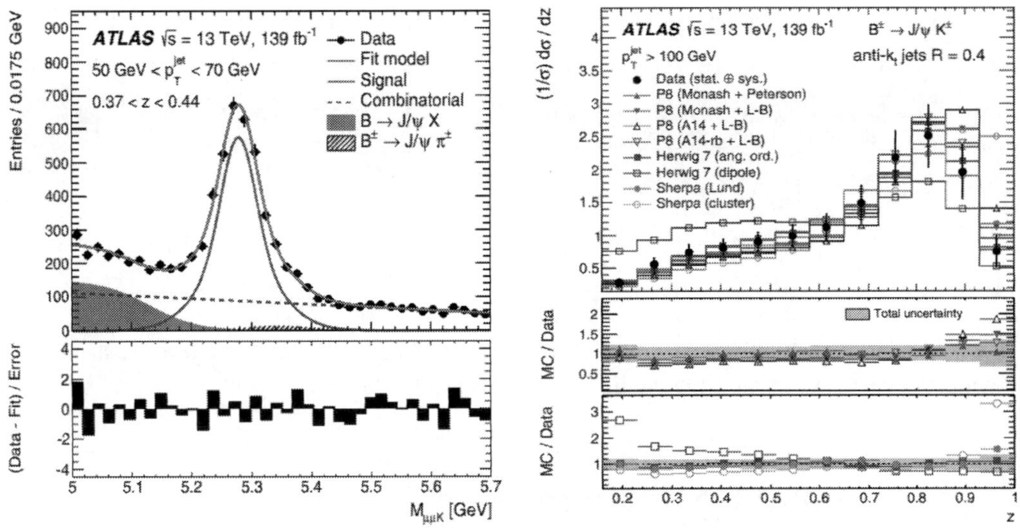

Figure 4. The invariant mass of the $B^{\pm}$ candidates reconstructed in a particular bin of jet p_T and z (left). The longitudinal profile distribution for jets with $p_T > 100$ GeV (right) [14].

longitudinal and transverse profiles of the momentum of $B^{\pm}$ mesons, $\vec{p}_B$, with respect to the momentum of the jets containing them, $\vec{p}_J$, defined as

$$z = \frac{\vec{p}_J \cdot \vec{p}_B}{\left|\vec{p}_J\right|^2}; \qquad p_T^{\mathrm{rel}} = \frac{\left|\vec{p}_J \times \vec{p}_B\right|}{\left|\vec{p}_J\right|}, \tag{2}$$

are measured. The results show a significant dependence of the studied variables on the modelling of gluon splittings $g \to b\bar{b}$, and an important spread of the different Monte Carlo predictions with respect to the data. Figure 4 shows the invariant mass of the reconstructed $B^{\pm}$ candidates, as well as the longitudinal profile for jets with $p_T > 100$ GeV.

Reference [15] presents a measurement of observables sensitive to the fragmentation functions of b-quarks in dileptonic $t\bar{t}$ events containing an electron and a muon. The measurement is performed using jets containing a displaced secondary vertex with at least four tracks originating from it, which are used to reconstruct the charged momentum of the B-hadron, $\vec{p}_b^{\,\mathrm{ch}}$. The charged momentum of the jet, $\vec{p}_{\mathrm{jet}}^{\,\mathrm{ch}}$, is reconstructed using all tracks geometrically associated to it.

The measured distributions, including the charged longitudinal profile as well as the number of charged particles originating from the B-hadron decay, are shown in Figure 5. The distributions are compared to different Monte Carlo expectations with the $t\bar{t}$ matrix elements calculated up to NLO, providing a generally reasonable agreement with the data.

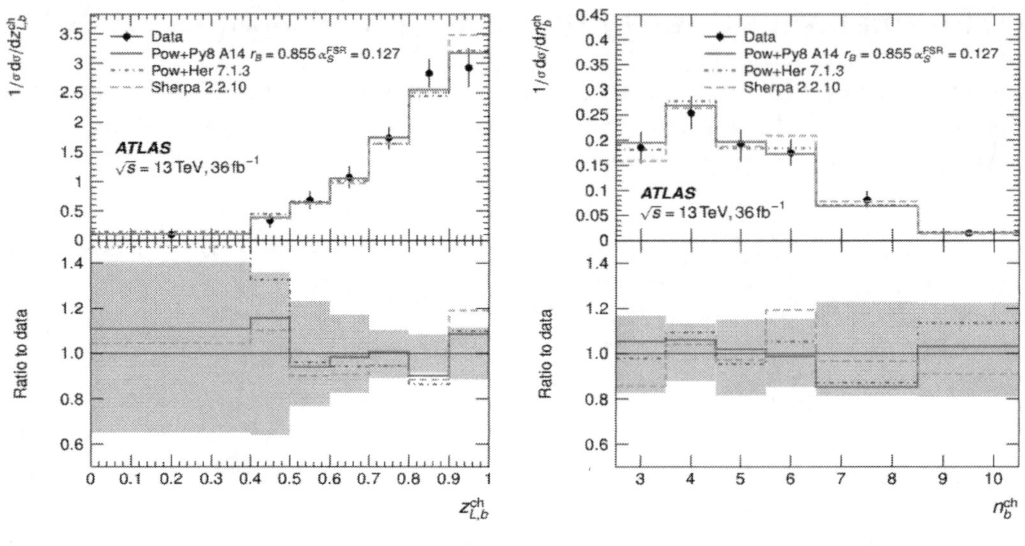

Figure 5. Distribution of the charged longitudinal profile of B hadrons inside jets (left). Distribution of the number of charged particles originating from the decay of the B hadron (right).

5 Summary and Conclusions

Results by the ATLAS Collaboration, sensitive to various aspects of the modelling of QCD, have been presented. The measurements include variables sensitive to the initial state, which are used for determining the Parton Distribution Functions, as well as to the hard-scattering process. This includes the production cross section of Z bosons in association with high-p_T jets as well as the production cross section of photon pairs.

The strong coupling constant α_s is determined from fits to the Transverse Energy-Energy Correlation distributions in different kinematic ranges. The running of $\alpha_s(Q)$ is obtained and compared to previous measurements and to the prediction from the Renormalisation Group Equation.

The fragmentation of heavy quarks has been probed in two different analyses. The first of them presents a measurement of the longitudinal and transverse profiles of the momentum of charged B mesons, decaying via $B^{\pm} \to J/\psi K^{\pm}$, with respect to the jet containing them. The second one presents different distributions sensitive to the fragmentation of b-quarks obtained in dileptonic $t\bar{t}$ events.

References

[1] J.C. Collins, D.E. Soper and G. Sterman, [Nucl. Phys. B. 261, 104 (1985)]
[2] ATLAS Collaboration, [Eur. Phys. J. C 82, 438 (2022)]
[3] H1 and ZEUS Collaborations, [Eur. Phys. J. C 75, 580 (2015)]
[4] T-J. Hou *et al.*, [Phys. Rev. D 103, 014013 (2021)]
[5] R.D. Ball *et al.*, [Eur. Phys. J. C 77, 663 (2017)]
[6] S. Bailey *et al.*, [Eur. Phys. J. C 81, 341 (2021)]
[7] ATLAS Collaboration, [arXiv:2205.02597 (hep-ex)]
[8] J. Alwall *et al.* [JHEP 07, 079 (2014)]
[9] E. Bothmann *et al.* [SciPost Phys. 7, 034 (2019)]
[10] ATLAS Collaboration [arXiv:2112.09588 (hep-ex)]
[11] ATLAS Collaboration, [ATLAS-CONF-2020-025]
[12] ATLAS Collaboration, [JHEP 11, 169 (2021)]
[13] T. Binoth, J. P. Guillet, E. Pilon and M. Werlen, [Eur. Phys. J. C 16, 311 (2000)]
[14] ATLAS Collaboration [JHEP 12, 131 (2021)]
[15] ATLAS Collaboration [Phys. Rev. D 106, 032008 (2022)]

Modified TMD Factorization and Sub-leading Power Corrections

Sergio Leal-Gomez[1,*] and *Massimiliano* Procura[1,**]

[1]University of Vienna, Faculty of Physics, Boltzmanngasse 5, A-1090 Vienna, Austria

Abstract. Collinear factorization and transverse-momentum-dependent (TMD) factorization are two complementary approaches to perform QCD calculations of Drell-Yan differential cross sections. The former is designed to correctly describe the behavior of the observable at large values of the gauge boson transverse momentum q_T, while the latter accounts for non-perturbative effects relevant at small q_T. We present basic features and first numerical results of a novel factorization formalism which is related to both previous frameworks and allows for an improved description of the intermediate-q_T region.

1 Introduction

The Drell-Yan process plays a central role in the study of vector gauge bosons and allows us also to extract information about the internal structure of the proton, like the distribution of quark spin and angular momentum. QCD calculations of Drell-Yan multi-differential cross sections have been performed using two different frameworks, namely collinear factorization [1–4] and TMD factorization [5–12], the former designed to describe physical cross sections at high values of the magnitude of the gauge-boson transverse momentum $q_T \equiv \sqrt{\mathbf{q}_T^2}$, and the latter accounting for features relevant at sufficiently small values of q_T. In the collinear factorization framework, the gauge boson transverse momentum has a purely perturbative origin rooted in the hard scattering process that involves purely collinear partons. In TMD factorization instead, $\mathbf{q}_T$ is due to the fact that the colliding partons have non-vanishing non-perturbative intrinsic transverse momentum inside the initial-state hadrons, as described by transverse-momentum-dependent parton distributions functions (TMDPDFs) [10, 13–15].

1.1 Collinear factorization

For two protons $h_{A,B}$ collinear to the z−axis and colliding in the center-of-mass (CM) frame, the following factorization formula for Drell-Yan holds at leading power [9, 10, 16] :

$$\frac{d\sigma_{h_A h_b \to X l\bar{l}}}{dQ^2\, dy\, d\mathbf{q}_T^2} = \sum_{a,b} \int_{x_A}^{1} \frac{dz_a}{z_a} \int_{x_B}^{1} \frac{dz_b}{z_b} H_{ab}\left(\alpha_s\left(\mu^2\right), Q^2, \mathbf{q}_T^2, \frac{x_A}{z_a}, \frac{x_A}{z_a}\right) f_{a\leftarrow h_A}\left(\alpha_s\left(\mu^2\right), z_a\right) f_{b\leftarrow h_B}\left(\alpha_s\left(\mu^2\right), z_b\right), \tag{1}$$

*e-mail: sergio.leal.gomez@univie.ac.at
**e-mail: massimiliano.procura@univie.ac.at

where y is the gauge boson rapidity and Q^2 is the invariant mass of the final lepton pair. The variables $x_{A(B)}$ are defined as $x_{A(B)} = \sqrt{Q^2/s}\, e^{\pm y}$ where s is the CM energy squared. The value of the renormalization scale μ^2 is typically set equal to Q^2. Non-perturbative information encoded in the parton distribution functions (PDFs) $f_{a(b)\leftarrow h_{A(B)}}$ is extracted from experimental data. Using perturbative QCD, the coefficients H_{ab} are given by

$$H_{ab}\left(\alpha_s\left(\mu^2\right), Q^2, \mathbf{q}_T^2, \frac{x_A}{z_a}, \frac{x_B}{z_b}\right) = \delta^{(2)}\left(\mathbf{q}_T\right)\delta\left(1-\frac{x_A}{z_a}\right)\delta\left(1-\frac{x_B}{z_b}\right)\sigma_{ab}^{\text{Born}}\delta_{ah_A}\delta_{bh_B} + \sum_n \alpha_S\left(\mu^2\right)^n H_{ab}^{[n]}\left(Q^2, \mathbf{q}_T^2, \frac{x_A}{z_a}, \frac{x_B}{z_b}\right), \tag{2}$$

with $H_{ab}^{[n]}$ determined from a partonic calculation [1, 4]. The last expression contains logarithmically enhanced terms of the form $\ln^k \mathbf{q}_T^2/Q^2$ which needs to be resummed to all orders to obtain reliable theory predictions [2, 3, 17, 18]. The resummed differential cross section in the limit $\mathbf{q}_T^2/Q^2 \ll 1$ is given by [1–4, 17–19]:

$$\left[\frac{d\sigma_{h_A h_B\to Xl\bar{l}}}{dQ^2\, dy\, d\mathbf{q}_T^2}\right]_{\text{Res}} = \sum_{c,a,b}\int_{x_A}^1 \frac{dz_a}{z_a}\int_{x_B}^1 \frac{dz_b}{z_b}\int \frac{d^2\mathbf{b}_T}{(2\pi)^2} e^{i\mathbf{q}_T\cdot\mathbf{b}_T}\sigma_{c\bar{c}}^{\text{Born}} C_{c\leftarrow a}\left(\alpha_s\left(b_0^2/\mathbf{b}_T^2\right), \frac{x_A}{z_a}\right) C_{\bar{c}\leftarrow b}\left(\alpha_s\left(b_0^2/\mathbf{b}_T^2\right), \frac{x_B}{z_b}\right)$$
$$\times S(Q^2, b_0^2/\mathbf{b}_T^2) f_{a\leftarrow h_A}\left(\alpha_s\left(b_0^2/\mathbf{b}_T^2\right), z_a\right) f_{b\leftarrow h_B}\left(\alpha_s\left(b_0^2/\mathbf{b}_T^2\right), z_b\right), \tag{3}$$

where $\mathbf{b}_T$ is the impact parameter, $b_0^2 = 4e^{-2\gamma_E}$, $C_{c(\bar{c})\leftarrow a(b)}$ denote perturbative coefficients and $S(Q^2, b_0^2/\mathbf{b}_T^2)$ is the Sudakov form factor [1–4, 8, 19]. The last expression allows us to define, order-by-order in QCD perturbation theory, the fixed-order (FO) contribution to the cross section as

$$\left[\frac{d\sigma_{h_A h_B\to Xl\bar{l}}}{dQ^2\, dy\, d\mathbf{q}_T^2}\right]_{\text{FO}} = \frac{d\sigma_{h_A h_B\to Xl\bar{l}}}{dQ^2\, dy\, d\mathbf{q}_T^2} - \left[\frac{d\sigma_{h_A h_B\to Xl\bar{l}}}{dQ^2\, dy\, d\mathbf{q}_T^2}\right]_{\text{Res}}, \tag{4}$$

where the most singular terms in the limit $\mathbf{q}_T^2/Q^2$ are removed via a subtraction of partonic collinear divergences using PDFs. Combining fixed-order and resummed contributions, the cross section derived in the collinear factorization framework describes well experimental data at sufficiently high q_T, but fails to do so in the small-q_T regime. The non-perturbative contributions that are responsible for this discrepancy are captured by TMD factorization.

1.2 TMD factorization

In TMD factorization [5–7, 9–13, 20, 21], the differential cross section at leading power is given by

$$\frac{d\sigma_{h_A h_B\to Xl\bar{l}}}{dQ^2\, dy\, d\mathbf{q}_T^2} = \sum_c \sigma_{c\bar{c}}^{\text{Born}} H\left(\alpha_s\left(\mu^2\right), Q^2/\mu^2\right)$$
$$\times \int \frac{d^2\mathbf{b}_T}{(2\pi)^2} e^{i\mathbf{b}_T\cdot\mathbf{q}_T} F_{c\leftarrow h_A}\left(\alpha_s\left(\mu^2\right), \mathbf{b}_T^2, x_A, \zeta_A\right) F_{\bar{c}\leftarrow h_B}\left(\alpha_s\left(\mu^2\right), \mathbf{b}_T^2, x_B, \zeta_B\right), \tag{5}$$

where the hard factor H is computed in perturbative QCD [11, 22, 23] and $F_{i\leftarrow h_{A,B}}$ describes the non-perturbative intrinsic transverse momentum distribution of the parton i inside the colliding hadrons. Here the renormalization scale μ^2 is typically set equal to Q^2 and the same choice is made for the rapidity renormalization scale $\zeta_{A(B)}$ [24–26]. The TMDPDFs can be matched onto collinear PDFs as [13, 15, 27]

$$F_{a\leftarrow h_A}\left(\alpha_s\left(\mu^2\right), \mathbf{b}_T^2, x_A, \zeta_A\right) = \sum_{n,r} (|\mathbf{b}_T|M)^n\, C_{a\leftarrow r}^{(n)}\left(\alpha_s\left(\mu^2\right), \mathbf{b}_T^2, \frac{x_A}{z_a}, \zeta_A\right) \otimes f_{r\leftarrow h_A}^{(n)}\left(\alpha_s\left(\mu^2\right), z_a\right), \tag{6}$$

where M is typically identified with the colliding hadron mass, n stands for the twist of the matching and $\otimes$ denotes the Mellin convolution in z_a.

The distribution obtained by combining fixed-order and resummed contributions in this framework provides a good description of experimental data at small q_T, as shown in fig. (1), but it undershoots experimental points at intermediate/large q_T, see also [28]. Since TMD and collinear factorizations achieve complementary descriptions, we study the possibility of combining them into a more general factorization framework that allows us to compute a set of power corrections in $\mathbf{q}_T^2/Q^2$ playing an important role in the intermediate-q_T region.

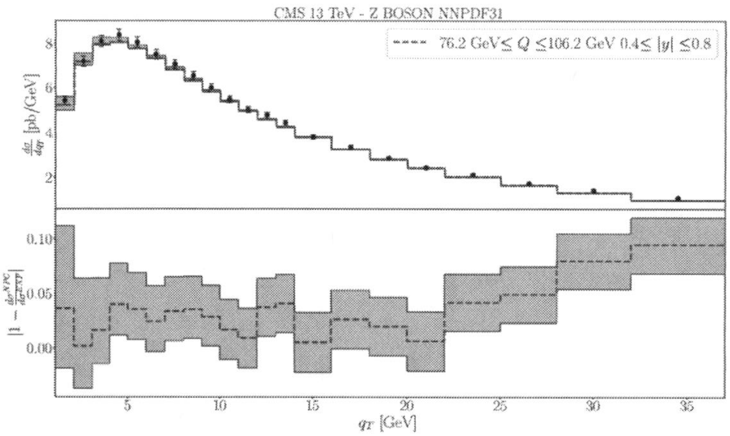

Figure 1. Drell-Yan Z-boson production at the LHC: comparison between the theoretical prediction in TMD factorization at NNLO+NNLL accuracy [23] and CMS experimental data [28].

2 Modified TMD factorization and power corrections

In the spirit of the work by Catani et al. [1, 3, 17, 18, 29], we worked out a phenomenological modification of the factorization formula in eq. (1) where the hard factor depends *both* on the collinear momentum components $z_{a(b)}$ *and* on the intrinsic transverse momenta $\mathbf{k}_{a(b)T}$ of the partons in the underlying scattering process. In our formalism, collinear divergences are subtracted using TMDPDFs instead of collinear PDFs, which enables the resummation of all logarithmically enhanced contributions in $\mathbf{q}_T^2/Q^2$ as well as the inclusion of a set of sub-leading power corrections. The Drell-Yan cross section in this formalism is given by

$$\frac{d\sigma_{h_A h_B \to X l\bar{l}}}{dQ^2\, dy\, d\mathbf{q}_T^2} =$$
$$\sum_{a,b} \int d^2\mathbf{k}_{aT} d^2\mathbf{k}_{bT} d^2\mathbf{q}'_T \int_{x_A}^1 \frac{dz_a}{z_a} \int_{x_B}^1 \frac{dz_b}{z_b} \delta^{(2)}\left(\mathbf{q}_T - \mathbf{q}'_T - \mathbf{k}_{aT} - \mathbf{k}_{bT}\right) \theta\left(\frac{(z_a - x_A)(z_b - x_B)}{x_A x_B} - \frac{\mathbf{q}'^2_T}{Q^2 + \mathbf{q}_T^2}\right)$$
$$\times \tilde{H}_{ab}\left(\alpha_s\left(\mu^2\right), Q^2, \mathbf{q}_T^2, \mathbf{q}'^2_T, \frac{x_A}{z_a}, \frac{x_B}{z_b}\right) F_{a\leftarrow h_A}\left(\alpha_s\left(\mu^2\right), \mathbf{k}_{aT}^2, z_a, \zeta_A\right) F_{b\leftarrow h_B}\left(\alpha_s\left(\mu^2\right), \mathbf{k}_{bT}^2, z_b, \zeta_B\right). \tag{7}$$

This expression shows that, according to the most general kinematic picture, $\mathbf{q}_T$ receives contributions from both perturbative parton emissions and non-perturbative intrinsic transverse-momentum parton distributions, related by four-momentum conservations. We explicitly checked up to NNLO in our framework that collinear divergences in the partonic calculations are taken care of by the TMDPDFs. The modified hard factor $\tilde{H}_{ab}$, which differs from both the hard coefficient in collinear factorization and the one in standard TMD factorization, is thus free of singularities in the small-q_T limit and can be determined through a fixed-order QCD calculation as the difference between left-hand side and right-hand side of eq. (7). This definition of $\tilde{H}_{ab}$ guarantees that in the small-q_T limit, the Drell-Yan distribution in this formalism reproduces the results from TMD factorization. Notice that in eq. (7), the TMDPDFs are written in momentum space, while in TMD factorization, eq. (5), the impact-parameter space notation is used.

We stress that our approach differs from previous attempts to combine TMD and collinear factorization, as the one in [30] where the following formula was proposed,

$$\frac{d\sigma_{h_A h_B \to X l\bar{l}}}{dQ^2\, dy\, d\mathbf{q}_T^2} = g_1\left(\frac{\mathbf{q}_T^2}{Q^2}\right)\left[\frac{d\sigma_{h_A h_B \to X l\bar{l}}}{dQ^2\, dy\, d\mathbf{q}_T^2}\right]_{\text{TMD}} + g_2\left(\frac{\mathbf{q}_T^2}{Q^2}\right)\left[\frac{d\sigma_{h_A h_B \to X l\bar{l}}}{dQ^2\, dy\, d\mathbf{q}_T^2}\right]_{\text{high}\, q_T}, \tag{8}$$

which combines TMD factorization and collinear factorization results in a weighted sum where $g_{1,2}$ are non-perturbative model functions such that $g_1 \to 1$, $g_2 \to 0$ as $q_T \to 0$, while $g_1 \to 0$, $g_2 \to 1$ as $q_T \to \infty$. In our approach instead, TMD and collinear factorization results are combined through the perturbative matching coefficient $\tilde{H}$.

We also point out that our formalism is different from the one presented in [6] to work out next-to-leading power corrections to the Drell-Yan cross section in TMD factorization. In that study, the first term on the right-hand side of eq. (4) was computed in the $q\bar{q} \to g l\bar{l}$ channel by first performing an expansion in $\mathbf{k}_T^2/Q^2$ up to next-to-leading power, where $\mathbf{k}_T$ denotes the transverse momentum of the emitted gluon, and then carrying out the phase space integrals. In our formalism no expansion in powers of $\mathbf{k}_T^2/Q^2$ is performed.

For a first numerical analysis within our modified TMD factorization formalism, we determined the hard factor $\tilde{H}_{ab}$ at NLO from the right-hand side of eq. (4) in a partonic calculation involving the channels $q\bar{q} \to X l\bar{l}$ where $X = \{g, gg, q\bar{q}, \cdots\}$ and $q\,(\bar{q})$ has collinear(anti-collinear) momentum with respect to the z-axis. No assumption about the hierarchy between $\mathbf{q}_T^2$ and Q^2 has been made. Fig. (2) shows our results where the phenomenological parameters in the TMDPDFs were fixed according to the outcome of the fits presented in [23]. Our modified TMD factorization framework leads to a better agreement with experimental results in the intermediate-q_T region. Our analysis allows us to quantify the impact of a set of power suppressed terms compared to standard TMD factorization: the contribution of these sub-leading effect turns out to be at least about 5% in the intermediate region, namely of comparable size or even larger than electroweak corrections that have recently been quantified to be around 1%-2% [31, 32].

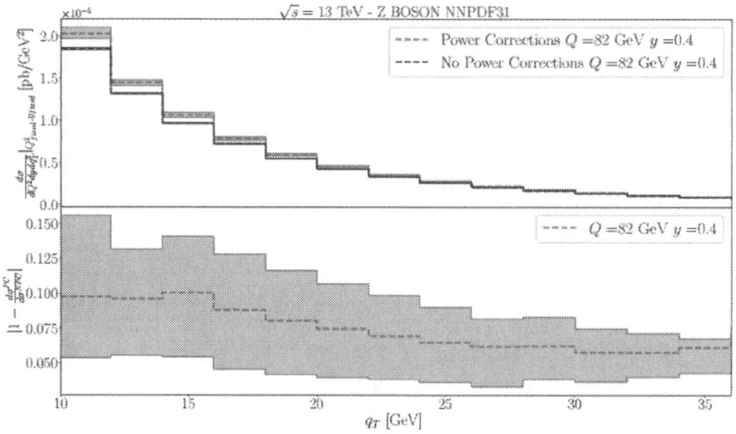

Figure 2. Comparison between theory predictions using our modified TMD factorization framework, with and without the power corrections that we can account for.

3 Conclusions

Modifying the collinear factorization formula for Drell-Yan by subtracting collinear divergences using TMDPDFs instead of standard PDFs allows us to include power suppressed terms in $\mathbf{q}_T^2/Q^2$ in the cross section. These terms provide additional positive contributions which improve the agreement between theory predictions and experimental data in the tail of the distribution. This enables future improved phenomenological extractions of TMDPDFs by fitting to enlarged sets experimental data. Furthermore, the fact that the hard factor in our modified TMD factorization formalism does not contain logarithmically enhanced terms makes this a smooth function in the limit $\mathbf{q}_T^2 \to 0$, opening up the possibility to reliably perform a numerical computation of $\tilde{H}$ using subtraction methods [17, 18, 29, 33], which can be automatized and implemented in Monte-Carlo event generators.

References

[1] S. Catani, D. de Florian, M. Grazzini, Nucl. Phys. B **596**, 299 (2001)
[2] S. Catani, D. de Florian, M. Grazzini, P. Nason, JHEP **07**, 028 (2003)
[3] D. de Florian, M. Grazzini, Phys. Rev. Lett. **85**, 4678 (2000)
[4] D. de Florian, M. Grazzini, Nucl. Phys. B **616**, 247 (2001)
[5] M.A. Ebert, F.J. Tackmann, JHEP **02**, 110 (2017)
[6] M.A. Ebert, I. Moult, I.W. Stewart, F.J. Tackmann, G. Vita, H.X. Zhu, JHEP **04**, 123 (2019)
[7] J.C. Collins, D.E. Soper, Nucl. Phys. B **194**, 445 (1982)
[8] J.C. Collins, D.E. Soper, G.F. Sterman, Nucl. Phys. B **250**, 199 (1985)
[9] J.C. Collins, D.E. Soper, G.F. Sterman, Adv. Ser. Direct. High Energy Phys. **5**, 1 (1989)
[10] J. Collins, Int. J. Mod. Phys. Conf. Ser. **4**, 85 (2011)
[11] M.G. Echevarría, A. Idilbi, I. Scimemi, Phys. Lett. B **726**, 795 (2013)
[12] M.G. Echevarria, A. Idilbi, I. Scimemi, JHEP **07**, 002 (2012)
[13] M.G. Echevarria, I. Scimemi, A. Vladimirov, JHEP **09**, 004 (2016)
[14] X.d. Ji, J.p. Ma, F. Yuan, Phys. Rev. D **71**, 034005 (2005)
[15] A. Bacchetta, D. Boer, M. Diehl, P.J. Mulders, JHEP **08**, 023 (2008)

[16] G. Altarelli, R.K. Ellis, M. Greco, G. Martinelli, Nucl. Phys. B **246**, 12 (1984)
[17] G. Bozzi, S. Catani, D. de Florian, M. Grazzini, Phys. Lett. B **564**, 65 (2003)
[18] G. Bozzi, S. Catani, D. de Florian, M. Grazzini, Nucl. Phys. B **737**, 73 (2006)
[19] C. Balazs, C.P. Yuan, Phys. Rev. D **56**, 5558 (1997)
[20] M.G. Echevarria, A. Idilbi, I. Scimemi, Phys. Rev. D **90**, 014003 (2014)
[21] V. Bertone, I. Scimemi, A. Vladimirov, JHEP **06**, 028 (2019)
[22] I. Scimemi, A. Vladimirov, Eur. Phys. J. C **78**, 89 (2018)
[23] I. Scimemi, A. Vladimirov, JHEP **06**, 137 (2020)
[24] I. Scimemi, A. Vladimirov, JHEP **08**, 003 (2018)
[25] J.Y. Chiu, A. Jain, D. Neill, I.Z. Rothstein, JHEP **05**, 084 (2012)
[26] M.G. Echevarria, I. Scimemi, A. Vladimirov, Phys. Rev. D **93**, 054004 (2016)
[27] M.G. Echevarria, T. Kasemets, P.J. Mulders, C. Pisano, JHEP **07**, 158 (2015), [Erratum: JHEP 05, 073 (2017)]
[28] A.M. Sirunyan et al. (CMS), JHEP **12**, 061 (2019)
[29] A. Broggio, C. Gnendiger, A. Signer, D. Stöckinger, A. Visconti, JHEP **01**, 078 (2016)
[30] J. Collins, L. Gamberg, A. Prokudin, T.C. Rogers, N. Sato, B. Wang, Phys. Rev. D **94**, 034014 (2016)
[31] R. Bonciani, L. Buonocore, M. Grazzini, S. Kallweit, N. Rana, F. Tramontano, A. Vicini, Phys. Rev. Lett. **128**, 012002 (2022)
[32] L. Cieri, G. Ferrera, G.F.R. Sborlini, JHEP **08**, 165 (2018)
[33] S. Catani, M.H. Seymour, Nucl. Phys. B **485**, 291 (1997), [Erratum: Nucl.Phys.B 510, 503–504 (1998)]

Heavy-flavor impact on CTEQ-TEA global QCD analyses

Marco Guzzi[1,*], *Alim* Ablat[2], *Sayipjamal* Dulat[2], *Tie-Jiun* Hou[3], *Pavel* Nadolsky[4], *Ibrahim* Sitiwaldi[2], *Keping* Xie[5], and *C.-P.* Yuan[6]

[1]Department of Physics, Kennesaw State University, 370 Paulding Ave., 30144 Kennesaw, GA, USA
[2]School of Physics Science and Technology, Xinjiang University, Urumqi, Xinjiang 830046, China
[3]School of Nuclear Science and Technology, Univ. of South China, Hengyang, Hunan 421001, China
[4]Department of Physics, Southern Methodist University, Dallas, TX 75275-0181, USA
[5]Pittsburgh Particle Physics, Astrophysics, and Cosmology Center, Department of Physics and Astronomy, University of Pittsburgh, Pittsburgh, PA 15260, USA
[6]Department of Physics and Astronomy, Michigan State University, East Lansing, MI 48824, USA

Abstract. We discuss heavy-flavor production at hadron colliders in recent global QCD analyses to determine parton distribution functions (PDFs) in the proton. We discuss heavy-flavor treatments in precision theory predictions at the LHC. In particular, we discuss factorization schemes in presence of heavy flavors in proton-proton collisions, as well as the impact of heavy-flavor production at the LHC on PDFs. We show results of recent updates beyond CT18, the latest global QCD analysis from the CTEQ-TEA group.

1 Introduction

Precise determination of parton distribution functions (PDFs) in the proton is critical for all current and future precision programs at the LHC. Proton PDFs are an essential ingredient of QCD factorization theorems and are therefore ubiquitous in theory predictions for standard candle observables in hadronic collisions at high perturbative order in QCD. Precise and accurate theory predictions for such observables are necessary to investigate properties of the Higgs boson and to explore the Electroweak (EW) sector of the Standard Model (SM). In addition, they are important to scrutinize and validate SM extensions and search for signals of new physics interactions. PDFs are obtained through global QCD analyses of experimental hadronic cross section measurements by using a variety of analytical and statistical methods [1–4]. They are "data-driven" quantities and they currently challenge precision in theory predictions at the LHC. In particular, precision measurements of heavy-quark (HQ) production observables in hadronic reactions play a significant role in PDF determinations because they can constrain PDFs in kinematic regions at intermediate and large parton momentum fraction x which are currently poorly constrained by data. In addition, they provide us with complementary information besides that obtained from jet production in global QCD analyses. HQ production at the LHC at small transverse momentum p_T and large rapidity y of the HQ is sensitive to PDFs at both small and large x, where $x \approx \left(\sqrt{p_T^2 + m_Q^2}/\sqrt{S}\right)e^{\pm y}$. This allows us to probe QCD dynamics simultaneously in these two regimes. This is especially true for charm and bottom (c/b) quark production. In fact, in the $4 < |y| < 4.5$ rapidity range

*e-mail: mguzzi@kennesaw.edu,presenter

at the LHC with a proton-proton collision energy of 13 TeV, it can probe $x \leq 10^{-5}$. When $p_T \geq 40$ GeV, it can probe $x \geq 0.2$. On the other hand, top-quark pair production at the LHC, at the same collision energy, can probe the gluon PDF already at $x \gtrsim 0.01$, while c/b production at HERA is sensitive to intermediate and small $x \lesssim 10^{-5}$. Probing these regimes (and beyond, at future facilities like the Electron Ion Collider (EIC) [5–7], the Future Circular Collider (FCC) [8–10], the Forward Physics Facility at the High-Luminosity (HL) LHC [11, 12], and the Super proton-proton Collider (SppC) [13]) will allow us to explore QCD factorization with an unprecedented level of accuracy. This will shed light on open questions like the intrinsic heavy-flavor content in the proton, and small-x dynamics. In this conference proceeding contribution, we shall report preliminary results of global PDF analyses which use recent high-precision HQ measurements from the LHC and HERA which are added on top of the CT18 [1] baseline. Moreover, treatments of QCD factorization in presence of HQ multiscale processes are also discussed.

2 HQ treatments in QCD factorization and the S-ACOT-MPS scheme

HQ production dynamics is nontrivial due to the interplay of massless and massive schemes, which essentially are different ways of organizing the perturbation series. In general, one can calculate cross sections by using massive schemes, that include HQs only in the final-state and apply at masses such that $p_T \lesssim m_Q$. In this case, the HQ p_T-spectrum can be obtained by keeping the number of flavors as fixed, *i.e.*, using a fixed-flavor number (FFN) scheme. That is, there is no HQ PDF in the proton; HQs are generated as massive final states, and m_Q acts as an infrared cutoff. Moreover, all power terms of the type $\left(p_T^2/m_Q^2\right)^p$, where p is a positive integer, are correctly accounted for in the perturbative series. Cross sections can also be computed by using massless schemes when $p_T \gg m_Q \gg m_P$, where m_P is the mass of the proton. In this case, large logarithmic terms of the type $\log^n\left(p_T^2/m_Q^2\right)$ spoil the convergence of the fixed-order expansion. Massless schemes are also known as zero-mass (ZM) schemes; the HQ is considered essentially massless everywhere and also enters the running of the strong coupling α_s. These large logarithms need to be resummed by using DGLAP evolution, that is, initial-state logs are resummed into a HQ PDF, while final-state logarithms are resummed into a fragmentation function (FF). Modern global QCD analyses determine proton PDFs using a variety of experimental data which span wide areas of the Q-x kinematic plane. Therefore, amendments to the factorization formula are necessary to calculate key cross sections across wide ranges of energy and momentum transfer. Interpolating schemes, such as General Mass Variable Flavor Number (GMVFN) schemes, introduce modifications in QCD collinear factorization so that they appear as composite schemes which retain key mass dependence and efficiently resum collinear logarithms in order to combine the FFN and ZM schemes together. They are crucial for a correct treatment of HQ in deep inelastic scattering (DIS) processes [14–22] and proton-proton (pp) collisions, and for accurate predictions of key scattering rates at the LHC in global QCD analyses to determine proton PDFs. Moreover, they provide us with the possibility of directly accessing HQ PDFs parametrized at the initial scale. This motivated the development of a general-mass (GM) factorization scheme for pp collisions [23, 24], the Simplified ACOT scheme with Massive Phase Space (S-ACOT-MPS), which is based on an amended version of the S-ACOT scheme, developed for DIS [14, 15, 18, 19, 22]. Similar GMVFN schemes are studied in [25–27] and differences are related to the treatment of the phase space. The main idea behind a GMVFN scheme such as S-ACOT-MPS is the introduction of subtraction terms order-by-order in perturbation theory which avoid double counting and cancel enhanced collinear contributions from flavor-creation (FC) terms when $\hat{s} \gg m_Q^2$, or $p_T \gg M_Q$. In S-ACOT-MPS, the subtraction and flavor excitation (FE) contributions are evaluated in one single step. This improves stability in calculating differential cross sections and facilitates the extension of this scheme to higher orders. We applied S-ACOT-MPS to prompt c and b production at NLO in QCD at the LHC, at $\sqrt{S} = 13$ TeV. In Figure 1 (left), we illustrate the c-rapidity distribution determined with the CT18 and CT18X PDFs at NLO. We explore the sensitivity of prompt c production to nonperturbative charm contributions in the proton by considering theory predictions in very forward regions with $y_c > 8$. This is relevant for future applications at the Forward Physics Facility at CERN [11] which will be able to clarify multiple aspects of QCD in such an extended forward region, in coordination with the HL-LHC and EIC. In Figure 1 (right), we compare theory predictions for the rapidity distribution of $B^{\pm}$ meson production at NLO in QCD with measurements [28] from the LHCb collaboration at $\sqrt{S} = 13$ TeV. Theory uncertainties at NLO are large ($O(50\%)$) and mainly ascribed to scale variation. This can be improved by including higher-order corrections which require an extension of the S-ACOT-MPS scheme to NNLO.

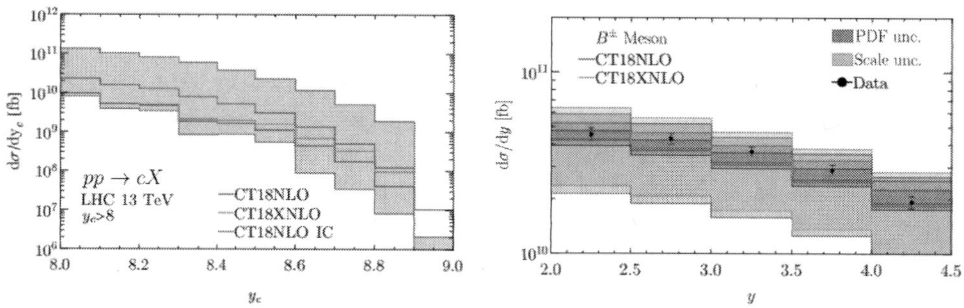

Figure 1. Left: Rapidity distributions of prompt charm at the LHC 13 TeV in the very forward region ($yc > 8$) [11]. The error band represents the CT18NLO induced PDF uncertainty at 68% CL.
Right: NLO theory predictions for the rapidity distributions obtained with CT18NLO and CT18XNLO PDFs compared to $B^{\pm}$ production data [28] from LHCb 13 TeV.

3 Impact on PDFs from high-precision $t\bar{t}$ production measurements

Top-quark pair production plays a significant role in setting constraints on PDFs, in particular the gluon, at intermediate to large x. In global PDF analyses, the impact on PDFs from differential cross section measurements of $t\bar{t}$ production at the LHC complements that from jet production measurements: $t\bar{t}$ and jet production overlap in the Q-x plane, but matrix elements and phase-space suppression are different. Therefore, constraints on the gluon may be placed at different values of x. On the other hand, challenges in the estimate of full impact from $t\bar{t}$ production on PDFs arise in presence of incompatibility of the $t\bar{t}$ measurements with other data sets in the baseline (see related discussion in [1]). More realistic estimates account for multiple PDF functional forms and some disagreements between the measurements. In this section, we explore the impact of high-precision $t\bar{t}$ measurements on proton PDFs, and illustrate preliminary results of a global QCD analysis in which $t\bar{t}$ single differential cross section measurements at $\sqrt{S} = 13$ TeV from the ATLAS [29] and CMS [30] collaborations are added on top of the CT18 data baseline. The measurements from ATLAS are obtained using events in the all-hadronic channel with 36.1 fb^{-1} of integrated luminosity (IL), while those from CMS are obtained with events containing two leptons with 35.9 fb^{-1} of IL. We consider the full phase space absolute measurements reported in Table 3, in which we specify the type of correlated systematic uncertainties published in the experimental papers. We added these

Table 1. Details of the single differential distributions for $t\bar{t}$ production considered in our analysis.

data type	N_{pt}	Exp	Corr Sys.
$d\sigma/dm_{t\bar{t}}$	9	ATLAS	nuisance par.
$d\sigma/dy_{t\bar{t}}$	12	ATLAS	nuisance par.
$d\sigma/dH_{T,t\bar{t}}$	11	ATLAS	nuisance par.
$d\sigma/dp_{T,t1}$	10	ATLAS	nuisance par.
$d\sigma/dp_{T,t2}$	8	ATLAS	nuisance par.
$d\sigma/dm_{t\bar{t}}$	7	CMS	Covariance matr.
$d\sigma/dp_{T,t}$	6	CMS	Covariance matr.
$d\sigma/dy_{t\bar{t}}$	10	CMS	Covariance matr.
$d\sigma/dy_t$	10	CMS	Covariance matr.

measurements one by one to the baseline because statistical correlations were not published for all data sets. The measurements from ATLAS and CMS are delivered by using a different binning for the same distribution. The theory predictions for the CMS distributions are obtained by using the FastNNLO grids [31], that are based on the calculation published in [32]. The theory predictions for the ATLAS distributions have been obtained in two steps. We have first generated QCD theory predictions at NLO by using in-house APPLGrid fast tables [33] from the public MCFM computer code [34, 35]. Then, we calculated bin-by-bin NNLO/NLO K-factors using the computer code MATRIX [36] that is based on the theory predictions calculated in [37]. The mass of the top quark, in the pole mass approximation, has been set as $m_t^{(pole)} = 172.5$ GeV. The factorization and renormalization scales μ_F and μ_R respectively, have been chosen according to [38] where for the $m_{t\bar{t}}$, $p_{T,t\bar{t}}$, $y_{t\bar{t}}$, and y_t distributions one uses $\mu_F = \mu_R = H_T/4$ and $H_T = \sqrt{m_t^2 + p_{T,t}^2} + \sqrt{m_t^2 + p_{T,\bar{t}}^2}$, for the $p_{T,t}$ distribution one uses $M_T = \sqrt{m_t^2 + p_{T,t}^2}$, and for the $p_{T,tavg}$ one uses $\mu_F = \mu_R = M_T/2$. EW corrections [39] are also included in this analysis, but their impact on the global fit is negligible. In Figure 2, we illustrate the pulls on the gluon PDF resulting from a global fit in which the y_{tt} distributions from ATLAS and CMS are added on top of the CT18 baseline. The pulls from these distributions from both experiments are in the same direction, that is, each of them prefers a softer gluon at large x ($x \gtrsim 0.2$). The individual impact of these measurements on the CT18 gluon is approximately the same as χ^2/N_{pt} for both distributions is of order 1. A more extensive analysis with complete results will be presented elsewhere [40].

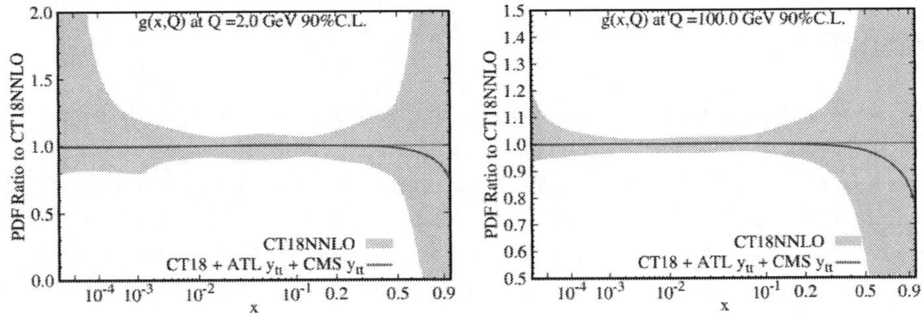

Figure 2. Left: PDF ratio to CT18NNLO at $Q = 2$ GeV for the gluon PDF. **Right**: same as left, but at $Q = 100$ GeV. The blue band represents the CT18NNLO PDF uncertainty at 90% CL.

4 CT18 and the new charm and bottom combination at HERA

To finalize our discussion, in this section we discuss the impact of the most recent c- and b-quark production combined measurements of semi-inclusive DIS at HERA from the H1 and ZEUS collaborations [41], on the CT18 gluon PDF. In particular, we illustrate the results of various fits in which these measurements replace their previous version [42, 43]. The new measurements exhibit an extended kinematic range of photon virtuality $2.5\ \text{GeV}^2 \leq Q^2 \leq 2000\ \text{GeV}^2$ and Bjorken scaling variable $3 \cdot 10^{-5} \leq x_{\text{Bj}} \leq 5 \cdot 10^{-2}$, and have reduced uncertainties due to a simultaneous combination of c and b cross-section measurements with reduced correlations between them. We find that the new c and b combination at HERA [41] cannot be well described in the CT18 global analysis, and χ^2/N_{pt} is never below 1.7. For the CT18NNLO fit, we obtained $\chi^2/N_{pt} = 1.98$ for c production ($N_{pt} = 47$), and $\chi^2/N_{pt} = 1.25$ for bottom production ($N_{pt} = 26$). For the CT18XNNLO fit, instead we obtained $\chi^2/N_{pt} = 1.71$ for c and 1.26 for b production. We observed tensions between these new combined data and several CT18 datasets such as the LHCb 7 and 8 TeV W/Z production data [44, 45], Z-rapidity data [46] at CDF run-II, CMS 8 TeV single inclusive jet production [47], and $t\bar{t}$ double differential p_T and y cross sections [48]. For this reason, these data were not included in the CT18 global analysis [1]. However, these measurements are very important in PDF determinations because they impose direct constraints on the gluon PDF at intermediate and small x. Therefore, we have re-analyzed these data by performing a series of attempts in which we tried to vary several parameters and settings in the fits. For example, we performed fits in which we varied the weight assigned to these data. We explored the impact of consistent initial-scale Q_0 variations, and tried different parametrizations for the gluon. We tried to vary the value of the c-quark mass m_c in both the pole and $\overline{\text{MS}}$ definitions, and we also analyzed the impact of scale variations in which the factorization scale is defined as $\mu_{\text{DIS}}(x) = A\sqrt{m_Q^2 + B^2/x^C}$ according to saturation models [49, 50]. This is expected to have similar impact to low-x resummation [51]. In addition, we explored phase-space suppression in fits in which we varied the S-ACOT-χ rescaling parameter $\chi = \zeta(1 + \zeta^\lambda m_Q^2/Q^2)$. A more detailed description of this study can be found in [52]. Here we only provide the results relative to two of the exercises that are described above. In Figure 3, we illustrate the impact on the CT18XNNLO gluon when a scan over the $\overline{\text{MS}}$ $\overline{m}_c(m_c)$ is performed (left panel), and the impact from scale variation where parameter B is varied in the $\mu_{\text{DIS}}(x)$ scale definition, while $A = 0.5$ and $C = 0.33$(right panel). The CT18X fit is a variant of CT18 where $\mu_{F,DIS} = \mu_{\text{DIS}}(x)$. Our preliminary findings indicate that the new c/b production measurements at HERA seem to prefer a harder gluon at intermediate and small x.

5 Conclusions

In this work we have studied HQ production at hadron colliders in recent global QCD analyses and analyzed their impact in collinear proton PDF determinations. We discussed S-ACOT-MPS, a new GMVFN scheme which we applied to pp collisions to describe c/b production at central and forward rapidities. In the near future, it will be technically possible to generate predictions within the S-ACOT-MPS scheme at NNLO with suitable K-factors (NNLO/NLO) at hand. Moreover, it is easy to extend S-ACOT-MPS to other HQ production processes. We explored the impact of $t\bar{t}$ single differential cross section measurements at the LHC at $\sqrt{S}$ =13 TeV on the CT18 PDFs. The impact of $t\bar{t}$ production at the LHC at $\sqrt{S}$ =13 TeV complements that of jet data on the gluon PDF, in particular in the large-x region. $t\bar{t}$ and jets overlap in the Q-x plane, but being matrix elements and phase-space suppression different, constraints on the gluon PDF may be placed at different values of x. Overall, the

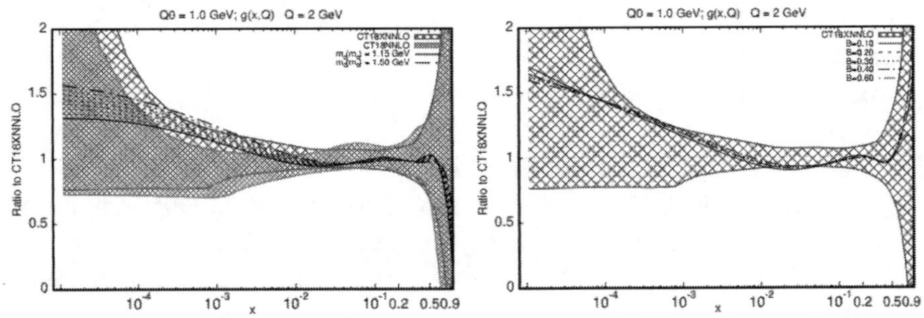

Figure 3. Ratio to the CT18XNNLO gluon PDF as a function of x at $Q = 2$ GeV. **Left**: scan over the c-quark mass $\overline{m}_c(m_c)$ while $\overline{m}_b(m_b) = 4.18$ GeV, and with $Q_0 = 1$ GeV. **Right**: scan over the B parameter in $\mu_{DIS}(x)$ with A = 0.5, C = 0.33, and $Q_0 = 1$ GeV. Error bands are shown at 90% confidence level for CT18NNLO (red) and CT18XNNLO (blue).

impact is found to be mild with a preference of softer gluon at large x. However, this may change when $t\bar{t}$ production in the lepton + jet channel at 13 TeV is included. We analyzed the most recent c/b combination at HERA. We find that these measurements are challenging and deserve more attention as they are very important for small-x dynamics. HQ production remains a critical process to constrain correlations between m_Q, α_s, and gluon-PDF.

Acknowledgments. The work of M.G. is supported by the NSF under Grant No. PHY-2112025. The work of K.X. at is supported in part by the U.S. DOE under Grant No. DE-SC0007914, the NSF under Grant No. PHY-1820760, and in part by PITT PACC. The work of P.N. is partially supported by the U.S. DOE under Grant No. DE-SC0010129. The work of C.P.Y. is partially supported by the NSF under Grant No. PHY-2013791. He is also grateful for the support from the Wu-Ki Tung endowed chair in particle physics.

References

[1] T.J. Hou et al., Phys. Rev. D **103**, 014013 (2021), `1912.10053`

[2] S. Bailey, T. Cridge, L.A. Harland-Lang, A.D. Martin, R.S. Thorne, Eur. Phys. J. C **81**, 341 (2021), `2012.04684`

[3] R.D. Ball et al. (NNPDF), Eur. Phys. J. C **82**, 428 (2022), `2109.02653`

[4] S. Alekhin, J. Blümlein, S. Moch, R. Placakyte, Phys. Rev. D **96**, 014011 (2017), `1701.05838`

[5] A. Accardi et al., Eur. Phys. J. A **52**, 268 (2016), `1212.1701`

[6] E.C. Aschenauer, S. Fazio, J.H. Lee, H. Mantysaari, B.S. Page, B. Schenke, T. Ullrich, R. Venugopalan, P. Zurita, Rept. Prog. Phys. **82**, 024301 (2019), `1708.01527`

[7] R. Abdul Khalek et al., Nucl. Phys. A **1026**, 122447 (2022), `2103.05419`

[8] A. Abada et al. (FCC), Eur. Phys. J. C **79**, 474 (2019)

[9] A. Abada et al. (FCC), Eur. Phys. J. ST **228**, 261 (2019)

[10] A. Abada et al. (FCC), Eur. Phys. J. ST **228**, 755 (2019)

[11] L.A. Anchordoqui et al., Phys. Rept. **968**, 1 (2022), `2109.10905`

[12] J.L. Feng et al. (2022), `2203.05090`

[13] J. Tang et al. (2015), `1507.03224`

[14] M.A.G. Aivazis, F.I. Olness, W.K. Tung, Phys. Rev. D **50**, 3085 (1994), `hep-ph/9312318`

[15] M.A.G. Aivazis, J.C. Collins, F.I. Olness, W.K. Tung, Phys. Rev. D **50**, 3102 (1994), hep-ph/9312319
[16] M. Buza, Y. Matiounine, J. Smith, W.L. van Neerven, Eur. Phys. J. C **1**, 301 (1998), hep-ph/9612398
[17] R.S. Thorne, R.G. Roberts, Phys. Rev. D **57**, 6871 (1998), hep-ph/9709442
[18] M. Krämer, F.I. Olness, D.E. Soper, Phys. Rev. D **62**, 096007 (2000), hep-ph/0003035
[19] W.K. Tung, S. Kretzer, C. Schmidt, J. Phys. G **28**, 983 (2002), hep-ph/0110247
[20] S.Alekhin, J.Blumlein, S.Klein, S.Moch, Phys. Rev. D **81**, 014032 (2010), 0908.2766
[21] S. Forte, E. Laenen, P. Nason, J. Rojo, Nucl. Phys. B **834**, 116 (2010), 1001.2312
[22] M.Guzzi, P. Nadolsky, H.L. Lai, C.P. Yuan, Phys.Rev. D **86**, 053005 (2012), 1108.5112
[23] K. Xie, Ph.D. thesis, Southern Methodist U. (2019)
[24] K. Xie, J.M. Campbell, P.M. Nadolsky, SciPost Phys. Proc. **8**, 084 (2022), 2108.03741
[25] I. Helenius, H. Paukkunen, JHEP **05**, 196 (2018), 1804.03557
[26] B.A. Kniehl, G. Kramer, I. Schienbein, H. Spiesberger, Phys. Rev. D **71**, 014018 (2005), hep-ph/0410289
[27] B.A. Kniehl, G. Kramer, I. Schienbein, H. Spiesberger, Eur. Phys. J. C **41**, 199 (2005), hep-ph/0502194
[28] R. Aaij et al. (LHCb), JHEP **12**, 026 (2017), 1710.04921
[29] G. Aad et al. (ATLAS), JHEP **01**, 033 (2021), 2006.09274
[30] A.M. Sirunyan et al. (CMS), JHEP **02**, 149 (2019), 1811.06625
[31] M. Czakon, D. Heymes, A. Mitov (2017), 1704.08551
[32] M. Czakon, D. Heymes, A. Mitov, Phys. Rev. Lett. **116**, 082003 (2016), 1511.00549
[33] T. Carli, D. Clements, A. Cooper-Sarkar, C. Gwenlan, G.P. Salam, F. Siegert, P. Starovoitov, M. Sutton, Eur. Phys. J. C **66**, 503 (2010), 0911.2985
[34] J.M. Campbell, R.K. Ellis, J. Phys. G **42**, 015005 (2015), 1204.1513
[35] J.M. Campbell, R.K. Ellis, W.T. Giele, Eur. Phys. J. C **75**, 246 (2015), 1503.06182
[36] M. Grazzini, S. Kallweit, M. Wiesemann, Eur. Phys. J. C **78**, 537 (2018), 1711.06631
[37] S. Catani, S. Devoto, M. Grazzini, S. Kallweit, J. Mazzitelli, JHEP **07**, 100 (2019), 1906.06535
[38] M. Czakon, D. Heymes, A. Mitov, JHEP **04**, 071 (2017), 1606.03350
[39] M. Czakon, A. Ferroglia, D. Heymes, A. Mitov, B.D. Pecjak, D.J. Scott, X. Wang, L.L. Yang, JHEP **05**, 149 (2018), 1803.07623
[40] A. Ablat et al., In preparation (2022)
[41] H. Abramowicz et al. (H1, ZEUS), Eur. Phys. J. C **78**, 473 (2018), 1804.01019
[42] A. Aktas et al. (H1), Eur. Phys. J. C **40**, 349 (2005), hep-ex/0411046
[43] H. Abramowicz et al. (H1, ZEUS), Eur. Phys. J. C **73**, 2311 (2013), 1211.1182
[44] R. Aaij et al. (LHCb), JHEP **08**, 039 (2015), 1505.07024
[45] R. Aaij et al. (LHCb), JHEP **05**, 109 (2015), 1503.00963
[46] T.A. Aaltonen et al. (CDF), Phys. Lett. B **692**, 232 (2010), 0908.3914
[47] V. Khachatryan et al. (CMS), JHEP **03**, 156 (2017), 1609.05331
[48] A.M. Sirunyan et al. (CMS), Eur. Phys. J. C **77**, 459 (2017), 1703.01630
[49] K.J. Golec-Biernat, M. Wusthoff, Phys. Rev. D **59**, 014017 (1998), hep-ph/9807513
[50] F. Caola, S. Forte, J. Rojo, Phys. Lett. B **686**, 127 (2010), 0910.3143
[51] R.D. Ball, V. Bertone, M. Bonvini, S. Marzani, J. Rojo, L. Rottoli, Eur. Phys. J. C **78**, 321 (2018), 1710.05935
[52] M. Guzzi, P. Nadolsky, K. Xie, SciPost Phys. Proc. **8**, 164 (2022), 2108.01791

Low-x behavior of gluon pseudo- and quasi-PDFs

Giovanni Antonio Chirilli[1,*]

[1]Institut für Theoretische Physik, Universität Regensburg, Universitätsstrasse 31, D-93040 Regensburg, Germany

Abstract. I will present the formalism one may use to study the behavior of the Ioffe-time distribution at large longitudinal distances and show that the pseudo-PDF and quasi-PDF are very different at low-x. Using light-ray operators, I will also show that the higher twist corrections of the quasi-PDF come in not as inverse powers of P but as inverse powers of x_BP.

1 Introduction

Lattice gauge theory, being formulated in Euclidean space, does not allow direct calculations of the PDFs. A way to overcome this difficulty is to consider equal-time correlators and to perform the Lattice analysis in coordinate space through the Ioffe-time distributions. From the Fourier transform of the Ioffe-time distribution, one may obtain the quasi-PDF [1] and the pseudo-PDF [2].

In references [1] it was suggested that to extract the PDFs from the quasi-PDFs, one has to take the infinite-momentum limit $P \to \infty$ with higher twist corrections expected to come in as inverse powers of the longitudinal momentum. Instead, from the pseudo-PDFs, instead, the PDFs are extracted in the short (longitudinal) distance limit [2]. Since lattice calculations provide values of the Ioffe-time distributions only for a limited range of the distance separating the bi-local operators, in order to perform the Fourier transform for the quasi-PDF or the pseudo-PDF, one has to extrapolate the large-distance behavior [3].

In Ref [4] we obtained the behavior of the Ioffe-time distribution at large longitudinal distances as well as the low-x_B behavior of the quasi-PDF and pseudo-PDF. We also calculated, using the light-ray operators, obtained as analytic continuation of local twist-two operators, [5–9], the behavior of the leading twist (LT) and next-to-leading twist (NLT) contributions for the gluon Ioffe-time distribution at large longitudinal distances as well as for the pseudo-PDF and quasi-PDF at low-x_B.

2 Pseudo-PDF and quasi-PDF from Ioffe-time distribution

The pseudo-PDF and the quasi-PDF are obtained from two different Fourier transforms of the Ioffe-time distributions. As we will see, these differences will lead to two completely different behaviors at small-x_B. To properly define the Ioffe-time distribution, we will make use of a Lorentz indexes

*e-mail: giovanni.chirilli@ur.de

decomposition of the gluon bi-local operator [10, 11] using the proton momentum P^μ, the coordinate z^μ and the antisymmetric tensor $g^{\mu\nu}$ as follow

$$\begin{aligned} M_{\mu\alpha;\lambda\beta} &\equiv \langle P|G_{\mu\alpha}(z)[z,0]G_{\lambda\beta}(0)|P\rangle \\ &= I_{1\mu\alpha;\lambda\beta}\mathcal{M}_{pp} + I_{2\mu\alpha;\lambda\beta}\mathcal{M}_{zz} + I_{3\mu\alpha;\lambda\beta}\mathcal{M}_{zp} + I_{4\mu\alpha;\lambda\beta}\mathcal{M}_{pz} + I_{5\mu\alpha;\lambda\beta}\mathcal{M}_{ppzz} + I_{6\mu\alpha;\lambda\beta}\mathcal{M}_{gg}\,. \end{aligned} \tag{1}$$

where the explicit expression of tensor structures I_i with $i = 1, \ldots, 6$ can be found in [10]. The amplitudes $\mathcal{M}$ are functions of the invariants z^2 and $z \cdot P$.

The light-cone gluon distribution is determined from $g_\perp^{\alpha\beta} M_{+\alpha;+\beta}(z^+, P)$ with z taken on the light-cone and proportional to the invariant amplitude $\mathcal{M}_{pp}$

$$g_\perp^{\alpha\beta} M_{+\alpha;+\beta}(z^+, P) = 2(P^-)^2\mathcal{M}_{pp}(z\cdot p, 0)\,. \tag{2}$$

As usual, we define the light-cone coordinates $x^\pm = \frac{x^0 \pm x^3}{\sqrt{2}}$, and light-cone vectors n^μ and n'^μ such that $n \cdot n' = 1, n \cdot x = x^-$ and $n' \cdot x = x^+$.

The relation between the gluon PDF $D_g(x_B)$ and the amplitude $\mathcal{M}_{pp}$ si given by

$$\mathcal{M}_{pp}(z\cdot P, 0) = \frac{1}{2}\int_{-1}^{1} dx_B\, e^{iz\cdot P\, x_B}\, x_B D_g(x_B)\,. \tag{3}$$

The distribution $\mathcal{M}_{pp}(z \cdot P, z^2)$ is the gluon Ioffe-time distribution with $z\cdot P$ the Ioffe-time. The gluon Ioffe-time distribution is given in terms of the zeroth and transverse components as

$$M_{0i;i0} + M_{ji;ij} = 2P_0^2\mathcal{M}_{pp}\,. \tag{4}$$

Since at high energy (Regge) limit the transverse components are suppressed while the 0th and 3rd components cannot be distinguished, calculating the behavior of the left-hand-side (LHS) of (2), will be equivalent, at high-energy, to calculating the behavior of LHS of (4).

The momentum-density pseudo-PDF is defined as the Fourier transform with respect to $z \cdot P$, that is a Fourier transform with respect to P keeping its orientation fixed. So, we define the gluon pseudo-PDF as

$$G_p(x_B, z^2) = \int \frac{d\varrho}{2\pi}\, e^{-i\varrho\, x_B}\mathcal{M}_{pp}(\varrho, z^2)\,, \tag{5}$$

where we defined $\varrho \equiv z \cdot P$.

The momentum-density quasi-PDF is defined, instead, as the Fourier transform with respect to z^μ keeping its orientation fixed. Let us define the vector $\xi^\mu = \frac{z^\mu}{|z|}$, and $P_\xi = P \cdot \xi$. The quasi-PDF is then defined as

$$G_q(x_B, P_\xi) = P_\xi \int \frac{d\varsigma}{2\pi}\, e^{-i\varsigma P_\xi x_B}\mathcal{M}_{pp}(\varsigma P_\xi, \varsigma^2)\,. \tag{6}$$

We will calculate the large distance (large ϱ) behavior of the Ioffe-time distribution $\mathcal{M}_{pp}(\varrho, z^2)$, and the low-$x_B$ behavior of the gluon pseudo-PDF (5), and of the gluon quasi-PDF (6).

3 Low-x_B behavior from high-energy OPE

To get the low-x_B behavior of the gluon pseudo- and quasi-PDF we should first consider the behavior of the Ioffe-time distribution for large values of the ϱ parameter using the high-energy operator product expansion (HE-OPE) (see [12] for review).

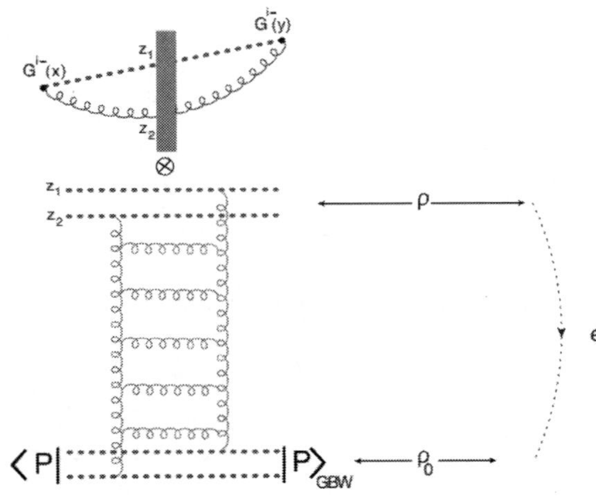

Figure 1. Diagrammatic representation of the HE-OPE applied to the gluon non-local operator with "quasi-PDF frame".

The Ioffe-time gluon distribution defined in the previous section is

$$\langle P|G^{ai-}(z)[z,0]G_i^{b-}(0)|P\rangle = 2(P^-)^2\mathcal{M}_{pp}(\varrho, z^2)\,. \tag{7}$$

In the high-energy (Regge) limit forward matrix elements of $U_xU_y^\dagger$ operator are divergent because they contain an unrestricted integration along the n^μ direction, so we have

$$\begin{aligned}&2\pi\delta(\epsilon^-)\langle P|G^{a\,i-}(x^+,x_\perp)[nx^+ + x_\perp, 0]^{ab}G_i^{b-}(0)|P\rangle\\ &= \int_{-\infty}^{+\infty} dy^+\langle P|G^{a\,i-}(x^+ + y^+, x_\perp)[n(x^+ + y^+) + x_\perp, ny^+]^{ab}G_i^{b-}(y^+)|P + \epsilon^- n\rangle\,.\end{aligned} \tag{8}$$

The HE-OPE procedure requires that one calculates the coefficient function (Impact Factor) first, and then the resummation of the relevant logarithms is obtained through the evolution with respect to the rapidity parameter of the resulting operators. The aforementioned procedure is schematically given in Fig 1.

Using the gluon propagator in the high-energy limit in coordinate space at $x^+ > 0 > y^+$ [13], the top part of Fig. 1, which represent the coefficient function of the HE-OPE, is

$$\begin{aligned}&\langle G^{a\,i-}(x^+,x_\perp)[nx^+ + x_\perp, ny^+ + y_\perp]^{ab}G_i^{b-}(y^+,y_\perp)\rangle_{\text{Fig.1}}\\ &\overset{y^+<0<x^+}{=} \int \frac{d^2z_2}{4\pi^3}\left[12\,\frac{-2x^+y^+(x-z_2,y-z_2)^2 + x^+y^+(x-z_2)_\perp^2(y-z_2)_\perp^2}{[x^+(y-z_2)_\perp^2 - y^+(x-z_2)_\perp^2]^4}\right]U^{ab}_{z_2}U^{ab}_{z_1}\,,\end{aligned} \tag{9}$$

Result (9) is proportional to two Wilson lines in the adjoint representation, one at point $z_{2\perp}$, as a result of the gluon propagation in the shock-wave external field, and the other one at point $z_{1\perp}$, which is the point at which the shock-wave intersects the gauge link $[nx^+ + x_\perp, ny^+ + y_\perp]^{ab}$. The point z_1 can actually be anywhere between point $nx^+ + x_\perp$ and $ny^+ + y_\perp$, so we parametrize the straight line between these two points as $x_u = ux_\perp + \bar{u}y_\perp = z_{1\perp}$, with $u = \frac{|y^+|}{\Delta^+}$, $\bar{u} = \frac{x^+}{\Delta^+}$ and $\Delta^+ = x^+ - y^+$, and at the end we will have to integrate over the parameter u. However, for technical reason, as it will be clear later, we will continue to call such a point $z_{1\perp}$.

We have to include the case $\theta(x^+y^+)$ and consider forward matrix elements performing the substitution $U^{ab}_{z_2}U^{ab}_{z_1}$ with $-2N_c^2\mathcal{U}(z_{12})$ where $\mathcal{U}(z_{12}) = 1 - \frac{1}{N_c}\mathrm{tr}\{U_{z_1}U^\dagger_{z_2}\}$, and tr denotes the trace in the fundamental representation. To perform the last step of the HE OPE we have to convolute the impact factor with the solution of the evolution equation of the Wilson-line operators, and use the GBW model for the initial condition to arrive at (see Ref. [4] for details)

$$\mathcal{M}_{pp}(\varrho, z^2) = \frac{3N_c^2}{4\pi^3}\frac{Q_s\sigma_0}{|z|\varrho}\int d\nu \left(\frac{2\varrho^2}{z^2M_N^2} + i\epsilon\right)^{\frac{\aleph(\gamma)}{2}} \frac{\gamma\Gamma^2(1-\gamma)\Gamma^3(1+\gamma)}{\Gamma(2+2\gamma)}\left(\frac{|z|Q_s}{2}\right)^{2i\nu}. \tag{10}$$

Using the saddle-point approximation technique we have

$$\mathcal{M}_{pp}(\varrho, z^2) \simeq \frac{3N_c^2}{128\,\varrho}\frac{Q_s\,\sigma_0}{|z|}\left(\frac{2\,\varrho^2}{z^2M_N^2} + i\epsilon\right)^{\bar{\alpha}_s 2\ln 2} \frac{e^{-\frac{\ln^2\frac{Q_s|z|}{2}}{7\zeta(3)\bar{\alpha}_s\ln\left(\frac{2\varrho^2}{z^2M_N^2}+i\epsilon\right)}}}{\sqrt{7\zeta(3)\bar{\alpha}_s\ln\left(\frac{2\varrho^2}{z^2M_N^2} + i\epsilon\right)}}. \tag{11}$$

In eq. (11), the first thing we should notice is that the logarithms resummed by BFKL are $\bar{\alpha}_s\ln\left(\frac{\sqrt{2}\varrho}{|z|M_M}\right)$ which, given that we are using $\bar{\alpha}_s = 0.2$, $|z| = 0.5\,\mathrm{GeV}^{-1}$, and $M_N = 1\,\mathrm{GeV}$, are of order 1 starting from $\varrho \sim 10$. The second is that the Ioffe-time ϱ acts like a rapidity parameter. We evolve the distribution in ϱ as long as $\bar{\alpha}_s\ln\left(\frac{\sqrt{2}\varrho}{|z|M_M}\right)$ is of order 1, i.e. we start the evolution at large values of ϱ and end at smaller ones. The dipole at the smallest value of ϱ is evaluated in the GBW model. We stopped the evolution at $\varrho = 10$ (see figure 1).

Considering the case $0 < \frac{Q_s^2|z|^2}{4} < 1$, which corresponds to the typical DIS region, taking the first two leading residues, we obtain the twist expansion as

$$\mathcal{M}_{pp}(\varrho, z^2) = \frac{N_c^2}{8\pi^2\bar{\alpha}_s}\frac{Q_s^2\sigma_0}{\varrho}\left(\frac{4\bar{\alpha}_s\left|\ln\frac{Q_s|z|}{2}\right|}{\ln\left(\frac{2\varrho^2}{z^2M_N^2} + i\epsilon\right)}\right)^{\frac{1}{2}} I_1(\tilde{t})\left(1 + \frac{Q_s^2|z|^2}{5}\right) + O\left(\frac{Q_s^4|z|^4}{16}\right), \tag{12}$$

with

$$\tilde{t} = \left[4\bar{\alpha}_s\left|\ln\frac{Q_s|z|}{2}\right|\ln\left(\frac{2\,\varrho^2}{z^2M_N^2} + i\epsilon\right)\right]^{\frac{1}{2}}. \tag{13}$$

In Fig. 2 we compare the coordinate space result of the high-energy (large ϱ) behavior of the gluon distribution with BFKL resummation, eq. (10), the LT term and LT plus NLT. We notice that the curves plotted in figure 2 are very slowly varying functions for large values of ϱ.

3.1 Gluon pseudo- and quasi-PDF with BFKL resummation

The pseudo-PDF is obtained from eq. (5) performing the Fourier transform of eq. (10) with respect to ϱ

$$G_p(x_B, z^2) = \frac{3N_c^2}{4\pi^3}\frac{Q_s\sigma_0}{|z|}\int\frac{d\varrho}{2\pi\varrho}e^{-i\varrho x_B}\int d\nu\left(\frac{2\varrho^2}{z^2M_N^2} + i\epsilon\right)^{\frac{\aleph(\gamma)}{2}}\frac{\gamma\Gamma^2(1-\gamma)\Gamma^3(1+\gamma)}{\Gamma(2+2\gamma)}\left(\frac{|z|Q_s^2}{2}\right)^{2i\nu}. \tag{14}$$

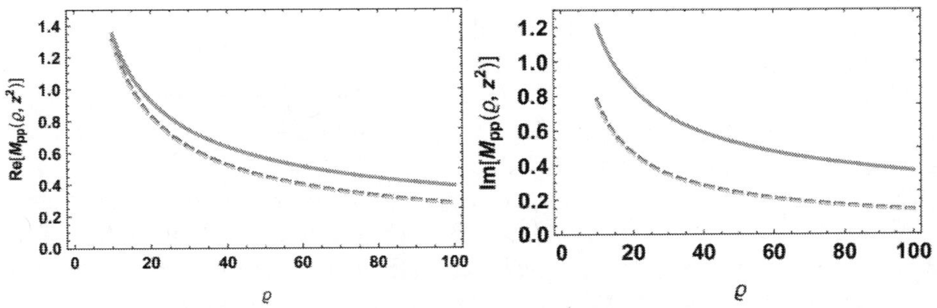

Figure 2. In the left panel, the orange curve is the numerical evaluation of the real part of eq. (10) (all twist resummed by BFKL), the green dashed curve is the real part of the LT term only, while the red dashed one is the real part of LT+NLT result. In the left panel we plot the imaginary parts. The range of ϱ is from 10 to 100.

Performing the Fourier transform first, and taking the saddle point approximation for the integration over the ν parameter at the end, we have

$$G_{\rm p}(x_B, z^2) \simeq -i\frac{3N_c^2}{128\pi}\frac{Q_s\sigma_0}{|z|}\frac{\Gamma(\bar{\alpha}_s 4\ln 2)\sin(\frac{\pi}{2}\bar{\alpha}_s 4\ln 2)}{\sqrt{7\zeta(3)\bar{\alpha}_s\ln\left(\frac{2}{x_B^2 z^2 M_N^2}+i\epsilon\right)}}\mathrm{sign}(x_B)$$
$$\times\ \exp\left\{\frac{-\ln^2\frac{Q_s|z|}{2}}{7\zeta(3)\bar{\alpha}_s\ln\left(\frac{2}{x_B^2 z^2 M_N^2}+i\epsilon\right)}\right\}\left(\frac{2}{x_B^2 z^2 M_N^2}+i\epsilon\right)^{\bar{\alpha}_s 2\ln 2}. \quad (15)$$

Next, we perform the Fourier transform to obtain the leading and next-to-leading twist corrections for the pseudo-PDF. The result is (see Ref. [4] for details)

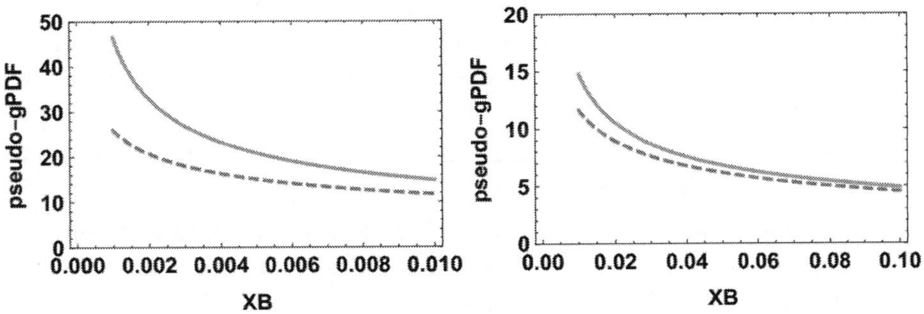

Figure 3. In the left panel we plot the pseudo-PDF with BFKL resummation (orange curve), and the LT (green dashed curve) and LT+NLT (red dashed curve) of the pseudo-PDF result; The value of x_B is between 0.001 to 0.02. In the right panel we plot the same curves in a different range of x_B i.e. between 0.01 to 0.1.

$$G_{\rm p}(x_B, z^2) = \frac{N_c^2\, Q_s^2\sigma_0}{16\bar{\alpha}_s\pi^3}\frac{1}{2\pi i}\int_{1-i\infty}^{1+i\infty}\frac{d\omega}{\omega}\left(\frac{2}{x_B^2|z|^2 M_N^2}\right)^{\frac{\omega}{2}}\left(\frac{4}{Q_s^2|z|^2}\right)^{\frac{\bar{\alpha}_s}{\omega}}\left(1+\frac{Q_s^2|z|^2}{5}\right)+O\left(\frac{Q_s^4|z|^4}{16}\right). \quad (16)$$

In Fig. 3, we plot the gluon pseudo-PDF with the BFKL resummation eq. (15) (orange curve), and the LT term and the LT plus NLT (green dashed and red dashed curve respectively). We observe that the BFKL resummation result agrees with the LT and LT+NLT result in the region of moderate x_B. When we move into the low-x_B region, we notice a strong disagreement which confirms the necessity of a $\ln \frac{1}{x_B}$ resummation represented by the BFKL eq.

For the the quasi-PDF case, we need to perform a Fourier transform of eq. (10) again, but with respect to a different parameter. Let us introduce the real parameter ς such that $-z^2 = \varsigma^2 > 0$, and the four-vector $\xi^\mu \equiv \frac{z^\mu}{|z|} = \frac{z^\mu}{|\varsigma|}$ with $|z| = \sqrt{-z^2}$. The quasi-PDF are defined as the Fourier transform of the coordinate-space gluon distributions (10) keeping, this time, the orientation of the vector z^μ fixed.

In the high-energy limit, where $x^+ \to \infty$ and $x^- \to 0$, we can not distinguish between the zeroth and the third component. We can then rewrite $LP^- = z \cdot P = \varsigma P_\xi$ with $P_\xi \equiv P \cdot \xi = P^-$ because the ξ^μ vector, in the limit we are considering, selects the minus component of the P^μ vector. Moreover, in coordinate space, in the high-energy limit, every fields depends only on x^+ and $x_\perp$, so, to restore the x^- components we make the substitution $(x-y)_\perp^2 = \Delta_\perp^2 \to -z^2 = \varsigma^2$. In the quasi-PDF notation the gluon bi-local operator becomes

$$\frac{\xi_\mu \xi_\nu}{2P_\xi^2}\langle P|G^{a\,\alpha\mu}(\varsigma)[\varsigma,0]^{ab}G^{b\,\nu}_{\alpha}(0)|P\rangle$$
$$= \frac{3N_c^2}{4\pi^3}\frac{Q_s\sigma_0}{\varsigma^2 P_\xi}\int d\nu \left(-\frac{2P_\xi^2}{M_N^2}+i\epsilon\right)^{\frac{\aleph(\gamma)}{2}} \frac{\gamma\Gamma^2(1-\gamma)\Gamma^3(1+\gamma)}{\Gamma(2+2\gamma)}\left(\frac{Q_s^2\varsigma^2}{4}\right)^{i\nu}. \tag{17}$$

Using the definition of the gluon quasi-PDF eq. (6), performing the integration over ς (recall that we are using $\gamma = \frac{1}{2}+i\nu$) and then taking the saddle point approximation for the integration over ν, we get

$$G_{\rm q}(x_B,P_\xi) \simeq -\frac{3N_c^2}{256}Q_s\sigma_0 P_\xi|x_B|\left(-\frac{2P_\xi^2}{M_N^2}+i\epsilon\right)^{\bar\alpha_s 2\ln 2}\frac{e^{-\frac{\ln^2\frac{Q_s}{2P_\xi|x_B|}}{7\bar\alpha_s\zeta(3)\ln\left(-\frac{2P_\xi^2}{M_N^2}+i\epsilon\right)}}}{\sqrt{7\zeta(3)\bar\alpha_s\ln\left(-\frac{2P_\xi^2}{M_N^2}+i\epsilon\right)}}. \tag{18}$$

We notice that, in the gluon quasi-PDF case, the usual exponentiation of the pomeron intercept (the LO BFKL eigenvalues $\aleph(\gamma)$), which indicates the resummation of large logarithms of $\frac{1}{x_B}$, is absent. The exponentiation of the pomeron intercept that we instead have in eq. (18) suggests that the logarithms resummed by BFKL equation are $\alpha_s \ln\left(\frac{2P_\xi^2}{M_N^2}\right)$ rather than the usual $\alpha_s \ln\frac{1}{x_B}$ as in the pseudo-PDF case.

To obtain the leading and next-to-leading twist gluon quasi-PDF we have to perform the Fourier ala quasi-PDF and make the inverse Mellin transform at the end. So, using the quasi-PDF variables ς and $P_\xi$and performing the integration over ς we get

$$G_{\rm q}(x_B,P_\xi) = -\frac{N_c^2 Q_s^2\sigma_0}{16\bar\alpha_s^2\pi^3}\frac{1}{2\pi i}\int_{1-i\infty}^{1+i\infty} d\omega\left(-\frac{2P_\xi^2}{M_N^2}+i\epsilon\right)^{\frac{\omega}{2}}\left(-\frac{4P_\xi^2 x_B^2}{Q_s^2}+i\epsilon\right)^{\frac{\bar\alpha_s}{\omega}}$$
$$\times\left(\omega + \frac{2\bar\alpha_s Q_s^2}{5P_\xi^2 x_B^2} + O\left(\frac{\bar\alpha_s}{\omega}\right)\right). \tag{19}$$

where we have used the approximation $\bar\alpha_s \ll \omega \ll 1$. Equation (19) is the gluon quasi-PDF up to next-to-leading twist contribution. What one should notice in result (19) is the strong enhancement of the NLT term with respect to the LT term due to the $\frac{1}{P_\xi^2 x_B^2}$ factor.

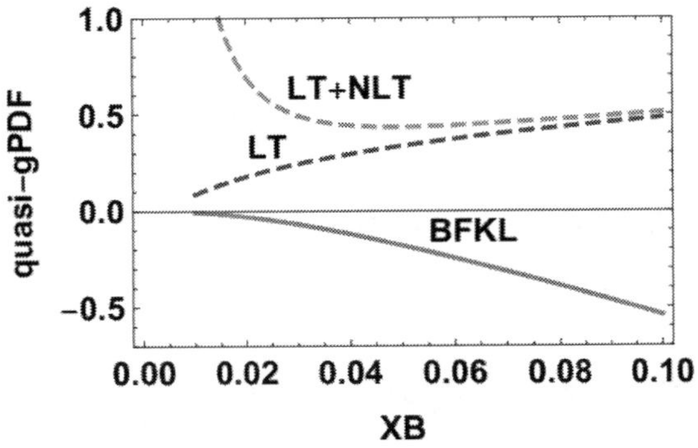

Figure 4. Here we plot the quasi-PDF with BFKL rsummation (real part), the LT and the LT+NLT (real part). The curves are plotted in the range $x_B \in [0.01, 0.1]$ with $P_\xi = 4\,\text{GeV}$.

In Fig. 4, using $P_\xi = 4\,\text{GeV}$, we plot the quasi-PDF with BFKL resummation (red curve), and the LT (blue dashed curve), and LT+NLT (magenta dashed curve) contributions. The behavior of the three curves is different than the usual low-x_B behavior of gluon distributions which we, instead, observed for the pseudo-PDF distribution in Fig. 3.

4 Conclusions

From the above analysis we may conclude that the pseudo-PDF and the quasi-PDF have a very different behavior at low-x_B. We can trace the physical origin of the difference between the two distributions to the two different Fourier transforms under which they are defined. Indeed, in the pseudo-PDF case, the scale is the resolution, i.e. the square of the length of the gauge link separating the bi-local operator. On the other hand, in the quasi-PDF case, the scale is the energy, i.e. the momentum of the hadronic target (the nucleon) projected along the direction of the gauge link. Therefore, if on one hand the pseudo-PDF has the typical behavior of the gluon distribution at low-x_B (see figure 3), on the other hand, the quasi-PDF has a rather unusual low-x_B behavior (see figure 4).

References

[1] X. Ji, Phys. Rev. Lett. **110**, 262002 (2013), 1305.1539
[2] A.V. Radyushkin, Phys. Rev. D **96**, 034025 (2017), 1705.01488
[3] V. Braun, D. Müller, Eur. Phys. J. C **55**, 349 (2008), 0709.1348
[4] G.A. Chirilli, JHEP **03**, 064 (2022), 2111.12709
[5] I. Balitsky, Int. J. Mod. Phys. Conf. Ser. **25**, 1460024 (2014)
[6] I. Balitsky, V. Kazakov, E. Sobko (2013), 1310.3752
[7] I. Balitsky, V. Kazakov, E. Sobko, Phys. Rev. D **93**, 061701 (2016), 1506.02038
[8] I. Balitsky, V. Kazakov, E. Sobko (2015), 1511.03625
[9] I. Balitsky, JHEP **04**, 042 (2019), 1812.07044
[10] I. Balitsky, W. Morris, A. Radyushkin, Phys. Lett. B **808**, 135621 (2020), 1910.13963
[11] I. Balitsky, W. Morris, A. Radyushkin (2021), 2111.06797
[12] I. Balitsky (2001), hep-ph/0101042
[13] I. Balitsky, G.A. Chirilli, Int. J. Mod. Phys. **A25**, 401 (2010), 0911.5192

Hard exclusive meson production in muon scattering at COMPASS

Andrzej Sandacz[1,*,**], on behalf of the COMPASS Collaboration

[1]National Institute for Nuclear Research, Division of High Energy Physics,
ul. Pasteura 7 Warsaw, PL 02-093, Poland

Abstract. We report on measurements of hard exclusive muoproduction of ρ^0, ω and π^0 mesons on the proton at COMPASS using 160 GeV/c polarized μ^+ and μ^- beams scattering off a liquid hydrogen target. The extracted Spin Density Matrix Elements for vector mesons indicate a violation of s-channel helicity conservation for transitions of transversely polarized virtual photons to longitudinally polarized vector mesons, $\gamma_T \rightarrow V_L$, where V denotes ρ^0 or ω meson. Additionally, for ρ^0 production we observe a dominant contribution of natural-parity-exchange transitions and a small contribution of unnatural-parity-exchange transitions. On the contrary, the contribution of unnatural-parity-exchange for ω production is significant and it decreases with increasing W, being still non-negligible at the largest W values accessible at COMPASS. Further, we present also the absolute value of the differential cross section for exclusive π^0 production and its dependence on squared four-momentum transfer t and azimuthal angle ϕ between the scattering plane and π^0 production plane. The results provide an important input for modeling Generalized Parton Distribution (GPDs), and in particular to constrain chiral-odd (transversity) GPDs.

1 Introduction

Exclusive meson production in lepton-nucleon scattering is a convenient tool for studying the production mechanism and, in a model-dependent way, the structure of the nucleon. Such process is an important tool to access Generalised Parton Dostributions (GPDs) [1–3]. The theoretical framework of GPDs offers a novel description of the nucleon's partonic structure and contain a wealth of new information. The GPDs provide a description of the nucleon as an extended object, sometimes referred to as a 3-dimensional 'nucleon tomography' [4], which correlates transverse spatial and longitudinal momentum degrees of freedom. The GPDs also allow access to such a fundamental property of the nucleon as the orbital angular momentum of quarks [2]. The mapping of the nucleon GPDs requires a comprehensive program of measuring various hard processes in a broad kinematic range, in particular Hard Exclusive Meson Production (HEMP).

The phenomenological GPD model that presently provides predictions for all types of experimental observables is the one of Goloskokov and Kroll [5–7] (GK model). It is a so-called handbag model that contains terms beyond the collinear factorization, which was

*Work supported by the Polish NCN Grant 2020/37/B/ST2/01547.
**e-mail: andrzej.sandacz@ncbj.gov.pl

proven rigorously only for longitudinally polarized virtual photons. The model is applicable for a comparison with results of existing measurements, also without transverse vs. longitudinal separation.

2 Experimental set-up

The data used for this analysis were recorded in 2012 during a four weeks pilot run for the COMPASS GPD program. COMPASS is a fixed-target experiment situated at the high-intensity M2 beam-line of the CERN SPS. It can deliver either hadron or naturally polarised muon beams of a given charge in the energy range between 50 and 280 GeV. For the GPD program the data are collected with the muon beam nominal energy of 160 GeV and with muons of both polarities: μ^+ (μ^-) with polarisation of about -0.8 (+0.8).

The COMPASS apparatus [8] consists of a two-stage forward spectrometer comprising various tracking detectors, electromagnetic and hadronic calorimeters, muon identification detectors, and a ring imagining Cherenkov counter grouped around two dipole magnets SM1 and SM2. For the GPD program the set-up was complemented by installing the recoil proton detector CAMERA around the 2.5 m long liquid hydrogen target and the large-angle electromagnetic calorimeter ECAL0. The recoil proton detection is based on the ToF measurement. The calorimeter ECAL0 that is situated just downstream of CAMERA allows to extend the accessible kinematic domain for Deeply Virtual Compton Scattering and exclusive π^0 production towards higher values of Bjorken x_{Bj} and to improve hermeticity for detection of exclusive events. By adding these detectors COMPASS has been converted into a facility measuring exclusive reactions within kinematic domain from $x \sim 0.01$ to ~ 0.15, which cannot be explored at any other existing or planned facility in the near future.

3 Results on exclusive vector meson production

3.1 Selections of events from exclusive vector meson production

In the following we present in more detail selections of events for ρ^0 data, which are preliminary, while analogous selections for published ω results can be found in Ref. [9]. An event to be accepted for the analysis is required to have an incident muon track, a scattered muon track and only two hadron tracks with opposite charges. Most of other applied event selections are either the same or similar to those used for the published COMPASS results on Spin Density Matrix Elements (SDMEs) for exclusive ω production [9]. As the recoil proton detector restricts the kinematic coverage towards low momentum transfers, it was not used in this analysis. Instead, in order to enhance the fraction of events with exclusively produced ρ^0 mesons, the missing energy

$$E_{\mathrm{miss}} = \frac{M_{\mathrm{X}}^2 - M^2}{2M} \tag{1}$$

is constrained by $-2.5\ \mathrm{GeV} < E_{\mathrm{miss}} < 2.5\ \mathrm{GeV}$ to take into account the experimental resolution. Here, M is the proton mass, $M_{\mathrm{X}}^2 = (p + q - p_{\pi^+} - p_{\pi^-})^2$ is the missing mass squared, and p, q, p_{π^+} and p_{π^-} are the four-momenta of target nucleon, virtual photon and each of two pions.

Selected kinematic distributions are shown in Fig. 1. These are the invariant mass of the $\pi^+\pi^-$ system and E_{miss} distributions. A clear ρ^0 signal is observed. The distribution for the data is compared to the result of the HEPGEN Monte Carlo (MC), which generates only exclusive production of ρ^0 resonance. An observed skewing in the invariant mass distribution for the data with respect to MC is due to a small contribution of exclusive non-resonant pion-pair production and its interference with the resonant production. Within the selected range

0.5 GeV/c^2 < $M_{\pi^+\pi^-}$ < 1.1 GeV/c^2 the contribution from non-resonant production, including the interference term, is equal to about 3% and thus is neglected.

In Fig. 1 (right) the $E_{\rm miss}$ distributions for the data and the background, which was evaluated using LEPTO Monte Carlo (MC), are shown. The amount of background in the signal window $|E_{\rm miss}|$ < 2.5 GeV depends on Q^2, W and p_T^2, and when averaged over the total kinematic range it amounts to 17%. The background-corrected data are represented by the shaded histogram.

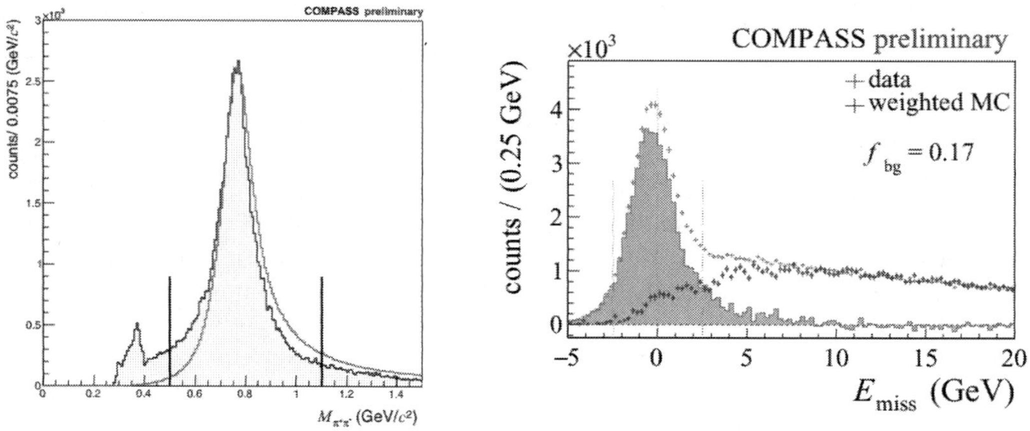

Figure 1. **Left:** Distribution of the $\pi^+\pi^-$ invariant mass for the experimental data (shaded histogram) and for the HEPGEN MC (open histogram). The invariant mass distribution from HEPGEN is normalized to the data in the region 0.75 GeV/c^2 < $M_{\pi\pi}$ < 0.77 GeV/c^2. The vertical lines indicate the applied limits. **Right:** The missing-energy distribution from experimental data compared to the distribution of SIDIS events from a LEPTO MC simulation. The MC distribution is normalized to the data in the region 7 GeV < $E_{\rm miss}$ < 20 GeV. The background-corrected distribution for the data is shown as shaded histogram. The vertical lines at $|E_{\rm miss}|$ = 2.5 GeV indicate the limits of the exclusive region. Each LEPTO MC event is reweighted by a $E_{\rm miss}$-dependent weight that is calculated using both experimental and simulated data with same-charge hadron pairs.

The selected ρ^0 sample consists of 23785 events for μ^+ and 28472 for μ^- beam. For comparison we quote the published total number of selected events for exclusive ω production [9], which is 3060 events. The lower number for the ω channel is due to the cross section being about an order of magnitude smaller and a lower acceptance than those for ρ^0 production. The measurements for both channels cover the kinematic region 5.0 GeV/c^2 < W < 17.0 GeV/c^2, 1.0 (GeV/c)^2 < Q^2 < 10.0 (GeV/c)^2 and 0.01 (GeV/c)^2 < $p_{\rm T}^2$ < 0.5 (GeV/c)^2.

3.2 Spin Density Matrix Elements

The vector meson spin density matrix elements are proportional to the bilinear combinations of helicity amplitudes $F_{\lambda_V\lambda'_N\lambda_\gamma\lambda_N}$ [10]. These amplitudes describe the transition of a virtual photon with helicity λ_γ to a vector meson with helicity λ_V, while λ_N and λ'_N are the helicities of the nucleon in the initial and final states, respectively.

The SDMEs can be determined from the fit of three-dimensional angular distribution describing the production and decay of a vector meson to the corresponding experimental distribution. Definitions of angles and reference frames are given in Refs. [9, 11].

SDMEs for exclusive ρ^0 production extracted for the entire kinematic region with $\langle Q^2 \rangle$ = 2.4 (GeV/c)^2, $\langle W \rangle$ = 9.9 GeV and $\langle p_T^2 \rangle$ = 0.18 (GeV/c)^2 are presented in Fig. 2 (left). These SDMEs are divided into five classes corresponding to different helicity transitions. The main terms in the expressions of SDMEs by helicity amplitudes of class A , correspond to the transitions from longitudinal virtual photons to longitudinal vector mesons, $\gamma_L^* \rightarrow V_L$, and from transverse virtual photons to transverse vector mesons, $\gamma_T^* \rightarrow V_T$. In class B the dominant terms correspond to the interference of the above two transitions. The main terms of SDMEs in classes C, D and E are proportional to the amplitudes describing $\gamma_T^* \rightarrow V_L$, $\gamma_L^* \rightarrow V_T$ and $\gamma_T^* \rightarrow V_{-T}$ transitions, respectively.

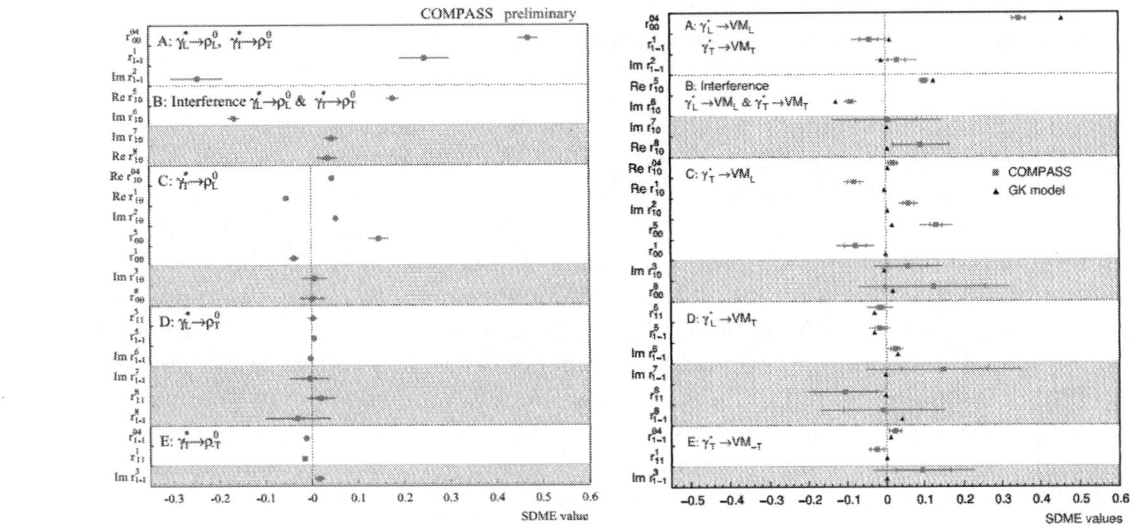

Figure 2. Left: The 23 SDMEs for exclusive ρ^0 leptoproduction extracted in the entire COMPASS kinematic region with $\langle Q^2 \rangle$ = 2.4 (GeV/c)^2, $\langle W \rangle$ = 9.9 GeVc^2, $\langle p_T^2 \rangle$ = 0.18 (GeV/c)^2. The inner error bars represent the statistical uncertainties, while the outer ones indicate the statistical and systematic uncertainties added in quadrature. Unpolarized (polarized) SDMEs are displayed in the unshaded (shaded) areas. **Right:** The measured SDMEs for exclusive ω production compared to predictions of the GPD model of Goloskokov and Kroll [7]. The calculations are obtained for Q^2 = 2.0 (GeV/c)^2, W = 7.5 GeV/c^2 and p_T^2 = 0.14 (GeV/c)^2.

In case of s-channel helicity conservation (SCHC) only the seven SDMEs of classes A and B may not vanish. However, for the transition $\gamma_T^* \rightarrow V_L$ (class C) one can observe a clear violation of SCHC. In the GK model [6] the SDMEs from class C are related to the contributions from transversity GPDs $\bar{E}_T$ and H_T coupled with the higher-twist wave function of the meson.

The SDME values for exclusive ω production [9] for the total kinematic region with mean values $\langle Q^2 \rangle$ = 2.13 (GeV/c)^2, $\langle W \rangle$ = 7.6 GeV/c^2 and $\langle p_T^2 \rangle$ = 0.16 (GeV/c)^2 are shown in Fig. 2 (right). They are compared to the predictions for COMPASS of the GPD model of Goloskokov and Kroll [6, 7]. The model was tuned to HERMES results on SDMEs and spin asymmetries for exclusive ρ^0 and ω production, which led to a satisfactory agreement between the model and the HERMES data.

The most noticeable differences between COMPASS results and the model predictions are following: i) the predicted value of SDME r_{00}^{04}, which represents the fraction of longitudinally polarized mesons in the produced sample, is significantly larger than the measured one; ii) the SDMEs sensitive to the transitions $\gamma_T^* \rightarrow \omega_L$ (class C) are in general close to zero in the model, while in the data several of them (r^5_{00}, $\mathrm{Re}\{r^1_{10}\}$) indicate a clear violation of the SCHC hypothesis. As the COMPASS results on SDMEs cover the kinematic range of W that extends considerably towards higher values of W compared to HERMES, it is expected that they may help to further constrain the model parameterizations. For constraining models the COMPASS results for the ω and ρ^0 channels are complementary. First, ω production is sensitive to unnatural parity exchange (UPE) amplitudes with large contribution from pion-pole exchange, while ρ^0 production is predominantly described by natural parity exchange (NPE) amplitudes (see Sec. 3.3). Next, for the helicity-flip NPE amplitudes each of the two channels has a different sensitivity to transversity GPDs H_T and $\bar{E}_T = 2\tilde{H}_T + E_T$.

3.3 Unnatural parity exchange contribution

The existence of UPE transitions in exclusive vector meson production can be tested by examining linear combination of certain SDMEs. The UPE fractional contribution Δ_{UPE} to the cross section for exclusive ρ^0 production at COMPASS is very small. The value of Δ_{UPE} for the entire kinematic range is $0.024 \pm 0.005 \pm 0.014$. This is in a sharp contrast with a significant UPE contribution in exclusive ω production observed by COMPASS [9] in the same kinematic range. For the latter one the value for the entire kinematic range is equal to 0.415 $\pm$ 0.037 $\pm$ 0.025. For ω production the UPE dominates at small W values and its contribution decreases with increasing W without vanishing towards the larges W values accessible in COMPASS. In the GK model, UPE is described by the GPDs $\widetilde{H}^f$ and $\widetilde{E}^f$ (non-pole), and by the pion-pole contribution treated as a one-boson exchange [7]. The difference between UPE for ω and ρ^0 is mostly explained by the different $\pi-\omega$ and $\pi-\rho^0$ transition form factors, with the former one being about 3 times larger than the latter. The difference comes from different quark-charge contributions in both cases.

Another observable that is sensitive to the relative contributions of UPE and NPE amplitudes is the NPE-to-UPE asymmetry of the transverse differential cross section for the transition $\gamma^*_T \to V_T$. It is defined [7] as

$$P = \frac{2r^1_{1-1}}{1 - r^{04}_{00} - 2r^{04}_{1-1}} \approx \frac{d\sigma^N_T(\gamma^*_T \to V_T) - d\sigma^U_T(\gamma^*_T \to V_T)}{d\sigma^N_T(\gamma^*_T \to V_T) + d\sigma^U_T(\gamma^*_T \to V_T)}, \tag{2}$$

where the superscripts N and U denote the parts of the cross section that are fed by NPE and UPE transitions, respectively.

The value of P for ρ^0 obtained in the total kinematic region is $0.887 \pm 0.016 \pm 0.029$, which indicates that NPE contribution dominates when averaged over the whole kinematic range of COMPASS. The kinematic dependences of the asymmetry P for ρ^0 are shown in the upper panel of Fig. 3. A possible small UPE contribution is observed only at small values of W and it becomes compatible with zero at larger W. No significant Q^2 and p^2_T dependences of the asymmetry are observed.

The COMPASS results for exclusive ω production [9] on the asymmetry P and its kinematic dependences exhibit a different behaviour as shown in the lower panel of Fig. 3. For the whole kinematic region the value of P is compatible with zero ($P = -0.007 \pm 0.076 \pm$ 0.125), which indicates that the UPE and NPE contributions averaged over the whole kinematic range are of similar size. The UPE contribution dominates at small values of W and decreases with increasing W. At large values of W the NPE contribution becomes dominant, while a non-negligible UPE contribution still remains.

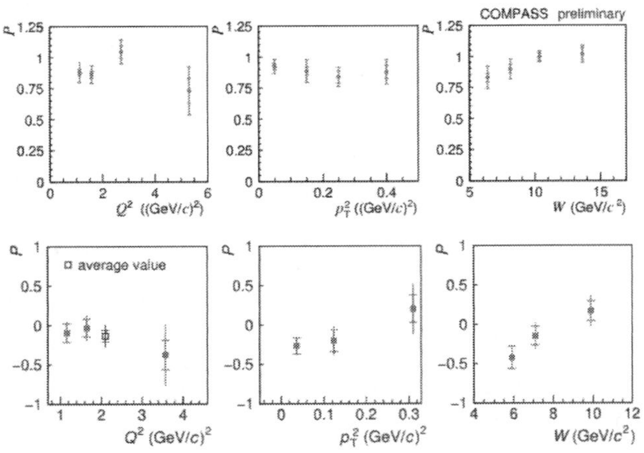

Figure 3. Q^2, p_T^2 and W dependences of the NPE-to-UPE asymmetry for transverse cross section for transitions $\gamma_T^* \to V_T$ in exclusive ρ^0 production (**top**) and exclusive ω production (**bottom**). The inner error bars represent the statistical uncertainties, while the outer ones indicate the statistical and systematic uncertainties added in quadrature.

4 Results on exclusive π^0 production

For the analysis of exclusive π^0 production, $\mu p \to \mu' p' \pi^0$, the use of the information from the CAMERA detector was essential. The procedure for event selection, subtraction of non-exclusive background and extraction of the absolute value of the differential cross section is described in Ref. [12]. Here, we only mention briefly that the four-fold differential μp cross sections was extracted separately for μ^+ and μ^- data. Then the unpolarised cross section was obtained as the weighted mean of the two. In this way the dependence on the beam polarisation cancels. In order to convert μp cross section into $\gamma^* p$ cross section we divide the former one by the transverse virtual photon flux.

The unpolarised $\gamma^* p$ cross section for exclusive π^0 production reads

$$\frac{d^2\sigma^{\gamma^* p}}{dt d\phi} = \frac{1}{2\pi}\left[\frac{d\sigma_T}{dt} + \epsilon\frac{d\sigma_L}{dt} + \epsilon\cos(2\phi)\frac{d\sigma_{TT}}{dt} + \sqrt{2\epsilon(1+\epsilon)}\cos(\phi)\frac{d\sigma_{LT}}{dt}\right], \tag{3}$$

where $\sigma_T, \sigma_L, \sigma_{TT}, \sigma_{LT}$ are the structure functions, ϵ is the virtual photon polarisation parameter and ϕ represents the azimuthal angle between lepton scattering plane and π^0 production plane. Here, the subscript $T(L)$ denotes the contribution from transversely (longitudinally) polarised virtual photons, while the subscripts TT and LT denote the contributions from the interference between transversely-transversely and longitudinally-transversely polarised photons.

After integration over ϕ, we extract the cross section in bins of $|t|$, which is displayed in Fig. 4 (left). It decreases exponentially with increasing $|t|$ at values of $|t|$ larger than about 0.25 (GeV/c)^2, while at smaller $|t|$ the t-dependence becomes weaker. Being the first measurement at energies significantly higher than previous results, it is worth to compare it to the model prediction. The COMPASS result is compared to the predictions of the Goloskokov-Kroll model [13, 14]. The dash-dotted and dotted curves represent the cross section from two versions of the model as a function of t, while the triangles correspond to the cross section averaged over $|t|$ bins of the data. The mean values of the cross section for the full t-range are compared at the rightmost part of the plot. We observe that the magnitude of the earlier prediction of the model [13] for the cross section overshoots our measurement by approximately a factor of two.

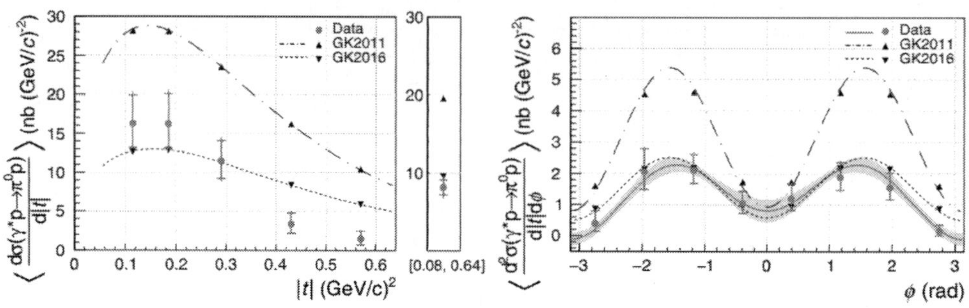

Figure 4. Left: The exclusive π^0 cross section as a function of $|t|$. COMPASS results are compared to predictions of the GPD model of Ref. [13, 14]. **Right:** The exclusive π^0 cross section as a function of ϕ. Solid line indicates the result of the fit to the data.

Furthermore, we extract the cross section as a function of ϕ in the range of measured t and eight bins of ϕ of equal width. The dots in Fig. 4 (right) show the measured cross section for each bin and a fit of the function given by Eq. (3), which is represented by the continuous curve. We observe a large contribution from σ_{TT} and a smaller positive one from σ_{LT}. This indicates a significant role of transversely polarised photons in exclusive π^0 production.

According to Refs. [13, 15] the structure functions for exclusive pseudo-scalar meson production, which appear in Eq. (3), depend on GPDs $\tilde{H}^q$, $\tilde{E}^q$, H_T^q and $\bar{E}_T^q = 2\tilde{H}_T^q + E_T^q$. For π^0 production a large contribution from transversely polarised virtual photons is expected, which is mainly generated by the chiral-odd ('transversity') GPDs $\bar{E}_T^q$. It manifests itself in a large contribution from σ_{TT} and a dip in the differential cross section $d\sigma/dt$ as $|t|$ decreases to zero.

Acknowledgements

We acknowledge the support by the European Union's Horizon 2020 research and innovation programme under grant agreement STRONG-2020 - No 824093.

References

[1] D. Müller et al., Fortschr. Phys. **42**, 101 (1994)
[2] X. Ji, Phys. Rev. Lett. **78**, 610 (1997); Phys. Rev. D **55**, 7114 (1997)
[3] A.V. Radyushkin, Phys. Lett. B **385**, 333 (1996); A.V. Radyushkin, Phys. Rev. D **56**, 5524 (1997)
[4] M. Burkardt, Phys. Rev. D **62** (2000) 071503; erratum-ibid. D **66** (2002) 119903; Int. J. Mod. Phys. A **18** (2003) 173; Phys. Lett. B **595** (2004) 245
[5] S.V. Goloskokov and P. Kroll, Eur. Phys. J. C **42**, 281 (2005); Eur. Phys. J. C **53** , 367 (2008); Eur. Phys. J. C **59** , 809 (2009)
[6] S.V. Goloskokov and P. Kroll, Eur. Phys. J. C **74** 2725 (2014)
[7] S.V. Goloskokov and P. Kroll, Eur. Phys. J. A **50** 146 (2014)
[8] P. Abbon et al. (COMPASS Collaboration), Nucl. Instr. Meth. A **577** 455 (2007); Nucl. Instr. Meth. A **779** 69 (2015)
[9] G.D. Alexeev et al. (COMPASS Collaboration), Eur. Phys. J. C **81**, 126 (2021)
[10] K. Schilling and G. Wolf, Nucl. Phys. B **61**, 381 (1973)
[11] A. Airapetian et al. (HERMES Collaboration), Eur. Phys. J. C **62**, 659 (2009)
[12] M.G. Alexeev et al. (COMPASS Collaboration), Phys. Lett. B **805**, 135454 (2020)
[13] S.V. Goloskokov and P. Kroll, Eur. Phys. J. A **47**, 112 (2011)
[14] S.V. Goloskokov and P. Kroll, private communications, 2016
[15] I. Bedlinsky et al. (CLAS Collaboration), Phys. Rev. C **90**, 025205 (2014)

EPJ Web of Conferences **270**, 00007 (2022) https://doi.org/10.1051/epjconf/202227000007
QCD@Work 2022

TOTEM, recent experimental review

Francesco S. Cafagna[1,*], on behalf of the TOTEM Collaboration[**]

[1]INFN Sezione di Bari, Italy

Abstract.
The TOTEM (TOTal cross section, Elastic scattering and diffraction dissociation Measurement at the LHC) experiment, located at the interaction point 5 of the LHC, has measured the total, elastic and inelastic proton-proton cross-sections, using a luminosity independent method, based on the optical theorem, in a center-of-mass energy range from 2.76 to 13 TeV. The elastic scattering was investigated in a wide range of the squared four-momentum transfer $|t|$ allowing study of the Coulomb-nuclear interference region down to $|t| \sim 8 \times 10^{-4}$ GeV^2. This made possible the first measurement of the ρ parameter at $\sqrt{s} = 13$ TeV, ρ being the ratio between the real and the imaginary part of the nuclear elastic scattering amplitude at $t = 0$. This measurement, combined with the total cross-section results, led to the exclusion of all the models classified and published by the COMPETE Collaboration. The results obtained by TOTEM are indeed compatible with predictions of a colourless 3-gluon bound state exchange in the t-channel of proton-proton elastic scattering, as postulated by alternative theoretical models both in the Regge-like framework and in the modern QCD framework. This result has been confirmed, with a significance of 5.4σ, also by the comparison with the $\bar{p}p$ data measured by the D0 collaboration at Fermilab. In this contribution the TOTEM experiment results will be described, along with the actual experiment status, the future physics program for the LHC Run 3.

1 Introduction

TOTEM measured the total pp cross-section at the energy of $\sqrt{s} = 2.76, 7, 8$ and 13 TeV [1–4], using a luminosity independent method and validating it by comparing several methods to determine the total cross-sections [5–10]. The total cross-section can be obtained via the optical theorem that combines the simultaneous measurements of the total inelastic rate, N_{inel}, and the total nuclear elastic rate, N_{el}, with its extrapolation to the optical point $t = 0$, $\mathrm{d}N_{el}/\mathrm{d}t|_{t=0}$:

$$\sigma_{tot} = \frac{16\pi}{1+\rho^2}\frac{(\mathrm{d}N_{el}/\mathrm{d}t)_{t=0}}{(N_{el}+N_{inel})} \tag{1}$$

where ρ is the ratio of the real to imaginary part of the hadronic scattering amplitude at $t = 0$. The method requires the simultaneous measurements of the inelastic and elastic rates, as well as the extrapolation of the latter in the invisible region down to $|t| = 0$. This is achieved with the experimental set-up of TOTEM, which consists of two inelastic telescopes, called T1 and

*e-mail: francesco.cafagna@ba.infn.it
**For collaboration list see http://cern.ch/totem

T2, to detect charged particles produced in inelastic pp collisions, and Roman Pot stations to detect elastically scattered protons at very small angles [11]. The measured ratio of the elastic and inelastic rates N_{el}/N_{inel} allows for the determination of the elastic and inelastic cross-sections as well.

The optical theorem can be used also to derive the total and fully inclusive inelastic pp cross-section, from the measurement of the elastic scattering one:

$$\sigma^2_{tot} = \frac{16\pi}{1+\rho^2}\frac{d\sigma_{el}}{dt}|_{t=0}, \quad \sigma_{inel} = \sigma_{tot} - \sigma_{el}. \tag{2}$$

To access the smaller $|t|$-value region, the colliding beams must have a beam divergence of not more than a few micro-radians [12]. This can be obtained by either increasing the beta function value, β^*, or by reducing the beam emittance, ϵ (beam divergence = $\sqrt{\epsilon/\beta^*}$). With special runs having dedicated beam optics configurations, β^*= 90, 1000, 2500 m, TOTEM extended the measurement to $|t|$-values as low as 8×10^{-4} GeV^2, measuring the differential elastic cross-section over a wide range of t [5, 8, 10, 13]. This made the extrapolation of the differential cross-section to the optical point at $t = 0$ possible, allowing, for the first time at the LHC, the determination of the elastic scattering cross-section, the total cross-section via the optical theorem as well as the measurement of the ρ parameter [10].

To increase robustness and understanding of the systematic uncertainties, the TOTEM Collaboration, cross-checked the results of its measurements by comparing several independent approaches to determine the total cross-section, proving the reliability of the luminosity independent method. Measurements at 7 TeV, were double checked using different methods under the same beam conditions, while the ones at 8 and 13 TeV where checked leveraging different beam and optics conditions. Moreover, with the data collected by the very forward telescopes T1 and T2, during special runs at $\sqrt{s}$ = 7 TeV, also the double diffractive (DD) cross-section was measured in the pseudorapidity region $4.7 < |\eta_{min}| < 6.5$ [26].

During the special runs, TOTEM provides minimum and zero bias triggers to the CMS experiment [14], making also possible physics studies in common between the two experiments. The collaboration with CMS gave birth to a joint project: CMS-TOTEM Precision Proton Spectrometer (CT-PPS) [15], designed to take data during the LHC Run 2 standard run with high luminosity. The project collected slightly less than 100 fb^{-1}, in the last two years of Run 2 data taken, proving the feasibility of the project [16] that has been integrated into CMS as PPS subsystem, eventually.

In particular the direct measurement of forward pseudorapidity distributions, that are also valuable in constraining theoretical models and Monte Carlo programs of the high energy air showers initiated by cosmic rays, have been measured by both experiments [27] and by TOTEM standalone [28].

The results obtained by TOTEM are indeed compatible with predictions of a colourless 3-gluon bound state exchange in the t-channel of proton-proton elastic scattering, as postulated by alternative theoretical models both in the Regge-like framework and in the modern QCD framework. This result has been confirmed, with a significance of 5.4σ, also by the comparison with the $\bar{p}p$ data measured by the D0 collaboration at Fermilab.

2 The Totem Experiment

The TOTEM inelastic telescopes[11], T1 and T2, consist of two arms located symmetrically on both sides of the LHC interaction point 5 (IP5): the T1 telescope with cathode strip chambers (CSC), placed at a distance of 9 m from IP5, covers the pseudorapidity range $3.1 \leq |\eta| \leq 4.7$. The T2 telescope with gas electron multiplier (GEM) chambers, located at ±13.5 m from IP5, covers the pseudorapidity range $5.3 \leq |\eta| \leq 6.5$. The pseudorapidity coverage of the two telescopes allowed the detection of 95% of the inelastic events, including events with diffractive mass down to 3.6 GeV.

For the LHC Run 3, an upgrade of the T2 detector has been scheduled. The new T2 will fit in the reduced space left by the new LHC beamline configuration; it will be based on scintillator slabs read out by silicon photomultipliers (SiPM) [17].

Roman Pot (RP) stations are located symmetrically on both sides of the IP5. During the LHC Run 1, the Roman Pot stations were located at 147 m (RP147) and 220 m (RP220) along the LHC beam line. Each station is composed of two units (near and far, in the TOTEM jargon) separated by a distance of about 5 m. A unit consists of 3 RPs, two approaching the outgoing beam vertically, from the top and the bottom, and one horizontally. During LHC Run 1, each RP was equipped with a stack of 10 silicon strip detectors, designed with the specific objective of reducing the insensitive area at the edge facing the beam to only a few tens of micrometers. This edgelessness design permits to measure the proton distance from the beam center, in both coordinates perpendicular to the beam, with a resolution of about 11 μm [18]. The movement and the alignment of all pots are monitored with a precision better than 20 μm based on track reconstruction and external alignment tools. The long lever arm between the near and the far units allows the determination of the track angle in both projections with a precision of about 5 μrad. A more detailed description of the TOTEM detector and performance, in this configuration, can be found in [11, 19].

The RP layout for the LHC Run 2, has been modified. The stations previously installed at ±147 m, from the interaction point, have been relocated at ±210 m, with one of the two units, the far one, tilted around the beam axis by 8° degree, to improve multi-track resolution[20]. Moreover, two newly designed horizontal RP, have been installed between the two units at ±220 m. All RP have been refurbished and upgraded to stand the increase in luminosity foreseen. This new layout has been proposed to improve the apparatus performances, adding timing detectors, to resolve the pileup of multiple events in the same bunch crossing, and operate the experiment at higher luminosities. This has been fulfilled by TOTEM, for the special runs with low luminosity, installing new timing detectors in a vertical RP [21, 23, 24]. The RP upgrade was a feat that boosted the CT-PPS project. In fact the CT-PPS collaboration installed, in two horizontal RP, new tracking detectors, based on the silicon pixels developed for the CMS tracker upgrade [15], while the diamond timing detectors, developed for the TOTEM upgrade, were installed in the new horizontal ones [23]. At the end, CT-PPS was leveraging four different detector technologies: silicon strips and pixels, and diamond and Ultra Fast Silicon Detector (UFSD) [24], for tracking and timing respectively.

For the LHC Run 3, a consolidation and refurbishment of the mechanics and electronics of the station is in progress. Moreover a new diamond based timing detector package will be installed inside an horizontal RP.

TOTEM will finish the cross section measurement campaign, during the LHC Run 3, when the machine will reach the maximum energy of $\sqrt{s} = 14\ TeV$. For this measurement an upgrade of the T2 detector has been scheduled. The new T2 will fit in the reduced space left by the new LHC beamline configuration; it will be based on scintillator slabs readout by silicon photomultipliers (SiPM).

3 The Cross Sections and ρ Measurements

The cross-sections obtained using the luminosity independent method of eq. 1, measured by TOTEM, are reported in Table 1 and shown in Figure 1, along with the ones obtained by other experiments at LHC and using cosmic ray measurements (see [4] and references therein). The

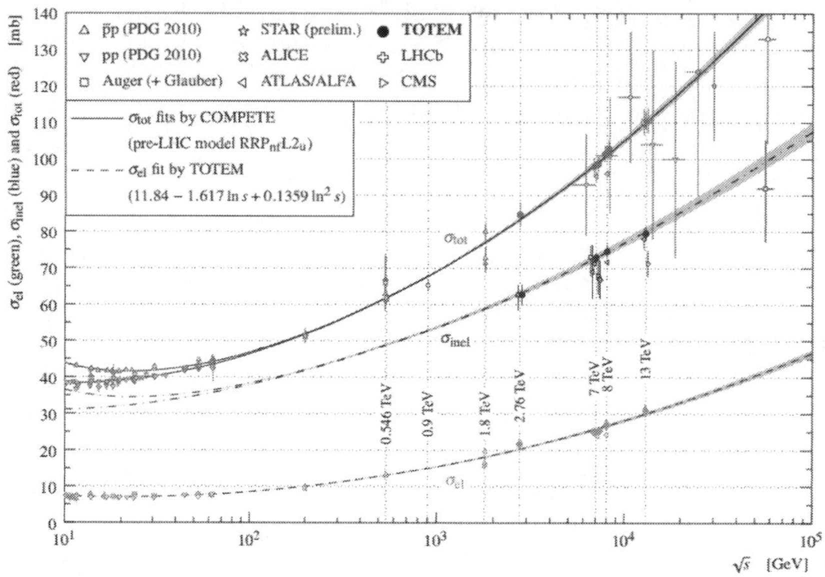

Figure 1. A compilation (see [4] and references therewith, of pp and $\bar{p}p$ total (σ_{tot}), inelastic (σ_{inel}) and elastic (σ_{el}) cross-sections measurements as a function of $\sqrt{s}$. The TOTEM measurements using the luminosity independent method are reported using full circle. The continuous lines (lower, for pp, upper for $\bar{p}p$) represent the best fits of the total cross-section data by the COMPETE collaboration [32]. The dashed line results from a fit of the elastic cross-section data. The dash-dotted lines refer to the inelastic cross-section and are obtained as the difference between the continuous and dashed fits.

total systematic uncertainty has been reported, obtained combining in quadrature all the systematic uncertainties, taking into account the correlations between contributions. For a more detailed description and discussion of the systematic uncertainties see [2–4, 8] and references therein. For the ρ parameter value the COMPETE [32] preferred-model extrapolation of 0.141 ± 0.007 was chosen.

Given the various methods of measuring elastic and inelastic rates, it was possible to perform completely different measurement of the three cross-section types, comparing the results for a better understanding of the systematic uncertainties and a double check of the results.

In particular at $\sqrt{s} = 7\ TeV$ the luminosity independent result [2] was shown to be consistent with the total cross section calculated by extrapolating the differential elastic cross-section to the optical point $|t| = 0$, using Equation 2 [5, 8], as well as with a luminosity dependent method [7]. While from the direct measurement of the inelastic rates the inelastic cross-section was determined and compared with the one deduced from the difference between the total and the elastic cross section [6].

Also results at $\sqrt{s} = 8\ TeV$ [3], has been double checked using the extrapolation of the differential elastic cross-section. This time, thanks to a very-high statistics collected at $\beta^{*} = 90$ m, the TOTEM experiment was able to exclude a pure exponential behavior of the elastic differential cross-section [13], so refined parametrizations have been used for the extrapolation to the optical point. The significance of the exclusion is greater than 7σ in the $|t|$ range from 0.027 to 0.2 GeV^2.

Thanks to the data collected at this energy, $\sqrt{s} = 8\ TeV$, with $\beta^{*} = 1000$ m, it was possible to explore the elastic scattering in the Coulomb Nuclear Interference (CNI) region. For the

Table 1. Summary of the cross-sections measured using a luminosity ($\mathcal{L}$) independent method. The full systematic uncertainty is reported for each measurements, see [2–4, 8] for a detailed description of this uncertainties.

	$\mathcal{L}$ independent using eq. 1			
	$\sqrt{2.76}\ TeV$ [1]	$\sqrt{7}\ TeV$ [2]	$\sqrt{8}\ TeV$ [3]	$\sqrt{13}\ TeV$ [4]
σ_{tot} (mb)	84.7 ± 3.3	98.0 ± 2.5	101.7 ± 2.9	110.6 ± 3.4
σ_{inel} (mb)	62.8 ± 2.9	72.9 ± 1.5	74.7 ± 1.7	79.5 ± 1.8
σ_{el} (mb)	21.8 ± 1.4	25.1 ± 1.1	27.1 ± 1.4	31.0 ± 1.7

first time at LHC, the ρ parameter was measured, via the CNI, and was found to be $\rho = 0.12 \pm 0.03$ [25]. The ρ parameter has been measured also at $\sqrt{s} = 13$ TeV, thanks to the data collected in a special run with $\beta^*= 2500$ m. The values obtained are: $\rho = 0.09 \pm 0.01$ and $\rho = 0.10 \pm 0.01$, depending on different physics assumptions and mathematical modelling [10]. This time the CNI region has been used to estimate the σ_{tot} using three different and independent approaches. This novel "Coulomb normalization" technique was applied for the first time to LHC data, finding a result of $\sigma_{tot} = 110.3 \pm 3.5$ mb, that, combined with the luminosity independent method calculation, yields $\sigma_{tot} = 110.5 \pm 2.4$ mb. A summary of this results is shown in Table 2.

Table 2. Summary of ρ and total cross-section results.

data	method	ρ	σ_{tot} [mb]
$\beta^* = 90$ m	Ref. [22]	-	110.6 ± 3.4
$\beta^* = 2500$ m	approach 1	0.09 ± 0.01	111.8 ± 3.2
	approach 2	0.09 ± 0.01	111.3 ± 3.2
	approach 3	0.08(5) ± 0.01	110.3 ± 3.5
	approach 3 (single fit)	0.10 ± 0.01	109.3 ± 3.5
$\beta^* = 90$ and 2500 m	Ref. [22] ⊕ approach 3		110.5 ± 2.4

The unprecedented precision of the ρ measurement, shown in Figure 2, combined with the TOTEM total cross-section measurements in an energy range from 2.76 to 13 TeV, has implied the exclusion of all the models classified and published by the COMPETE Collaboration [32]. The ρ results obtained by TOTEM are compatible with the predictions, from alternative theoretical models both in the Regge-like framework and in the QCD framework, of a colourless 3-gluon bound state exchange in the t-channel of the proton-proton elastic scattering.

4 Comparison between TOTEM and D0 results

At high energies, the scattering amplitudes of both pp and $\bar{p}p$ elastic collisions are dominated by t-channel exchanges that carry the quantum numbers of the vacuum. Out of these amplitudes, a subdominant crossing- odd combination can be formed and contribute to the elastic collision at the TeV energy scale. The differences between these cross sections are due to physics that couples differently to particles and their charge conjugates that result from the exchange of the so called *odderon*, in QCD, a colourless three-gluon state at leading order [33–35]. A quantitative analysis has been carried out to compare the $\bar{p}p$ elastic cross section as measured by the D0 collaboration at a center-of-mass energy of 1.96 TeV

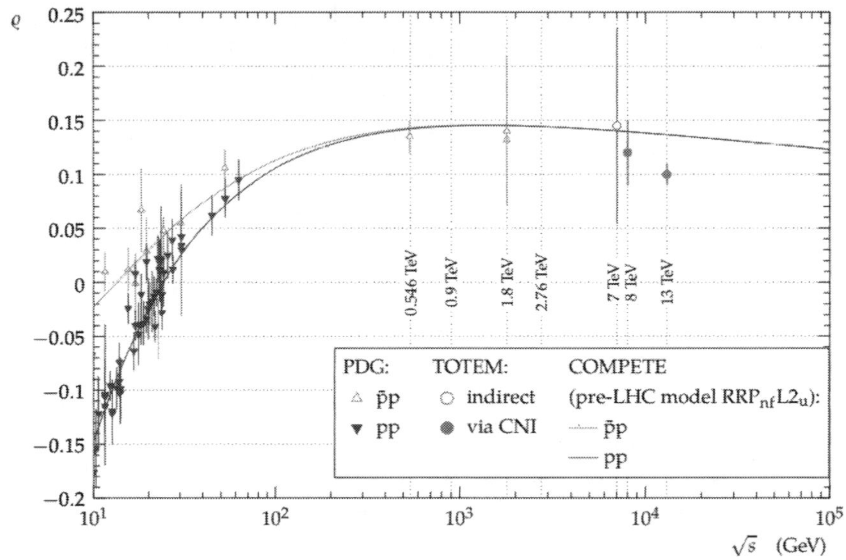

Figure 2. Dependence of the ρ parameter on $\sqrt{s}$. The pp (blue) and $p\bar{p}$ (green) data are taken from PDG [31], TOTEM measurements [10, 25] are marked in red. The continuous lines (lower, in blue,for pp, upper, in red, for $\bar{p}p$) represent the best fits of the total cross-section data by the COMPETE collaboration [32]. The dashed line.

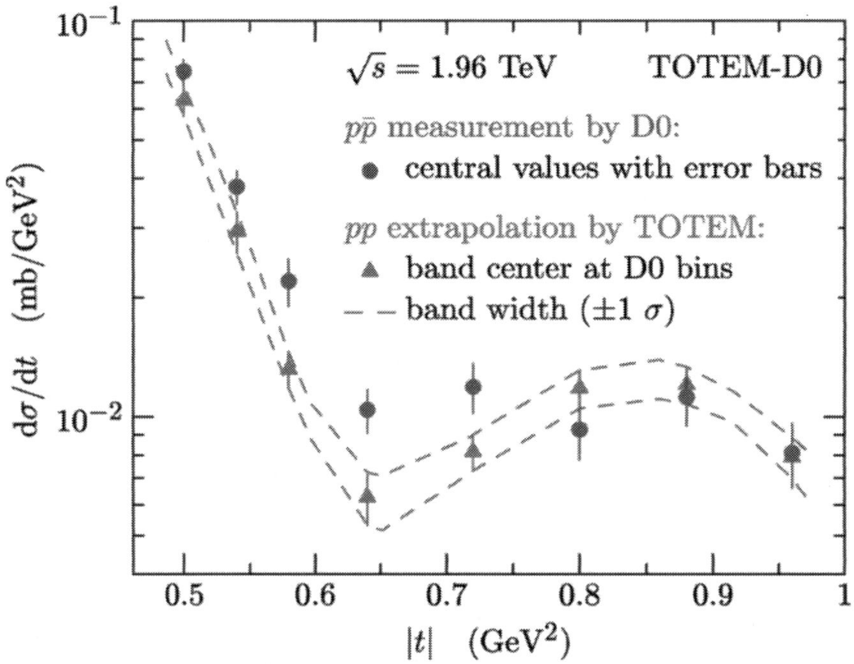

Figure 3. Comparison between the D0 $\bar{p}p$ measurement at 1.96 TeV and the extrapolated TOTEM pp cross section, rescaled to match the optical point ($d\sigma/dt|_{t=0}$) of the D0 measurement. The dashed lines show the 1σ uncertainty band on the extrapolated pp cross section.

to that in pp collisions as measured by the TOTEM collaboration at 2.76, 7, 8, and 13 TeV using a model-independent approach. The TOTEM cross sections have been extrapolated to a center-of-mass energy of 1.96 TeV and compared with the D0 measurement in the region of the diffractive minimum and second maximum of the pp cross section. The two data sets disagree with a significance os 3.4σ and thus provide a further evidence that an exchange of an *odderon* is needed to descibe elastic scattering at high energies. When combined with the result of [10], the significance is in the range 5.2 to 5.7σ and thus constitutes the first experimental observation of the *odderon*.

5 Conclusions

TOTEM has measured, for the first time at the LHC, the total, inelastic and elastic proton-proton cross sections at $\sqrt{s} = 2.76, 7, 8$ and 13 TeV using a luminosity independent method. The method was validated comparing the elastic and inelastic cross sections measured in independent ways and using a novel "Coulomb normalization" technique based on QED data. Furthermore, for the first time at LHC, the ρ parameter has been estimated studying the CNI region.

This measurement, combined with the total cross-section results, led to the exclusion of all the models published by the COMPETE Collaboration, and shows compatibility with predictions of a colourless 3-gluon bound state exchange in the t-channel of proton-proton elastic scattering.

The evidence for an *odderon* exchange is further confirmed by the comparison of the pp and $\bar{p}p$ cross secionts as measured by TOTEM and D0. The combined significance of the differences between the two measurement, is larger than 5σ and is interpreted as the first observation of the exchange of a colorless, C-odd gluonic compound: the *odderon*.

Acknowledgment: We are grateful to the beam optics development team for the design and the successful commissioning of the high β^* optics and to the LHC machine coordinators for scheduling the dedicated fills.

This work was supported by the institutions listed on the front page and also by the Magnus Ehrnrooth foundation (Finland), the Waldemar von Frenckell foundation (Finland), the Academy of Finland, the Finnish Academy of Science and Letters (The Vilho, Yrjö and Kalle Väisälä Fund), the OTKA NK 101438 and the EFOP-3.6.1-16-2016-00001 grants (Hungary). Individuals have received support from Nylands nation vid Helsingfors universitet (Finland), from the MŠMT ČR (Czech Republic) and the János Bolyai Research Scholarship of the Hungarian Academy of Sciences and the NKP-17-4 New National Excellence Program of the Hungarian Ministry of Human Capacities.

References

[1] G. Antchev *et al.* (TOTEM Collaboration), *Eur. Phys. J. C*, **80** (2020) 91, doi:10.1140/epjc/s10052-020-7654-y

[2] G. Antchev *et al.* (TOTEM Collaboration), *EPL*, **101** (2013) 21004, doi:10.1209/0295-5075/101/21004

[3] G. Antchev *et al.*, *Phys. Rev. Lett.* **111** (2013) 012001, doi: 10.1103/PhysRevLett.111.012001

[4] G. Antchev *et al.* (TOTEM Collaboration), *Eur. Phys. J. C*, **79** (2019) 103, doi:10.1140/epjc/s10052-019-6567-0

[5] G. Antchev *et al.* (TOTEM Collaboration), *EPL*, **101** (2013) 21002, doi:10.1209/0295-5075/101/21002

[6] G. Antchev *et al.* (TOTEM Collaboration), *EPL*, **101** (2013) 21003, doi:10.1209/0295-5075/101/21003

[7] G. Antchev *et al.* (TOTEM Collaboration), *EPL*, **96** (2011) 21002, doi: 10.1209/0295-5075/96/21002

[8] G. Antchev *et al.* (TOTEM Collaboration), *EPL*, **95** (2011) 41001, doi: 10.1209/0295-5075/95/41001
[9] G. Antchev *et al.* (TOTEM Collaboration), *Eur. Phys. J. C*, **79** (2019) 861, doi:10.1140/epjc/s10052-019-7346-7
[10] G. Antchev *et al.* (TOTEM Collaboration), *Eur. Phys. J. C*, **80** (2020) 91, doi:10.1140/epjc/s10052-020-7654-y
[11] G. Anelli *et al.* (TOTEM Collaboration), *JINST*, 3, S08007 (2008), doi:10.1088/1748-0221/3/08/S08007
[12] G. Antchev *et al.* (TOTEM Collaboration), *New J. Phys.* **16** (2014) 103041, doi:10.1088/1367-2630/16/10/103041
[13] G. Antchev *et al.* (TOTEM Collaboration), *Nucl. Phys. B* **899** (2015) 527, doi:10.1016/j.nuclphysb.2015.08.010
[14] CMS Collaboration, *JINST* 3, S08004 (2008). doi:10.1088/1748-0221/3/08/S08004
[15] M. Albrow *et al.* (CMS-TOTEM Collaboration), *CMS-TOTEM Precision Proton Spectrometer, CERN-LHCC-2014-021*, (2014), url: https://cds.cern.ch/record/1753795
[16] A. M. Sirunyan, A. Tumasyan *et al.*, *J. High Energ. Phys.*, **2018**, 153 (2018), doi:10.1007/JHEP07(2018)153
[17] TOTEM Collaboration, *Upgrade of the TOTEM T2 Telescope, Technical Design Report, CERN-LHCC-2019-007*, (2019), url: https://cds.cern.ch/record/2677468
[18] E. Alagoz *et al.*, *JINST*, 8, P06009 (2013), doi:10.1088/1748-0221/8/06/P06009
[19] G. Antchev *et al.*, *Int. J. Mod. Phys.* **A 28**, 1330046 (2013), doi: 10.1142/S0217751X13300469
[20] TOTEM Collaboration, *TOTEM Upgrade Proposal, CERN-LHCC-2013-009*, (2013), url: https://cds.cern.ch/record/1554299
[21] TOTEM Collaboration, *Timing Measurements in the Vertical Roman Pots of the TOTEM Experiment, CERN-LHCC-2014-020*, (2014), url: https://cds.cern.ch/record/1753189
[22] G. Antchev *et al.*, *CERN-EP-2017-321*, (2017), url: https://cds.cern.ch/record/2296409
[23] G. Antchev *et al.*, *JINST*, **12**, P03007 (2017), doi:10.1088/1748-0221/12/03/P03007
[24] M. Berretti *et al.*, *JINST*, **12**, P03024 (2017), doi:10.1088/1748-0221/12/03/P03024
[25] G. Antchev *et al.* (TOTEM Collaboration), *Eur. Phys. J. C* **76** (2016) 661, doi:10.1140/epjc/s10052-016-4399-8
[26] G. Antchev *et al.*, *Phys. Rev. Lett.*, **111** (2013) 262001, doi:10.1103/Phys-RevLett. 111.262001
[27] The CMS and TOTEM Collaborations, *EPJC* **74** (2014) 3053, doi:10.1140/epjc/s10052-014-3053-6
[28] G. Antchev *et al.* (TOTEM COllaboration), *EPJC* **75** (2015) 126, doi:10.1140/epjc/s10052-015-3343-7
[29] V. M. Abazov *et al.*, *Phys. Rev. Lett.*, **127** (2021) 062003, doi:10.1103/Phys RevLett.127.062003
[30] G. Antchev *et al.* (TOTEM Collaboration), *Eur. Phys. J. C*, **82** (2022) 263, doi:10.1140/epjc/s10052-022-10065-x
[31] C. Patrignani *et al.*, *(Particle Data Group), Chin. Phys.*, **C40**, 100001 (2016)
[32] J. R. Cudell *et al.* (COMPETE Collaboration), *Phys. Rev. Lett.* **89**, 201801 (2002)
[33] A. Martin, *Phys. Rev.* **129**, 1432 (1963)
[34] L. Lukaszuk, A. Martin, *Nuovo Cimento* **A 52**, 122 (1967)
[35] L.Baksay *et al.*, Nucl. Phys. **B141**, 1(1978); Erratum, **B148,** 538 (1979)

EPJ Web of Conferences **270**, 00008 (2022)
QCD@Work 2022
https://doi.org/10.1051/epjconf/202227000008

Latest results from NA62 and NA48/2 experiments

Michele Corvino[1,*], on behalf of the NA48/2 and NA62 Collaborations

[1]European Organization for Nuclear Research (CERN), Geneva, Switzerland

Abstract. Kaon physics at CERN has a long tradition, started in 1980s by the NA31 experiment and continued by the current NA48/2 and NA62 collaborations. New results have been presented by both, namely a new measurement of the branching ratio of $K^{\pm} \to \pi^0\pi^0\mu^{\pm}\nu$ from NA48/2 and a set of new results from the broad physics programme of NA62.

1 Introduction

The NA48/2 and NA62 experiments at CERN are fixed target experiments located in the same experimental area. NA48/2 collected data in 2003 and 2004 with simultaneous K^+/K^- beams, then in 2016 the NA62 experiment started to take data with a positively charged particle beam and a new experimental apparatus. The layout of the two experiments is shown in fig. 1. A detailed description of the detectors can be found in [1] and in [2].

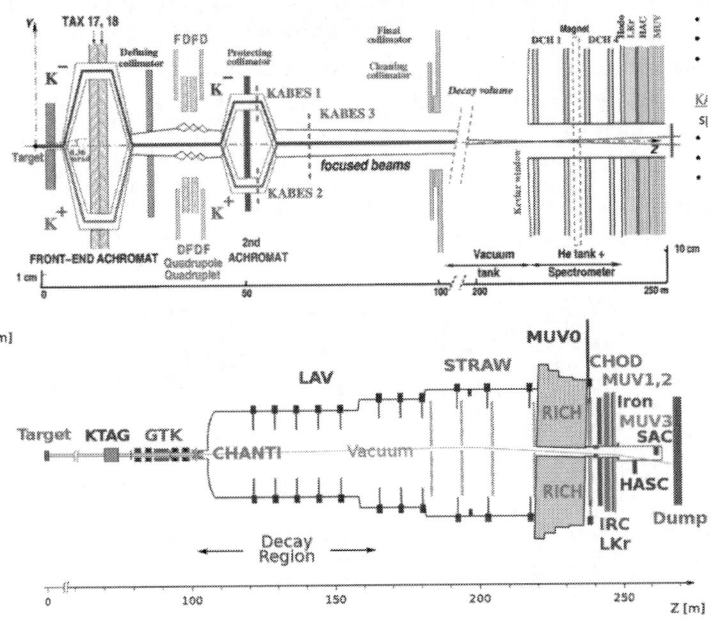

Figure 1. Experimental layout of the NA48/2 (top) and NA62 (bottom) experiments.

*e-mail: michele.corvino@cern.ch

2 Latest results from NA48/2 experiment

2.1 Measurement of the branching ratio for $K^{\pm} \rightarrow \pi^0\pi^0\mu^{\pm}\nu$ ($BR(K^{00}_{\mu 4})$)

The semileptonic decays $K \rightarrow \pi\pi l\nu$ (K_{l4}) are described by the Cabibbo-Maksymowicz variables S_π (dipion mass), S_l (dilepton mass) and the angles θ_π, θ_l and ϕ. In general, the decay amplitudes depends on the form factors F, R, G and H; in the case of $K^{00}_{\mu 4}$ decay, assuming S-wave state for the dipion system, there is no dependance on $cos\theta_\pi$ and ϕ and therefore only F and R contribute to its branching ratio. The NA48/2 experiment already provided the best measurements for the branching ratios of K^{00}_{e4} and $K^{\pm}_{e4}$ decays, while $K^{00}_{\mu 4}$ has never been observed. In the muon modes, the largest source of background is represented by the $K \rightarrow \pi\pi\pi$ decays with subsequent decay in flight of a charged pion into a muon and a neutrino. For this measurement, the parametrization $F(S_\pi, S_l)$ from K^{00}_{e4} has been used profiting from lepton universality, while $R(S_\pi, S_l)$ is estimated by Chiral Perturbation Theory (ChPT) [3].

The signal is normalized to the $K^{\pm} \rightarrow \pi^0\pi^0\pi^{\pm}$ decay, using a common selection: 4 isolated photons, compatible with a pair of pi0s are required in the electromagnetic calorimeter. The photons must be associated in time with a downstream track and the momentum of the total final state must be compatible with the decay of a beam kaon. For the signal the muon is identified by associating the track with hits in the muon system. Background events are rejected by cuts on the 3-pion invariant mass, p_t, $cos\theta_l$ and the missing mass. A data-driven background estimate is performed using the signal sidebands in the missing mass distribution.

The branching ratio is measured in the restricted phace space $S_l > 0.03$ GeV^2/c^4 and then extrapolated to the full phase space. The result is obtained from

$$BR(K^{00}_{\mu 4}) = \frac{N_S}{N_N} \cdot \frac{A_N}{A_S} \cdot K_{trig} \cdot BR(K^{00}_{3\pi}) \tag{1}$$

where N_S and N_N are the number of signal and background normalization events, A_N and A_S the selection acceptances, K_{trig} is a trigger correction factor and $BR(K^{00}_{3\pi})$ the BR of the normalization channel. After the background subtraction, $2083 \pm 59_{stat}$ signal events are observed, resulting in a measurement in the full phase space of:

$$BR(K^{00}_{\mu 4}) = (3.45 \pm 0.17) \times 10^{-6} \tag{2}$$

The result is compatible with the theoretical prediction when a non-zero contribution from the R form factor is taken into account (fig. 2).

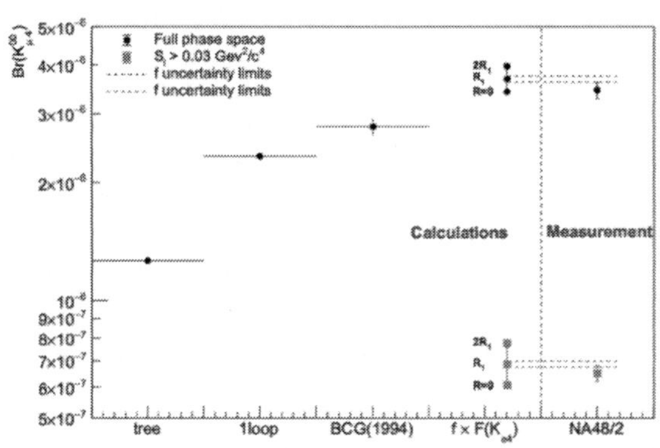

Figure 2. Result for $BR(K^{00}_{\mu 4})$ in the restricted (red) and full (black) phase space. The teoretical predictions in case of different contributions from the R form factor are also shown.

3 Latest results from NA62 experiment

3.1 Measurement of $BR(K^+ \to \pi^+ \nu\bar{\nu})$

The $K^+ \to \pi^+ \nu\bar{\nu}$ decay is a golden candidate to search for new physics in the kaon sector. It is a Flavour Changing Neutral Current (FCNC) process, suppressed in the Standard Model (SM) by the GIM mechanism, and therefore very sensitive to new physics effects. The decay amplitude is dominated by short-distance contributions, in particular from top quark loops, and the hadronic matrix element is linked by the isospin symmetry to those of the semileptonic kaon decays that are well measured. As a consequence, the theoretical prediction for its BR is very precise: $BR(K^+ \to \pi^+ \nu\bar{\nu})_{SM} = (8.4 \pm 1.0) \times 10^{-11}$.

The experimental signature is represented by one track in the initial state and one track in the final states, as the neutrinos are not detected. The main variable for the analysis is $m^2_{miss} = (P_K - P_\pi)^2$; from the m^2_{miss} distribution, two signal regions have been identified to discriminate the signal from the backgrounds. A blind analysis technique has been adopted, with the signal regions kept masked until the background evaluation was assessed using appropriate control regions. The experiment analyzed the data collected in 2016, 2017 and 2018: the data from 2018 have been divided into two subsamples S1 and S2, respectively before and after the installation of a new kaon beam collimator before the entrance of the decay volume, which reduced the background contribution from beam interactions in the last plane of the beam tracker. The sample S2, corresponding to $\sim$ 80% of the 2018 sample, has been split into 6 pion momentum categories, from 15 to 45 GeV/c with a bin width of 5 GeV/c. Finally, data from 2016 and 2017 have been treated as two separate categories, for a total of 9.

The single event sensitivity (SES) is given by:

$$SES = \frac{BR(K^+ \to \pi^+\pi^0)}{D \cdot N_{\pi\pi} \cdot \epsilon_{trig} \cdot \epsilon_{RV}} \frac{A_{\pi\pi}}{A_{\pi\nu\nu}} \tag{3}$$

where $N_{\pi\pi}$ are the number of events in the normalization channel, $A_{\pi\nu\nu}$ and $A_{\pi\pi}$ the acceptances of signal and normalization selections respectively, D is the downscaling factor between the signal and normalization trigger streams, ϵ_{trig} is the trigger efficiency and ϵ_{RV} is the random veto efficiency, namely the probability of not discarding an event because of accidental activity in the veto system. For the samples S1 and S2 the SES are $(0.54 \pm 0.04) \times 10^{-10}$ and $(0.14 \pm 0.01) \times 10^{-10}$ respectively. The backgrounds from the main kaon decays have been evaluated with dedicated background regions as $N^{exp}_{decay} = N_{bkg} \cdot f_{kin}$, where N_{bkg} is the number of events in each background region and f_{kin} is the fraction of events entering the signal region for that specific decay, estimated from MonteCarlo simulations. Beam-induced background ("upstream background") has been estimated using a data-driven technique, and the evaluation has been validated using a sample that is orthogonal to the signal sample.

After the background validation, the signal regions have been unblinded: 17 events have been found, which combined with the events found from the 2016 and 2017 analyses gives a total of 20 signal candidates. Taking into account the signal and background expectations, the resulting measurement of the branching ratio is [4]

$$BR(K^+ \to \pi^+ \nu\bar{\nu}) = (10.6^{+4.0}_{-3.5}|_{stat} \pm 0.3_{syst}) \times 10^{-11} \tag{4}$$

which corresponds to an observation of the decay at a significance of 3.4 standard deviations and is the most precise measurement of the branching ratio to date.

3.2 Measurement of $BR(K^+ \to \pi^0 e^+ \nu\gamma)$

The decay $K^+ \to \pi^0 e^+ \nu\gamma$ ($K_{e3\gamma}$) provides a good test of the Chiral Perturbation Theory (ChPT). In particular, its branching ratio is expressed in terms of the branching ratio of the non-radiative K_{e3} decay in different regions of the phase space defined by E_γ and $\theta_{e,\gamma}$, namely the energy of the radiative photon and the angle between the positron and the photon:

$$R_j = \frac{BR(K_{e3\gamma}|E^j_\gamma, \theta^j_{e,\gamma})}{BR(K_{e3})} \tag{5}$$

The most common definition of the restricted phase space regions are reported in table 1, together with the theoretical expectations from ChPT and the most recent results obtained by the ISTRA+ and OKA experiments.

Table 1. Definitions of R_j, preditions from ChPT and result from ISTRA+ and OKA experiments.

	E^j_γ	$\theta^j_{e,\gamma}$	$O(p^6)$ *ChPT*	ISTRA+	OKA
$R_1 \times 10^2$	$E_\gamma > 10$ MeV	$\theta_{e,\gamma} > 10°$	1.804 ± 0.021	$1.81 \pm 0.03 \pm 0.07$	$1.990 \pm 0.017 \pm 0.021$
$R_2 \times 10^2$	$E_\gamma > 30$ MeV	$\theta_{e,\gamma} > 20°$	0.640 ± 0.008	$0.63 \pm 0.02 \pm 0.03$	$0.587 \pm 0.010 \pm 0.015$
$R_3 \times 10^2$	$E_\gamma > 10$ MeV	$0.6 < \cos(\theta_{e,\gamma}) < 0.9$	0.559 ± 0.006	$0.47 \pm 0.02 \pm 0.03$	$0.532 \pm 0.010 \pm 0.012$

The process can be used to study possible violations of T-parity by measuring the T-odd observable ξ and the relative asymmetry:

$$\xi = \frac{\vec{p}_\gamma \cdot (\vec{p}_e \times \vec{p}_\pi)}{M_K^3} \quad ; \quad A = \frac{N_+ - N_-}{N_+ + N_-} \tag{6}$$

In the second equation, N_+ and N_- correspond to the number of events with positive and negative values of ξ respectively.

The analysis is performed by using the $K(e3)$ decay as a normalization channel, in order to cancel out most of the systematic uncertainties. The main background is represented by K_{e3} events with accidental activity in the electromagnetic calorimeter, that can mimic the radiative photon. The background contribution has been evaluated on data using the signal sidebands.

The preliminary results for the ratios R^j are:

$$R_1 = 1.684 \pm 0.005 \pm 0.010) \times 10^{-2} \tag{7}$$

$$R_2 = 0.599 \pm 0.003 \pm 0.005) \times 10^{-2} \tag{8}$$

$$R_3 = 0.523 \pm 0.003 \pm 0.003) \times 10^{-2} \tag{9}$$

In all cases, the results improved the experimental precision by a factor between 2 and 3 compared to the previous results. The asymmetries have been computed using (6) and corrected for the offset induced by the reconstruction and the selection, estimated using a MonteCarlo simulation. The results are reported in fig. 3. No statistically relevant differences with zero have been observed.

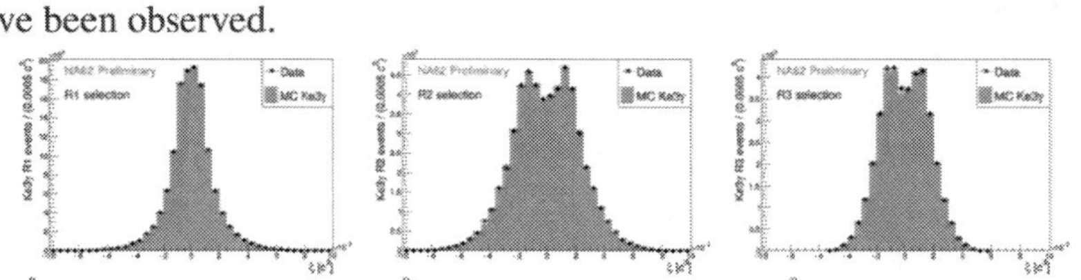

$A_\xi^{R_1} = (-0.1 \pm 0.3_{stat} \pm 0.2_{MC}) \times 10^2$ $A_\xi^{R_2} = (-0.3 \pm 0.4_{stat} \pm 0.3_{MC}) \times 10^2$ $A_\xi^{R_3} = (-0.9 \pm 0.5_{stat} \pm 0.4_{MC}) \times 10^2$

Figure 3. Data and MonteCarlo distribution for the ξ observables in the three restricted phase space regions. The corresponding results for the asymmetries are reported.

3.3 Measurement of $BR(K^+ \to \pi^+\mu^+\mu^-)$

$K^\pm \to \pi^\pm l^+ l^-$ decays are FCNC processes described through the exchange of a virtual photon, $K^\pm \to \pi^\pm \gamma^* \to \pi^\pm l^+ l^-$ involving long-distance hadronic effects. Much theoretical effort has been put into completely describing the processes involved out [7–9]. Precision measurements of the ($K_{\pi ll}$) branching ratios can provide important tests of Lepton Flavour Universality (LFU). The form factors are parametrized by the coefficients a_+ and b_+; NA62 has made a new measurement of the branching ratio for the $K^+ \to \pi^+\mu^+\mu^-$ ($K_{\pi\mu\mu}$) decay and of the two form factor parameters using data collected in 2017 and 2018.

The normalization channel is the $K^+ \to \pi^+\pi^+\pi^-$ decay and a common selection has been adopted, based on the reconstruction of a three-track vertex in the final state, matched with an upstream kaon, and on the absence of positrons in the downstream detectors. The signal and normalization channels have particle identification requirements. The muons in the signal channel are identified by activity in the MUV3 detector, compatible with a track in space and time. A track with no associated MUV3 activity is identified as a pion. This selection leads to a $K_{\pi\mu\mu}$ sample of 27679 events, with a signal acceptance of $(8.688 \pm 0.009)\%$. The invariant mass for the event passing the signal selection is reported in fig. 4.

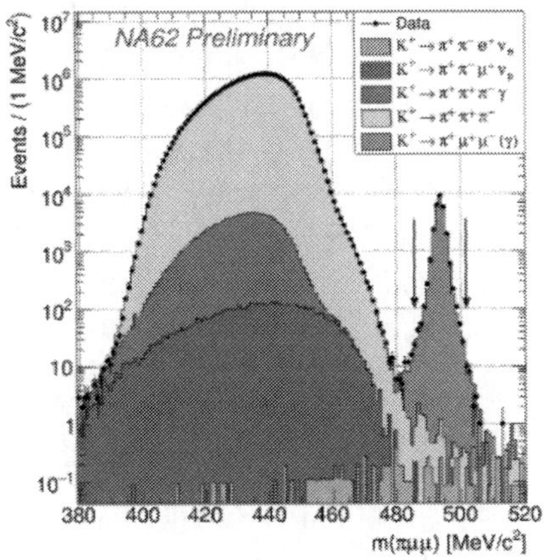

Figure 4. Invariant mass distribution for the event passing the $K_{\pi\mu\mu}$ selection. MC samples of the signal and the main background are shown.

The results for the branching ratio and for the form factor coefficients are reported in Table 2.

Table 2. Results obtained for the form factor coefficient and for $BR(K_{\pi\mu\mu})$.

	a_+	b_+	$BR(K_{\pi\mu\mu}) \times 10^8$
Best fit	-0.592	-0.699	9.27
Errors	δa_+	δb_+	$\delta BR(K_{\pi\mu\mu}) \times 10^8$
Statistical	0.013	0.046	0.07
Systematic			
Reconstruction efficiency	0.005	0.026	0.06
Beam and pileup simulation	0.005	0.024	0.05
Trigger efficiency	0.001	0.005	0.005
Background	-	0.001	0.004
Total systematic	0.007	0.035	0.08
External			
PDG error on $BR(K_{3\pi})$	0.001	0.003	0.04
Total	0.015	0.058	0.11

3.4 Searches for lepton number and lepton flavour violations

In the SM, Lepton Number (LN) and Lepton Flavour (LF) are conserved quantities not related to a symmetry principle of the theory and therefore considered as accidental symmetries. For this reason, many models of new physics beyond the SM contains LN and LF violating processes, mediated for instance by Majorana neutrinos, Z' bosons or leptoquarks.

Thanks to the large sample of K^+ decays collected, the NA62 experiment is able to perform many different searches for LNV and LFV in kaon and neutral pion decays. One of the most recent results obtained by NA62 in this context is the search for the LNV decays $K^+ \to \pi^-(\pi^0)e^+e^+$.

The signal rates have been measured normalizing to the SM decay $K^+ \to \pi^+e^+e^-$, which means most systematic uncertainties coming from detector or trigger inefficiencies and pileup activity in the experimental setup cancel. The common selection for both signal and normalization is based on the reconstruction of a 3-track vertex in the the final state, compatible with the decay of a kaon and made by one $\pi^\pm$ and two $e^\pm$; particles are identified by the measurement of the ratio E/p between the energy deposited in the electromagnetic calorimeter and the track momentum measured by the spectrometer. To suppress background from Dalitz decays of the $\pi 0$ events with in-time activity in the Large angle veto are rejected. Further e^+ identification requirements, using the RICH detector, are applied to reduce the background to the signal. Finally, a requirement on the invariant mass is performed and the signal region has been kept blind until the validation of the background estimate.

The main background contribution comes from $\pi^\pm$ mis-identification as $e^\pm$ and vice versa. Pion and positron mis-identification probabilities and identification efficiencies using the ratio E/p and the RICH have been estimated by data as a function of the particle momentum with pure $K^+ \to \pi^+\pi^+\pi^-$ and $K^+ \to \pi^0e^+\nu$ control samples respectively. The probabilities obtained from the control samples have been used as models in the MonteCarlo simulation.

The total number of K^+ decays selected is 10^{12}, obtained from 11041 SM $K^+ \to \pi^+e^+e^-$ events, which is the largest sample of the decay mode ever collected. After the validation of the background estimate, performed using control regions adjacent to the signal regions, no events have been found for both $K^+ \to \pi^-e^+e^+$ and $K^+ \to \pi^-\pi^0e^+e^+$ (fig. 5). As a result, the following 90% CL upper limits on the branching ratios of the LNV decay modes are obtained [10]:

$$BR(K^+ \to \pi^-e^+e^+) < 5.3 \times 10^{-11} \quad (10)$$

$$BR(K^+ \to \pi^-\pi^0e^+e^+) < 8.5 \times 10^{-10} \quad (11)$$

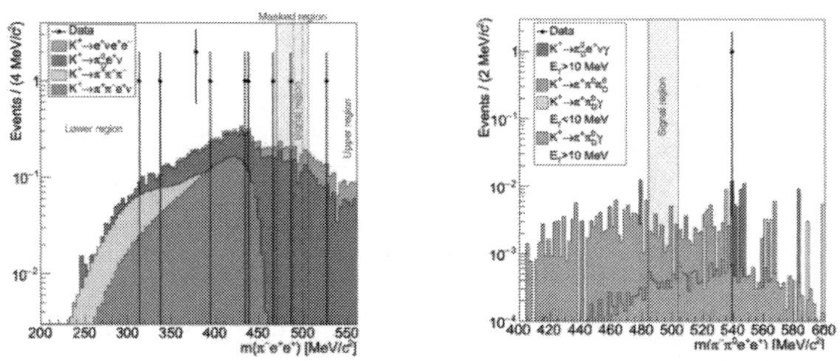

Figure 5. Invariant mass for $K^+ \to \pi^-e^+e^+$ (left) and $K^+ \to \pi^-\pi^0e^+e^+$ (right) decay modes. Signal regions and main background contributions are shown.

4 Conclusions

The first measurement of the $BR(K^{00}_{\mu 4})$ has been performed by the NA48/2 collaboration, showing an agreement with the theoretical expectation from ChPT. The NA62 experiment performed the most precise measurement of $BR(K^+ \to \pi^+ \nu\bar{\nu})$, based on the full Run1 sample (2016-2018) with a 3.4σ significance on the observation of the decay. The NA62 physics programme, however, is quite broad: and important results have been presented in the context of precision measurement of branching ratios of kaon decays ($K_{e3\gamma}$ and $K_{\pi\mu\mu}$), with improved precision in both cases with respect to previous results. In the searches for LNV decays from kaons and neutral pions, improved 90% CL upper limits have been obtained. Other LNV and LFV searches performed by NA62 can be found in [11].

The experiment restarted its data taking in 2021 and will continue until 2026, with the aim of measuring $BR(K^+ \to \pi^+ \nu\bar{\nu})$ to 10% precision.

References

[1] V. Fanti et al. [NA48 Collaboration], Nucl. Instrum. Methods A **574** 443 (2007)
[2] E. Cortina Gil et al., JINST **12** P05025 (2017)
[3] J.Bijnens, G.Colangelo, J.Gasser, Nucl.Phys. B **427** (1994)
[4] The NA62 collaboration., Cortina Gil, E., Kleimenova, A. et al., J. High Energ. Phys. **2021**, 93 (2021)
[5] Akimenko, S.A., Bolotov, V.N., Britvich, G.I. et al., Phys. Atom. Nuclei **70**, 702–708 (2007)
[6] Polyarush, A.Y., Akimenko, S.A., Artamonov, A.V. et al., Eur. Phys. J. C **81**, 161 (2021)
[7] G. Ecker, A. Pich, and E. de Rafael, Nucl. Phys., **B291** 692 (1987)
[8] G. D'Ambrosio, G. Ecker, G. Isidori, and J. Portoles, JHEP **08** 004 (1998)
[9] A. Dubnickova, S. Dubnicka, E. Goudzovski, V. Pervushin and M. Secansky, Phys. Part. Nucl. Lett. **5** 76 (2008)
[10] E. Cortina Gil et al., Phys. Lett. B **830**, 137172 (2022)
[11] R. Aliberti et al., Phys. Rev. Lett. **127**, 131802 (2021)

Computing one-loop radiative corrections in $\tau \to \pi(K)\nu[\gamma]$ and testing new physics

Ignasi Rosell[1,*]

[1]Departamento de Matemáticas, Física y Ciencias Tecnológicas, Universidad Cardenal Herrera-CEU, CEU Universities, 46115 Alfara del Patriarca, València, Spain

Abstract. Without a doubt the ratios $R_{\tau/P} \equiv \Gamma(\tau \to P\nu_\tau[\gamma])/\Gamma\big(P \to \mu\nu_\mu[\gamma]\big)$ $(P = \pi, K)$ give a convenient scenario to test the lepton universality, the CKM unitarity and to search for non-standard interactions in τ decays. Moreover, the previous theoretical estimation of these observables is more than twenty-five years old and some assumptions of that estimation are unrealistic. Therefore, an update of $R_{\tau/P}$ was timely. The one-loop radiative corrections have been computed by considering an expansion of Chiral Perturbation Theory including the lightest spin-one resonances and respecting the short-distance behavior dictated by QCD. We have reported $\delta R_{\tau/\pi} = (0.18 \pm 0.57)\%$ and $\delta R_{\tau/K} = (0.97 \pm 0.58)\%$, where the uncertainties have been induced fundamentally by the estimation of the counterterms. We have tested the lepton universality, the CKM unitarity and have searched for new physics in τ decays. As a by-product, we have also determined the theoretical radiative corrections to the $\tau \to P\nu_\tau[\gamma]$ decay rates, $\delta_{\tau\pi} = -(0.24 \pm 0.56)\%$ and $\delta_{\tau K} = -(0.15 \pm 0.57)\%$.

1 Introduction

Although within the Standard Model the three lepton families are expected to couple to the electroweak gauge bosons with the same intensity g_ℓ ($\ell = e, \mu, \tau$), the so-called lepton universality (LU), and this has been tested experimentally and most of the results have been compatible with LU, a few anomalies observed in semileptonic B meson decays [1] seem to indicate a small deviation from LU. Therefore, an interesting task is the study of all the different observables testing LU, including low-energy precision probes using pion, kaons and tau leptons, which currently provide the most precise tests of LU [2]. Following this idea, we have tested LU between the second and third families through the ratio ($P = \pi, K$) [3–5]

$$R_{\tau/P} \equiv \frac{\Gamma(\tau \to P\nu_\tau[\gamma])}{\Gamma\big(P \to \mu\nu_\mu[\gamma]\big)} = \left|\frac{g_\tau}{g_\mu}\right|_P^2 R^0_{\tau/P}(1 + \delta R_{\tau/P}), \tag{1}$$

where $g_\mu = g_\tau$ according to LU, $\delta R_{\tau/P}$ denotes the radiative corrections, and $R^{(0)}_{\tau/P}$ is the tree-level contribution given by

$$R^{(0)}_{\tau/P} = \frac{1}{2}\frac{M_\tau^3}{m_\mu^2 m_P}\frac{(1 - m_P^2/M_\tau^2)^2}{(1 - m_\mu^2/m_P^2)^2}, \tag{2}$$

*e-mail: rosell@uchceu.es

which is free from quark mixing angles and hadronic couplings.

In Ref. [4], more than twenty-five years ago, the first complete estimation for $\delta R_{\tau/P}$ was given: $\delta R_{\tau/\pi} = (0.16 \pm 0.14)\%$ and $\delta R_{\tau/K} = (0.90 \pm 0.22)\%$. However, there were important reasons to analysis this observable again: the hadronic form factors of Ref. [4] are different for virtual- and real-photon contributions, do not follow the correct QCD high-energy behavior, violate analyticity, unitarity, and the chiral limit at leading non-trivial orders; moreover, a cutoff is implemented in order to regulate loop integrals, splitting unphysically short- and long-distance contributions. Besides, errors reported in Ref. [4] are not realistic, since they are of the order of an expected $O(e^2 p^2)$ Chiral Perturbation Theory (ChPT) calculation, whose only model-dependence is usually the estimation of the counterterms, but it does not allow the inclusion of the τ lepton.

Depending on the process at hand, different values of $\left|g_\tau/g_\mu\right|$ can be found in the literature:

1. $\Gamma(\tau \to P\nu_\tau[\gamma])/\Gamma(P \to \mu\nu_\mu[\gamma])$ $(P = \pi, K)$. Considering the values of $\delta R_{\tau/P}$ given in Ref. [4], the 2018 HFLAV analysis [6] reported $\left|g_\tau/g_\mu\right|_\pi = 0.9958 \pm 0.0026$ and $\left|g_\tau/g_\mu\right|_K = 0.9879 \pm 0.0063$, at 1.6σ and 1.9σ of LU.

2. $\Gamma(\tau \to e\bar{\nu}_e\nu_\tau[\gamma])/\Gamma(\mu \to e\bar{\nu}_e\nu_\mu[\gamma])$. This purely leptonic extraction gives $\left|g_\tau/g_\mu\right| = 1.0010 \pm 0.0014$ [6], at 0.7σ of LU.

3. $\Gamma(W \to \tau\nu_\tau)/\Gamma(W \to \mu\nu_\mu)$. The weighted average of the W-boson decay determinations yields $\left|g_\tau/g_\mu\right| = 0.995 \pm 0.006$ [7, 8], at 0.8σ of LU.

Thus, a new estimation of $\delta R_{\tau/P}$ was interesting to solve these discrepancies, being the improvement of the theoretical Ansätze an important ingredient.

In Ref. [5] we have presented a new next-to-leading calculation of $\delta R_{\tau/P}$ considering a large-N_C effective approach including the lightest resonances [9] and overcoming all the aforementioned theoretical difficulties. Note that whereas P decays can be analyzed unambiguously by using Chiral Perturbation Theory (the Standard Model), being the estimation of the local counterterms the only model dependence, τ decays must be studied considering a model-dependent effective approach encoding the hadronization of the QCD currents: this is the reason why we have considered in Ref. [5] the large-N_C approach quoted previously [9].

The new calculation of $\delta R_{\tau/P}$ has been used not only to analyze the LU in $\left|g_\tau/g_\mu\right|$, but also to report the theoretical radiative corrections in individual $\tau \to P\nu_\tau[\gamma]$ decay rates, to test the CKM unitarity via $|V_{us}/V_{ud}|$ in $\Gamma(\tau \to K\nu_\tau[\gamma])/\Gamma(\tau \to \pi\nu_\tau[\gamma])$ or through $|V_{us}|$ in $\Gamma(\tau \to K\nu_\tau[\gamma])$ and to search for non-standard interactions on these decays. It is interesting to stress that our results have been used in the recent 2021 HFLAV analysis [10] instead of Ref. [4].

2 The calculation

Despite their different approaches, $P \to \mu\nu_\mu[\gamma]$ and $\tau \to P\nu_\tau[\gamma]$ decay rates can be organized similarly [3, 5, 11]:

$$\Gamma_{P_{\mu2[\gamma]}} = \Gamma^{(0)}_{P_{\mu2}} S_{\rm EW}\left\{1 + \frac{\alpha}{\pi}F(m_\mu^2/m_P^2)\right\}\left\{1 - \frac{\alpha}{\pi}\left[\frac{3}{2}\log\frac{m_\rho}{m_P} + c_1^{(P)} + \frac{m_\mu^2}{m_\rho^2}\left(c_2^{(P)}\log\frac{m_\rho^2}{m_\mu^2} + c_3^{(P)} + c_4^{(P)}(m_\mu/m_P)\right) - \frac{m_P^2}{m_\rho^2}\tilde{c}_2^{(P)}\log\frac{m_\rho^2}{m_\mu^2}\right]\right\}, \quad (3)$$

$$\Gamma_{\tau_{P2[\gamma]}} = \Gamma^{(0)}_{\tau_{P2}} S_{\rm EW}\left\{1 + \frac{\alpha}{\pi}G(m_P^2/M_\tau^2)\right\}\left\{1 - \frac{3\alpha}{2\pi}\log\frac{m_\rho}{M_\tau} + \delta_{\tau P}\big|_{\rm rSD} + \delta_{\tau P}\big|_{\rm vSD}\right\}, \quad (4)$$

Table 1. Numerical values of the different contributions to $\delta R_{\tau/P}$: Structure Independent (SI), real-photon Structure Dependent (rSD) and virtual-photon Structure Dependent (vSD). Uncertainties are not given if they are negligible for the level of accuracy of this analysis, that is, lower than 0.01%.

Contribution	$\delta R_{\tau/\pi}$	$\delta R_{\tau/K}$	Ref.
SI	+1.05%	+1.67%	[4]
rSD	+0.15%	+(0.18 ± 0.05)%	[11, 13]
vSD	−(1.02 ± 0.57)%	−(0.88 ± 0.58)%	[5, 11]
Total	+(0.18 ± 0.57)%	+(0.97 ± 0.58)%	[5]

where $S_{\rm EW} = 1.0232 \simeq 1 + \frac{2\alpha}{\pi}\log\frac{m_Z}{m_\rho}$ corresponds to the (universal) leading short-distance electroweak correction [3] (canceling in the ratio $R_{\tau/P}$), the first bracketed terms are the universal long-distance or point-like correction [4, 5, 12], the second bracketed terms include the structure-dependent (SD) contributions and $\Gamma^{(0)}_{P_{\mu2}}$ and $\Gamma^{(0)}_{\tau_{P2}}$ are the decay rate at leading order ($F_\pi \sim 92$ MeV),

$$\Gamma^{(0)}_{P_{\mu2}} = \frac{G_F^2|V_{uD}|^2F_P^2}{4\pi}\, m_P\, m_\mu^2\left(1-\frac{m_\mu^2}{m_P^2}\right)^2, \qquad \Gamma^{(0)}_{\tau_{P2}} = \frac{G_F^2|V_{uD}|^2F_P^2}{8\pi} M_\tau^3\left(1-\frac{m_P^2}{M_\tau^2}\right)^2, \tag{5}$$

being $D = d, s$ for $P = \pi, K$, respectively. Whereas for structure-dependent contributions in $P_{\mu2}$ the notation of Ref. [3] has been used and the numerical values for $c_n^{(P)}$ in Ref. [11] have been considered, we have separated the structure-dependent contributions in τ_{P2} into real-photon (rSD) and virtual-photon (vSD) corrections: we have taken rSD corrections from Ref. [13] and have performed a new calculation for vSD ones.

The technicalities of the calculation are given in Ref. [5] and the different photonic contributions are showed in Table 1 . The final result reads [5]

$$\delta R_{\tau/\pi} = (0.18 \pm 0.57)\%\,, \qquad \delta R_{\tau/K} = (0.97 \pm 0.58)\%\,. \tag{6}$$

The most relevant source of uncertainty comes from the estimation of the counterterms appearing in vSD contributions of $\tau \to P\nu_\tau$ (±0.57% and ±0.58% for the pion and kaon case, respectively, see Table 1). This uncertainty has been estimated by considering the running of the counterterms between 0.5 and 1.0 GeV (±0.52%, in a similar way to Ref. [3]) and considering the effect of assuming a less general resonance effective Lagrangian.[1]

Although the final result reported in (6) is compatible with Ref. [4], $\delta R_{\tau/\pi} = (0.16 \pm 0.14)\%$ and $\delta R_{\tau/K} = (0.90 \pm 0.22)\%$, the comparison needs to be done carefully. Firstly, in our view uncertainties were underestimated in Ref. [4], for they have approximately the expected size in a purely Chiral Perturbation Theory computation, a much more model-independent scenario. And secondly, we would like to stress again the inconsistencies in Ref. [4]: the hadronic form factors are different for real- and virtual-photon corrections, do not satisfy the asymptotic behavior dictated by QCD, violate analyticity, unitarity and the chiral limit at leading non-trivial orders, and a cutoff is used, separating for no reason short- and long-distance contributions. Moreover, as it is explained in Appendix B of Ref. [5], despite the

[1]Let us stress the conservative estimation followed here. Due to the different scales, counterterms in effective approaches with resonances are expected to be lower than in ChPT, without resonances. Nevertheless, with our estimation the counterterms affecting the vSD corrections in $P_{\mu2}$ and τ_{P2} are assumed to be of similar size.

Table 2. Comparison of photonic contributions to $\delta R_{\tau/\pi}$ [5]. The asterisk indicates that the figure at hand depends on the cutoff, which has been established at $\mu_{\rm cut} = 1.5$ GeV [4].

Contribution	Ref. [4] [$\mu_{\rm cut} = 1.5$ GeV]	Ref. [5]
SI	$+0.84\%^*$	$+1.05\%$
rSD	$+0.05\%$	$+0.15\%$
vSD	$-0.49\%^*$	$-(1.02 \pm 0.57)\%$
short-distance	$-0.25\%^*$	0
Total	$+(0.16 \pm 0.14)\%^*$	$+(0.18 \pm 0.57)\%$

fact that central values of (6) are very similar to the ones reported in Ref. [4], central values of the different SD corrections are very different within both analysis, as it can be seen in Table 2:

1. Virtual corrections by Ref. [4] are cutoff-dependent, since, as it has been spotlighted previously, long- and short-distance photonic contributions are separated unphysically: figures with an asterisk are cutoff-dependent.

2. The reported error in the radiative correction of Ref. [4] arises from uncertainties in hadronic parameters of SD contributions and from variations in $\mu_{\rm cut}$.

3. For the SI contribution in Ref. [4] we have added to the result obtained in the point-like approximation (1.05%), the term coming from cutting off the loops at $\mu_{\rm cut}$ (−0.21%).

4. Different contributions of $\delta R_{\tau/K}$ are not provided in Ref. [4], preventing a comparison.

Consequently, the agreement between the central values seems to be only a coincidence.

3 Applications

As it has been explained in the introduction, there are different interesting applications of our result [5]:

1. Radiative corrections in $\tau \to P\nu_\tau[\gamma]$ decay rates. We have used our results to estimate the radiative corrections in the individual $\tau_{P2[\gamma]}$ decays, $\Gamma_{\tau_{P2[\gamma]}} = \Gamma^{(0)}_{\tau_{P2}} S_{\rm EW}\,(1+\delta_{\tau P})$, where $\delta_{\tau P}$ includes all SI and SD radiative corrections of (4) and we have reported [5]:

$$\delta_{\tau\pi} = -(0.24 \pm 0.56)\%\,, \qquad \delta_{\tau K} = -(0.15 \pm 0.57)\%\,. \tag{7}$$

2. Lepton universality test. Considering (1) and (6), the LU can be tested [5],

$$\begin{aligned} \left|\frac{g_\tau}{g_\mu}\right|_\pi &= 0.9964 \pm 0.0028_{\rm th} \pm 0.0025_{\rm exp} = 0.9964 \pm 0.0038\,, \\ \left|\frac{g_\tau}{g_\mu}\right|_K &= 0.9857 \pm 0.0028_{\rm th} \pm 0.0072_{\rm exp} = 0.9857 \pm 0.0078\,, \end{aligned} \tag{8}$$

at 0.9σ and 1.8σ of LU ($g_\tau = g_\mu$), respectively. Note that while experimental and theoretical uncertainties are of similar size in the pion case, experimental ones dominate in the kaon case. We can compare these results with the 2018 HFLAV analysis [6], $|g_\tau/g_\mu|_\pi = 0.9958 \pm 0.0026$ and $|g_\tau/g_\mu|_K = 0.9879 \pm 0.0063$ (at 1.6σ and

1.9σ of LU), where $\delta R_{\tau/P}$ was taken from Ref. [4]. In the recent HFLAV analysis of Ref. [10], where our results of (6) have been considered, the agreement is larger: $\left|g_\tau/g_\mu\right|_\pi = 0.9959 \pm 0.0038$ and $\left|g_\tau/g_\mu\right|_K = 0.9855 \pm 0.0075$ (at 1.1σ and 1.9σ of LU).

3. CKM unitarity test via $|V_{us}/V_{ud}|$. One can use the ratio

$$\frac{\Gamma(\tau \to K\nu_\tau[\gamma])}{\Gamma(\tau \to \pi\nu_\tau[\gamma])} = \frac{|V_{us}|^2 F_K^2}{|V_{ud}|^2 F_\pi^2} \frac{(1-m_K^2/M_\tau^2)^2}{(1-m_\pi^2/M_\tau^2)^2} (1+\delta), \tag{9}$$

in order to extract $|V_{us}/V_{ud}|$. Considering that the radiative correction in (9) can be extracted from (7), $\delta = \delta_{\tau K} - \delta_{\tau\pi} = (0.10 \pm 0.80)\%$, we have found [5]:

$$\left|\frac{V_{us}}{V_{ud}}\right| = 0.2288 \pm 0.0010_{\text{th}} \pm 0.0017_{\text{exp}} = 0.2288 \pm 0.0020\,, \tag{10}$$

at 2.1σ of unitarity, being dominant the experimental uncertainties. Despite the fact that this analysis was absent in the 2018 HFLAV analysis [6], it has been included in the 2021 version [10], where our result of (6) has been considered and it has been reported $|V_{us}/V_{ud}| = 0.2289 \pm 0.0019$, very similar to (10). Our result is also consistent with Ref. [14], $|V_{us}/V_{ud}| = 0.2291 \pm 0.0009$, obtained in the context of kaon semileptonic decays, where the larger statistics allows lower uncertainties.

4. CKM unitarity test via $|V_{us}|$. Alternatively, $|V_{us}|$ can be obtained directly from the $\tau \to K\nu_\tau[\gamma]$ decay rate, $\Gamma_{\tau_{K2[\gamma]}} = \Gamma^{(0)}_{\tau_{K2}} S_{\text{EW}}\,(1 + \delta_{\tau K})$. We have found [5]:

$$|V_{us}| = 0.2220 \pm 0.0008_{\text{th}} \pm 0.0016_{\text{exp}} = 0.2220 \pm 0.0018\,, \tag{11}$$

at 2.6σ of unitarity and again being dominant the experimental uncertainties. This result can be compared with the 2018 HFLAV analysis [6], $|V_{us}| = 0.2234 \pm 0.0015$ and using Ref. [4], or the recent 2021 HFLAV analysis [10], $|V_{us}| = 0.2219 \pm 0.0017$ and considering our results. Again (11) is compatible with Ref. [14], $|V_{us}| = 0.2231 \pm 0.0006$, obtained using kaon semileptonic decays and, consequently, with a better precision.

5. New physics in $\tau \to P\nu_\tau[\gamma]$ decay rates. We have also used our results to constrain non-SM interactions in $\tau \to P\nu_\tau[\gamma]$ decays:

$$\Gamma(\tau \to P\nu_\tau[\gamma]) = \Gamma^{(0)}_{\tau_{P2}} \left|\frac{\widetilde{V}_{uD}}{V_{uD}}\right|^2 S_{\text{EW}} \left(1 + \delta_{\tau P} + 2\Delta^{\tau P}\right), \tag{12}$$

being $D = d, s$ for $P = \pi, K$, respectively. $\Delta^{\tau P}$ contains the leading-order new-physics corrections not present in $\widetilde{V}_{uD} = (1 + \epsilon_L^e + \epsilon_R^e)V_{uD}$, directly incorporated by considering V_{uD} from nuclear β decays [15, 16]:

$$\Delta^{\tau P} = \epsilon_L^\tau - \epsilon_L^e - \epsilon_R^\tau - \epsilon_R^e - \frac{m_P^2}{M_\tau(m_u + m_D)}\epsilon_P^\tau\,. \tag{13}$$

Considering our estimations of $\delta_{\tau P}$ and $|V_{us}/V_{ud}|$ given in (7) and (10), respectively, we have estimated [5]:

$$\Delta^{\tau\pi} = -(0.15 \pm 0.72)\%\,, \qquad \Delta^{\tau K} = -(0.36 \pm 1.18)\%\,, \tag{14}$$

to be compared to $\Delta^{\tau\pi} = -(0.15\pm0.67)\%$ in Ref. [15], $\Delta^{\tau\pi} = -(0.12\pm0.68)\%$ and $\Delta^{\tau K} = -(0.41 \pm 0.93)\%$ in Ref. [16], and $\Delta^{\tau\pi} = -(0.09 \pm 0.73)\%$ and $\Delta^{\tau K} = -(0.2 \pm 1.0)\%$ in Ref. [17]. All these values are reported at a scale of $\mu = 2$ GeV in the $\overline{\text{MS}}$-scheme. Note that all these determinations are consistent with each other and compatible with the SM, $\Delta^{\tau P} = 0$.

For all these determinations branching ratios and masses have been obtained from the PDG [18], $|V_{ud}| = 0.97373 \pm 0.00031$ from Ref.[19], $S_{\rm EW} = 1.0232$ from Ref. [3] and meson decay constants from the FLAG analysis [20]: $\sqrt{2}F_\pi = (130.2 \pm 0.8)$ MeV, $\sqrt{2}F_K = (155.7 \pm 0.3)$ MeV and $F_K/F_\pi = 1.1932 \pm 0.0019$, .

I wish to thank the organizers of the workshop for the pleasant conference. I am also very grateful to M. A. Arroyo-Ureña, G. Hernández-Tomé, G. López-Castro and P. Roig, since the work presented here has been done with them and also for their helpful comments to prepare these proceedings. This work has been supported in part by the Universidad Cardenal Herrera-CEU [INDI21/15]; by the Generalitat Valenciana [PROMETEU/2021/071] and by the Spanish Government [MCIN/AEI/10.13039/501100011033, Grant No. PID2020-114473GB-I00].

References

[1] J. Albrecht, D. van Dyk and C. Langenbruch, "Flavour anomalies in heavy quark decays," Prog. Part. Nucl. Phys. **120**, 103885 (2021) [arXiv:2107.04822 [hep-ex]]

[2] D. Bryman, V. Cirigliano, A. Crivellin and G. Inguglia, "Testing Lepton Flavor Universality with Pion, Kaon, Tau, and Beta Decays," [arXiv:2111.05338 [hep-ph]]

[3] W. J. Marciano and A. Sirlin, "Radiative corrections to pi(lepton 2) decays," Phys. Rev. Lett. **71** (1993), 3629-3632

[4] R. Decker and M. Finkemeier, "Radiative corrections to the decay tau → pi (K) tau-neutrino," Phys. Lett. B **316** (1993), 403-406 [arXiv:hep-ph/9307372 [hep-ph]]; "Radiative corrections to the decay tau → pi (K) tau-neutrino. 2," Phys. Lett. B **334** (1994), 199-202; "Short and long distance effects in the decay tau → pi tau-neutrino (gamma)," Nucl. Phys. B **438** (1995), 17-53 [arXiv:hep-ph/9403385 [hep-ph]]; "Radiative corrections to the decay tau → pi tau-neutrino," Nucl. Phys. B Proc. Suppl. **40** (1995), 453-461 [arXiv:hep-ph/9411316 [hep-ph]]

[5] M. A. Arroyo-Ureña, G. Hernández-Tomé, G. López-Castro, P. Roig and I. Rosell, "Radiative corrections to $\tau \to \pi(K)\nu_\tau[\gamma]$: a reliable new physics test," Phys.Rev.D 104 (2021) 9, L091502 [arXiv:2107.04603 [hep-ph]]; "One-loop determination of $\tau \to \pi(K)\nu_\tau[\gamma]$ branching ratios and new physics tests," JHEP **02** (2022), 173 [arXiv:2112.01859 [hep-ph]]

[6] Y. S. Amhis *et al.* [HFLAV], "Averages of b-hadron, c-hadron, and τ-lepton properties as of 2018," Eur.Phys.J.C 81 (2021) 3, 226, [arXiv:1909.12524 [hep-ex]]

[7] G. Aad *et al.* [ATLAS], "Test of the universality of τ and μ lepton couplings in W-boson decays with the ATLAS detector," Nature Phys. **17** (2021) no.7, 813-818 [arXiv:2007.14040 [hep-ex]]

[8] CMS Collaboration, "A precision measurement of the W boson decay branching fractions in pp collisions at $\sqrt{s} = 13$ TeV," CMS-PAS-SMP-18-011

[9] G. Ecker, J. Gasser, A. Pich and E. de Rafael, "The Role of Resonances in Chiral Perturbation Theory," Nucl. Phys. B **321** (1989), 311-342;
G. Ecker, J. Gasser, H. Leutwyler, A. Pich and E. de Rafael, "Chiral Lagrangians for Massive Spin 1 Fields," Phys. Lett. B **223** (1989), 425-432;
V. Cirigliano, G. Ecker, M. Eidemüller, R. Kaiser, A. Pich and J. Portolés, "Towards a consistent estimate of the chiral low-energy constants," Nucl. Phys. B **753** (2006), 139-177 [arXiv:hep-ph/0603205 [hep-ph]]

[10] Y. Amhis *et al.* [HFLAV], "Averages of b-hadron, c-hadron, and τ-lepton properties as of 2021," [arXiv:2206.07501 [hep-ex]]

[11] V. Cirigliano and I. Rosell, "Two-loop effective theory analysis of pi (K) → e anti-nu/e [gamma] branching ratios," Phys. Rev. Lett. **99** (2007), 231801 [arXiv:0707.3439 [hep-ph]]; "pi/K → e anti-nu(e) branching ratios to O(e^2 p^4) in Chiral Perturbation Theory," JHEP **10** (2007), 005 [arXiv:0707.4464 [hep-ph]]

[12] T. Kinoshita, "Radiative corrections to pi - e decay," Phys. Rev. Lett. **2** (1959), 477

[13] Z. H. Guo and P. Roig, "One meson radiative tau decays," Phys. Rev. D **82** (2010), 113016 [arXiv:1009.2542 [hep-ph]]

[14] C. Y. Seng, D. Galviz, W. J. Marciano and U. G. Meißner, "An update on $|V_{us}|$ and $|V_{us}/V_{ud}|$ from semileptonic kaon and pion decays," [arXiv:2107.14708 [hep-ph]]

[15] V. Cirigliano, A. Falkowski, M. González-Alonso and A. Rodríguez-Sánchez, "Hadronic τ Decays as New Physics Probes in the LHC Era," Phys. Rev. Lett. **122** (2019) no.22, 221801 [arXiv:1809.01161 [hep-ph]]

[16] S. Gonzàlez-Solís, A. Miranda, J. Rendón and P. Roig, "Exclusive hadronic tau decays as probes of non-SM interactions," Phys. Lett. B **804** (2020), 135371 [arXiv:1912.08725 [hep-ph]]

[17] V. Cirigliano, D. Díaz-Calderón, A. Falkowski, M. González-Alonso and A. Rodríguez-Sánchez, "Semileptonic tau decays beyond the Standard Model," JHEP **04** (2022), 152 [arXiv:2112.02087 [hep-ph]]

[18] P. A. Zyla *et al.* [Particle Data Group], "Review of Particle Physics," PTEP **2020** (2020) no.8, 083C01

[19] J. C. Hardy and I. S. Towner, "Superallowed $0^+ \to 0^+$ nuclear β decays: 2020 critical survey, with implications for V_{ud} and CKM unitarity," Phys. Rev. C **102** (2020) no.4, 045501

[20] S. Aoki *et al.* [Flavour Lattice Averaging Group], "FLAG Review 2019: Flavour Lattice Averaging Group (FLAG)," Eur. Phys. J. C **80** (2020) no.2, 113 [arXiv:1902.08191 [hep-lat]]

Overview of most recent BESIII results

Marco Scodeggio[1,2,*], on behalf of the BESIII Collaboration

[1]Istituto Nazionale di Fisica Nucleare (INFN), Sezione di Ferrara, Ferrara, Italy
[2]Dipartimento di Fisica e Scienze della Terra, Università degli Studi di Ferrara, Ferrara, Italy

Abstract. Since 2009, the BESIII experiment has been colliding e^+e^- beams in the energy range between 2.0 and 4.9 GeV. This allowed the BESIII Collaboration to collect three of the largest data sets of charmonium resonances (J/ψ, $\psi(2S)$, and $\psi(3770)$) in the world as well as other data sets at the centre-of-mass energies above 3.8 GeV. These datasets allow BESIII to deepen the knowledge of the properties of charmonia; investigate the light hadrons spectra; probe the decay and production mechanism of charmed mesons; search for the exotic XYZ hadrons and to understand their nature; perform precision Quantum ChromoDynamics measurements (such as R-value measurements and Form Factors estimations). Some of the most recent and noteworthy results from the BESIII collaboration are presented.

1 Introduction

The BESIII (BEijing Spectrometer III) detector records symmetric e^+e^- collisions provided by the BEPCII (Beijing Electron Positron Collider II) storage ring, which operates in the centre-of-mass energy ($\sqrt{s}$) range from 2.0 to 4.9 GeV. The machine is hosted at the Institute of High Energy Physics (IHEP) of the Chinese Academy of Sciences (CAS) in Beijing, People's Republic of China. Details on the sub-detectors and their performance can be found in Ref. [1]. Typically referred to as a τ-charm factory, BESIII physics programme [2] covers charmonium(-like) and light hadrons spectroscopy, charmed mesons and baryons decays, τ studies, Quantum ChromoDynamics (QCD) measurements, and new physics searches.

Being BEPCII a leptonic collider, BESIII can profit from the direct production of vector states ($J^{PC} = 1^{--}$). This allowed the collaboration to gather the biggest datasets of the J/ψ, $\psi(2S)$, and $\psi(3770)$ resonances in the world. Owing to the statistics of these data sets and those at $\sqrt{s} > 3.8$ GeV, BESIII can probe a plethora of QCD features, ranging from the charmonium potential model to precision tests of QCD fundamental properties.

A compendium of the latest results of the BESIII collaboration is presented to give a glimpse of what the same can provide.

*e-mail: mscodegg@fe.infn.it

2 Charmonium(-like) States

Charmonium ($c\bar{c}$) resonances are located in the transition region of perturbative and non-perturbative QCD and are thus an optimal benchmark for probing strong interaction phenomenological features. Despite conventional charmonia fitting fairly well potential model predictions, non-vector states are still not entirely known (e.g., the $h_c(1^1P_1)$) [3]. Moreover, the $c\bar{c}$ spectrum features supernumerary (*XYZ*) states, which do not fit potential model predictions, show strong couplings to hidden charm states, and can exhibit a non-zero charge [4].

2.1 The Latest Addition... the $Z_{cs}(3985)^0$ Resonance

Using 5 data sets with an integrated luminosity ($\mathcal{L}_{\text{Int}}$) of 3.8 fb^{-1} at $\sqrt{s}$ = [4.628, 4.699] GeV, Ref. [5] reports the evidence of a near-threshold structure with a significance of 4.6σ in the $e^+e^- \to K^0_S(D^-_s D^{*+} + D^{*-}_s D^+)$ reaction.
To increase the final event tagging efficiency, a partial reconstruction technique is employed and only the neutral kaon and the non-excited charmed meson (either the D^-_s or the D^+) are identified. So, to tag the event, the $KK\pi$, $K^0_S K$, $KK\pi\pi^0$, $K^0_S K\pi\pi$, and $\eta'\pi$ decays are used to reconstruct the D^-_s mesons; while the $K\pi\pi$, $K^0_S\pi$, and $K^0_S\, 3\pi$ channels are used to form the D^+ states. All the K^0_S states are reconstructed through their charged pions decay. Moreover, to select the signal events, each D-meson signal candidate is requested to have its reconstructed mass within 15 MeV/c^2 of its established [3] mass. Before analysing the K^0_S recoil mass, the RQ(DK^0_S) variable[1] is requested not to lay beyond 20 MeV/c^2 (10 MeV/c^2) of the D^{*+} (D^{*-}_s). Finally, the K^0_S recoil mass distributions for the 5 energy points are obtained and fitted with a simultaneous unbinned maximum likelihood fit. The signal is described by a combination of two Breit-Wigner functions (taking into account the two possible decay modes, $D^-_s D^{*+}/D^{*-}_s D^+$) properly smeared taking into account the detector resolution and the mass-dependent efficiency; the background is modelled from MC samples (for the non-resonant process and higher excited mesonic channels) and using signal sidebands (for simulating the combinatorial background).

The $Z_{cs}(3985)^0$ coupling to $D^-_s D^{*+}/D^{*-}_s D^+$ suggests a minimum quark content of $c\bar{c}s\bar{d}$. Following Ref. [6], which predicts the Z^+_{cs} [7] to have a neutral isospin partner, the $Z_{cs}(3985)^0$ mass is found to be consistent with predictions (m($Z_{cs}(3985)^0$) > m(Z^+_{cs})) and the Born cross-sections (σ^{Born}) multiplied by the decays branching fraction found to be compatible with the charged partner.

2.2 Y and ψ States

2.2.1 Y Resonances

Using 28 energy points at $\sqrt{s}$ = [4.127, 4.600] GeV with a $\mathcal{L}_{\text{Int}}$ = 15.6 fb^{-1}, Ref. [8] studies the $e^+e^- \to K^+K^-J/\psi$ line-shape employing a partial reconstruction technique to increase the event selection efficiency. Only the most energetic kaon and the J/ψ ($\to \ell\ell$) resonance are identified. Signal yields (i.e., number of $e^+e^- \to K^+K^-J/\psi$ events) are then obtained from the 28 background-subtracted invariant J/ψ distributions. A maximum likelihood fit is then applied to the dressed cross-sections (i.e., σ^{Born} but including vacuum polarization effects), with the fit function being parameterized as a coherent sum of two relativistic

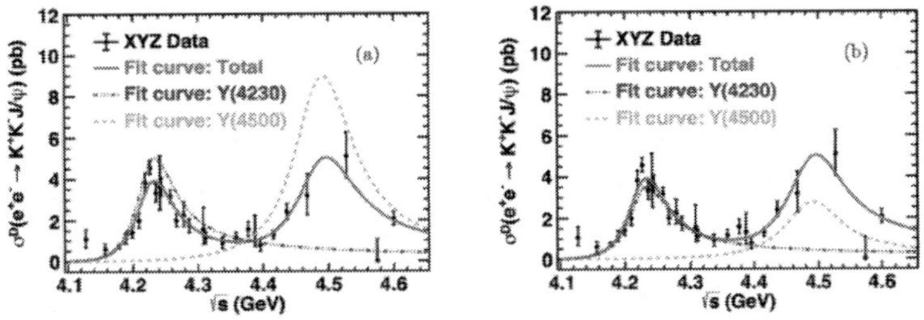

Figure 1. Fit solutions to the dressed cross-section of the $e^+e^- \to K^+K^-J/\psi$ channel. (a) corresponds to Solution I, (b) corresponds to Solution II.

Breit-Wigner distributions.

As Fig. 1 shows, both the solutions of the fit predict two resonances. The low-mass state is consistent with the $Y(4230)$ one, making it the first time this state is observed (at a significance of 29σ) in the K^+K^-J/ψ channel. The other resonance has no corresponding observed state, as such it is named $Y(4500)$. It is the first observation of this exotic Y-state with a significance of 8σ. The mass $Y(4500)$ is such that this state can be consistent with the 5S-4D mixing scheme [9], with the hadronic molecule model [10], or with a hidden-charm hidden-strange state [11].

2.2.2 ψ States

Similarly to the analysis presented in Sec. 2.2.1, Ref. [12] studies the $e^+e^- \to \pi^+\pi^-\psi_2(3823)$ cross-section using 20 energy points at $\sqrt{s}$ = [4.230, 4.700] GeV ($\mathcal{L}_{\text{Int}}$ = 11.3 fb^{-1}). To increase the signal events yield, the $\psi_2(3823)$ is partially reconstructed through its $\gamma\chi_{c1,c2} \to \gamma\ (\gamma J/\psi) \to \gamma\ (\gamma\ (\ell\ell)\ )$ decay, allowing one-missing-photon events. For each energy point, the $\pi^+\pi^-$ recoil mass (RM($\pi^+\pi^-$)) is then fitted to extrapolate the number of $e^+e^- \to \pi^+\pi^-\psi_2(3823) \to \pi^+\pi^-\gamma\chi_{c1,c2}$ events and the $\psi_2(3823)$ mass. This strategy allows estimating the most precise $\psi_2(3823)$ mass value up to date 3823.12 ± 0.43 ± 0.13 MeV/c^2 (cf., Ref. [3]).

No signal above the 3σ significance is found for the $\gamma\chi_{c1,c2} \to$ channel, hence, only the $e^+e^- \to \pi^+\pi^-\psi_2(3823) \to \pi^+\pi^-\gamma\chi_{c1}$ cross-section is fitted. Two structures are found, one corresponding to the $Y(4360)$ state and the other to the $Y(4660)$ resonance. With a significance greater than 5σ, this is the first observation of Y-states decaying to a D-wave charmonium resonance. Moreover, the sizeable observation of the $Y(4660)$ in this channel challenges both its $f_0(980)\psi(2S)$ hadron molecule interpretation [13] and the extended baryonium one [14], as the absence or at most a weak coupling to other non-$\psi(2S)$ charmonium states is essential for these two models.

2.3 Back to the Origins... Rediscovering the $h_c(1^1P_1)$

Despite extensive studies of the charmonium system, knowledge of the singlet state $h_c(1^1P_1)$ is sparse. Only nine decay modes have been observed, with the electromagnetic $h_c \to \gamma\eta_c$

[1]Essentially DK^0_S recoil mass, corrected by the difference between the reconstructed invariant mass of the signal D candidate, and its known PDG mass.

(E1) decay being the predominant one [3]. Refs. [15, 16] pursued the search for this state's non-E1 decays.

Using a $\mathcal{L}_{\text{Int}} = 11.3$ fb^{-1} ($\sqrt{s}$ = [4.189, 4.437] GeV) via the $e^+e^- \to \pi^+\pi^- h_c$ production channel, Ref. [15] searched for the hadronic transition $h_c \to \pi^0 J/\psi$ without finding any significant signal.

On the other hand, Ref. [16] searched for pure hadronic decay channels using the BESIII's 448 million of $\psi(2S)$ events via the $\psi(2S) \to \pi^0 h_c$. This study found evidence (4.9σ) for the $h_c \to p\bar{p}\pi^+\pi^-\pi^0$ decay and observed (5.1σ) the $h_c(1^1P_1)$ state decaying into $p\bar{p}\eta$ ($\eta \to \pi^+\pi^-\pi^0/\gamma\gamma$).

Using the $h_c(1^1P_1)$ state E1 decay and exploiting BESIII's $\psi(2S)$ data set, it is possible to provide another measurement of this state's mass and width. Ref. [17], using the $\psi(2S) \to \pi^0 h_c \to \pi^0$ $(\gamma\eta_c)$, gives, with an increased precision with respect to Ref. [3], the second measurement only of this state's width. This work increased the precision of the production and the decay branching fraction estimates, by a factor of two and three times concerning the PDG measurement [3]. Furthermore, with the $h_c(1^1P_1)$ state mass estimate, Ref. [17] provides the mass splitting between the pseudo-vector state and the centre-of-gravity mass of the three $\chi_{cJ}(1^3P_J)$ ones, finding it to be consistent to 0 at leading-order [18].

3 Light Hadrons

Despite the Naïve Quark Model describing conventional hadrons containing two or three quarks, QCD predicts an additional plethora of exotic states (hybrids, glueballs, mesonic molecules, etc.).
Indeed, also below the $c\bar{c}$ spectrum many supernumerary XYZ-like states are experimentally observed. Also here, in the light quarks regime, the nature of these exotic states is hard to determine, as they have overlapping masses with "regular" states.
Radiative J/ψ (i.e., $\to \gamma$ + had) decays are gluon-rich processes, ideal for studying light glueballs and hybrids. Owing to its statistics, the BESIII collaboration plays a fundamental role in the searches for these states.

3.1 A New Incognita... The $X(2600)$ State

Using the 10 billion J/ψ data set, Ref. [19] studies the radiative $J/\psi \to \gamma\pi^+\pi^-\eta'$ decay, reconstructing the η' meson from its $\gamma\pi^+\pi^-$ and η ($\to \gamma\gamma$) $\pi^+\pi^-$ main [3] decays. In said analysis, the invariant $\pi^+\pi^-\eta'$ mass (M($\pi^+\pi^-\eta'$)) is obtained and confronted to the $\pi^+\pi^-$ one (M($\pi^+\pi^-$)). In the M($\pi^+\pi^-\eta'$) distribution, apart from the already observed $X(1835)$, $X(2120)$, $X(2370)$, and η_c states, a new structure around 2.6 GeV/c^2 is found and is seen to be correlated with another one in the M($\pi^+\pi^-$) spectrum at 1.5 GeV/c^2.
To study the parameters and significance of this structure, a simultaneous fit, to the invariant $\pi^+\pi^-\eta'$ and $\pi^+\pi^-$ mass distributions is performed. The $X(2600)$, as this structure is called by Ref. [19], is found for the first time with a significance of more than 20σ, and its partner structure in the M($\pi^+\pi^-$) spectrum is described as an interference between the $f_0(1500)$ resonance and an $X(1540)$ state. Considering the $X(2600) \to f_0(1500)/X(1540)$ decay, the spin parity of the $X(2600)$ state is 0^{-+} or 2^{-+}. With these spin-parity assignments, the $X(2600)$ state can be a radial excitation of the η meson [20] or an exotic hadron.

3.2 Partial Wave Analyses on J/ψ Decays

Owing to its huge statistics (the aforementioned 10 billion J/ψ data set), the BESIII Collaboration can study with high precision gluon-rich processes (such as $J/\psi \to \gamma\eta\eta'$ and $J/\psi \to \gamma\eta'\eta'$) also via the Partial Wave Analysis (PWA) technique. Following the isobar model [21], the total amplitude of a decay is parameterised as a sum of sequential quasi-two-body processes and its intermediate resonances contributions are studied.

Ref. [22] studies the $J/\psi \to \gamma\eta\eta'$ decay. After event selection, the four-momenta of the reconstructed $\gamma\eta\eta'$ are used to perform the PWA fit by constructing the quasi-two-body decay amplitudes in the sequential $J/\psi \to \gamma X, X \to \eta\eta'$ and $J/\psi \to \eta' X, X \to \gamma\eta$ processes. Hence, the $\eta\eta'$ and $\gamma\eta$ spectra are first fitted using all the allowed resonances from Refs. [3, 23, 24]; then, the fit is re-performed using all those resonances which have a statistical significance larger than 5σ from the first iteration. Finally, a search is performed for additional resonances by individually adding them to the second iteration solution. It is found that only a resonance with $J^{PC} = 1^{-+}$ around 1.9 GeV/c^2 in the $\eta\eta'$ spectrum has a significant contribution (i.e., bigger than 20σ). This state, observed for the first time and called $\eta_1(1855)$, is an isoscalar resonance with exotic quantum numbers, and its parameters are consistent with LQCD calculations for the 1^{-+} hybrid [25]. Moreover, the absence of any $f_0(1710)$ resonance contribution in the PWA fit, supports the hypothesis that the $f_0(1710)$ meson overlaps with the ground state scalar glueball [26].

Ref. [27] focuses on the $J/\psi \to \gamma\eta'\eta'$ channel. Similarly to Ref. [22] , a PWA fit is performed to the $J/\psi \to \gamma\eta'\eta'$ radiative transition. In this study, the $f_0(2480)$ meson, a new scalar resonance (i.e., with $J^{PC} = 0^{++}$), is observed for the first time with a significance bigger than 5σ.

4 Charm Mesons for QCD

The BESIII collaboration can also probe the light hadron spectrum via the hadronic and semi-leptonic D_s decays. On these channels, Amplitude Analyses (AA) [21] are performed to study the intermediate resonance contributions to the decays. Moreover, to measure branching fractions, Single Tag (ST) - Double Tag (DT) analysis methods are employed. In the branching fraction estimation, the ST sample is used as a normalising factor, for this data set one of two D_s mesons is reconstructed from a specific hadronic decay; on the other hand, to ensure high signal purity, in the DT sample the D_s "signal" meson is reconstructed through the searched decay, while the other D_s (so-called "tag"), is reconstructed through several hadronic decays. The synergy of these two techniques allows to estimate SU(3) multiplets couplings and shine a light on their nature.

4.1 Back to the $f_0(1710)$

Using a $\mathcal{L}_{\text{Int}} = 6.32$ fb^{-1} ($\sqrt{s} = [4.178, 4.226]$ GeV), Ref. [28] searches for the $D_s^+ \to K_S^0 K_S^0 \pi^+$ decay in the $e^+e^- \to D_s^* D_s \to \gamma D_s D_s$ process. An AA is performed on the $K_S^0 K_S^0$ and $K_S^0 \pi^+$ invariant masses, reconstructed through a DT sample. The AA allows identifying a structure around 1.7 GeV/c^2 in the $K_S^0 K_S^0$ spectrum; this structure, referred to as $S(1710)$, is modelled as an admixture of the $f_0(1710)$ meson and its $a_0(1710)$ isovector partner. Using the ST sample, it is possible to estimate the $D_s^+ \to S(1710)\pi^+ \to K_S^0 K_S^0 \pi^+$ branching fraction, which is found consistent with Ref. [29], implying the existence of an isovector partner [30] of the $f_0(1710)$ meson and a constructive interference between them when decaying to neutral kaons.

4.2 The $f_0(1710)$... Glueball, Molecule, or Scalar Meson?

Using the same data set and process as Ref. [28], Ref [31] studies the $D_s^+ \to K_S^0 K^+ \pi^0$ decay. Using a DT sample, an AA is performed on the $K_S^0 K^+$, $K_S^0 \pi^0$, and $K^+ \pi^0$ invariant masses. The AA allows identifying a structure around 1.7 GeV/c^2 in the $K_S^0 K^+$ spectrum. The mass and width of this structure are measured and it is identified in the charged partner of the isovector state $a_0(1710)$ (i.e., $a_0(1710)^+$). Despite this, the mass value of the $a_0(1710)^+$ meson is found to be in tension (at a significance of 5σ) with the BaBar collaboration's one [32]. Using the ST sample, it is possible to estimate the $D_s^+ \to a_0(1710)^+$ branching fraction, which is found consistent with Ref. [33] predicting the $f_0(1710)$ being a $K\bar{K}^*$ molecule. Moreover, another possible interpretation comes from Ref. [34] suggesting this structure to be the isovector partner of the $X(1812)$ [35] scalar observed in $\omega\phi$ decay channel by the BES collaboration.

5 Baryon Studies

Finally, the BESIII collaboration can probe the perturbative regime of QCD by performing high-precision measurements such as R-value and Form Factors estimations.

5.1 R-Value Estimation

The R-value, which is the ratio between the $e^+e^- \to$ *hadrons* cross-section and the $e^+e^- \to \mu^+\mu^-$ one, allows the BESIII collaboration to probe the physics beyond the standard model, as it is fundamental for the determination of the anomalous muon magnetic moment, and to test quantum electrodynamics, as it enters in the running coupling constant calculations.

Using 14 energy points centre-of-mass energies between 2.2324 GeV and 3.6710 GeV, Ref. [36] provides the respective R-values. For such a measurement, good charged hadronic tracks are selected, while rejecting QED-related background. The trigger efficiency and the event selection are the least contributing uncertainties, thus, granting the most precise up-to-date measurement of the R-value for the aforementioned 14 energy points. Reaching an accuracy of 2.6% at $\sqrt{s} < 3.1$ GeV and 3.0% at $\sqrt{s} > 3.1$ GeV, the results are consistent with KEDR ones [37] and QCD predictions [38].

5.2 ElectroMagnetic Form Factors

An example of how the BESIII collaboration can perform regarding the form factors is represented in Ref. [39]. In this study, the $e^+e^- \to \Sigma^0\bar{\Sigma}^0$ Born cross-section at 7 centre-of-mass energies ($\sqrt{s}$ = [2.3864, 3.0200] GeV) is measured, providing the 7 corresponding Σ^0 effective form factors [Rev. Mod. Phys. 35, 335 (1963)]. For the data sets at $\sqrt{s} > 2.3960$ GeV, the process is reconstructed via Single Tag method, in which the Σ^0 baryon is completely reconstructed from its $\Sigma^0 \to \gamma\Lambda^0$, $\Lambda^0 \to p\pi^-$ decay chain. For the two energy points below 2.3960 GeV (i.e., near the $\Sigma^0\bar{\Sigma}^0$ production threshold), the soft-$\pi^\pm$ and the $\bar{p}$ annihilation products are reconstructed to identify the process.
The results are in agreement with the BaBar collaboration's ones [40], but with improved precision of a factor of 2. Moreover, following Ref. [41], an asymmetry in the effective form factors (G_{Eff}) of Σ-triplet is observed confirming that G_{Eff} is proportional to the squared sum of the valence quarks' charge.

6 Summary

The BESIII collaboration started taking data in 2008, and, since then, its physics reach has spanned a plethora of topics. The main focus of this review is the most recent results of the BESIII collaboration in the QCD field, ranging from charmonium(-like) studies, in which both conventional and exotic states are under investigation; to light hadrons spectroscopy and decays, used to probe the nature of the light hybrids, glueballs, and mesonic molecules; to open charm physics, which in synergy with light hadrons studies, investigate the light exotic states; to the precision QCD measurements of which the R-value measurements or the EMFFs are just two examples.
The BESIII collaboration can also provide useful insights outside of the QCD environment such as probing BSM physics or studying the electroweak sector.
Finally, new data sets (e.g., $\sim 3 \times 10^9$ of $\psi(2S)$ or $\sim$20 fb^{-1} of $\psi(3770)$) are currently being taken and analysed, hence more studies and analyses will help the community to shed light on the mysteries of QCD.

References

[1] M. Ablikim et al. (BESIII Collaboration), Nucl. Instrum. Methods Phys. Res. Sec. A **614**, 345 (2010)
[2] M. Ablikim et al. (BESIII Collaboration), Chin. Phys. C **44**, 040001 (2020)
[3] P. Zyla et al. (Particle Data Group), PTEP **2020**, 083C01 (2020)
[4] N. Brambilla, S. Eidelman, C. Hanhart, A. Nefediev, C.P. Shen, C.E. Thomas, A. Vairo, C.Z. Yuan, Physics Reports **873**, 1 (2020), the XYZ states: experimental and theoretical status and perspectives
[5] M. Ablikim et al. (BESIII Collaboration), Phys. Rev. Lett. **129**, 112003 (2022)
[6] B.D. Wan, C.F. Qiao, Nucl. Phys. B **968**, 115450 (2021), `2011.08747`
[7] M. Ablikim et al. (BESIII Collaboration), Phys. Rev. Lett. **126**, 102001 (2021), `2011.07855`
[8] M. Ablikim et al. (BESIII Collaboration) (2022), `2204.07800`
[9] J.Z. Wang, D.Y. Chen, X. Liu, T. Matsuki, Phys. Rev. D **99**, 114003 (2019), `1903.07115`
[10] X.K. Dong, F.K. Guo, B.S. Zou, Progr. Phys. **41**, 65 (2021), `2101.01021`
[11] T.W. Chiu, T.H. Hsieh (TWQCD), Phys. Rev. D **73**, 094510 (2006), `hep-lat/0512029`
[12] M. Ablikim et al. (BESIII Collaboration), Phys. Rev. Lett. **129**, 102003 (2022)
[13] F.K. Guo, C. Hanhart, U.G. Meissner, Phys. Lett. B **665**, 26 (2008), `0803.1392`
[14] C.F. Qiao, J. Phys. G **35**, 075008 (2008), `0709.4066`
[15] M. Ablikim et al. (BESIII Collaboration), JHEP **05**, 003 (2022), `2111.13915`
[16] M. Ablikim et al. (BESIII Collaboration), JHEP **05**, 108 (2022), `2203.10439`
[17] M. Ablikim et al. (BESIII Collaboration) (2022), `2204.09413`
[18] W. Kwong, J.L. Rosner, C. Quigg, Ann. Rev. Nucl. Part. Sci. **37**, 325 (1987)
[19] M. Ablikim et al. (BESIII Collaboration), Phys. Rev. Lett. **129**, 042001 (2022)
[20] L.M. Wang, Q.S. Zhou, C.Q. Pang, X. Liu, Phys. Rev. D **102**, 114034 (2020)
[21] B.S. Zou, D.V. Bugg, Eur. Phys. J. A **16**, 537 (2003)
[22] M. Ablikim et al. (BESIII Collaboration) (2022), `2202.00623`
[23] D.V. Bugg, Phys. Rept. **397**, 257 (2004), `hep-ex/0412045`
[24] M. Ablikim et al. (BESIII), Phys. Rev. D **87**, 032008 (2013), `1211.5668`

[25] J.J. Dudek, R.G. Edwards, P. Guo, C.E. Thomas (for Hadron Spectrum Collaboration), Phys. Rev. D **88**, 094505 (2013)

[26] J.M. Frère, J. Heeck, Phys. Rev. D **92**, 114035 (2015)

[27] M. Ablikim et al. (BESIII Collaboration), Phys. Rev. D **105**, 072002 (2022)

[28] M. Ablikim et al. (BESIII Collaboration), Phys. Rev. D **105**, L051103 (2022)

[29] M. Ablikim et al. (BESIII Collaboration), Phys. Rev. D **104**, 012016 (2021)

[30] J.P. Lees et al. (BaBar Collaboration), Phys. Rev. D **104**, 072002 (2021), 2106.05157

[31] M. Ablikim et al. (BESIII Collaboration) (2022), 2204.09614

[32] J.P. Lees, V. Poireau, V. Tisserand, E. Grauges, A. Palano, G. Eigen, D.N. Brown, Y.G. Kolomensky, M. Fritsch, H. Koch et al. (BABAR Collaboration), Phys. Rev. D **104**, 072002 (2021)

[33] Dai, L. R., Oset, E., Geng, L. S., Eur. Phys. J. C **82**, 225 (2022)

[34] D. Guo, W. Chen, H.X. Chen, X. Liu, S.L. Zhu, Phys. Rev. D **105**, 114014 (2022), 2204.13092

[35] M. Ablikim et al. (BES Collaboration), Phys. Rev. Lett. **96**, 162002 (2006)

[36] M. Ablikim et al. (BESIII Collaboration), Phys. Rev. Lett. **128**, 062004 (2022)

[37] V. Anashin, O. Anchugov, V. Aulchenko, E. Baldin, G. Baranov, A. Barladyan, A. Barnyakov, M. Barnyakov, S. Baru, I. Basok et al., Physics Letters B **788**, 42 (2019)

[38] P. Baikov, K. Chetyrkin, J. Kühn, J. Rittinger, Physics Letters B **714**, 62 (2012)

[39] M. Ablikim, M. Achasov, P. Adlarson, S. Ahmed, M. Albrecht, A. Amoroso, Q. An, Y. Bai, O. Bakina, R. Baldini Ferroli et al., Physics Letters B **831**, 137187 (2022)

[40] B. Aubert, M. Bona, D. Boutigny, Y. Karyotakis, J.P. Lees, V. Poireau, X. Prudent, V. Tisserand, A. Zghiche, J.G. Tico et al. (BABAR Collaboration), Phys. Rev. D **76**, 092006 (2007)

[41] S. Pacetti, R. Baldini Ferroli, E. Tomasi-Gustafsson, Physics Reports **550-551**, 1 (2015), proton electromagnetic form factors: Basic notions, present achievements and future perspectives

Impact of clustering inside compact tetraquarks

Hagop Sazdjian[1,*]

[1]Université Paris-Saclay, CNRS/IN2P3, IJCLab, 91405 Orsay, France

Abstract. Due to the reducibility of tetraquark operators into mesonic clusters, a strong interplay exists in tetraquarks between compact and molecular structures. This issue is studied within an effective field theory approach, where the compact tetraquark is treated as an elementary particle. Under the influence of the coupling to the mesonic clusters, an initially formed compact tetraquark bound state is deformed towards a new structure of the molecular type, having the attributes of a shallow bound state.

1 Introduction

There has been growing experimental evidence, during the last two decades, about the existence of exotic hadrons, also called multiquark states, containing more valence quarks than ordinary mesons ($\overline{q}q$) and baryons (qqq). Prototypes are tetraquarks with valence quark structure $\bar{q}\bar{q}qq$, pentaquarks, with structure $\overline{q}qqqq$, hexaquarks, with structure $qqqqqq$ or $\bar{q}\bar{q}\bar{q}qqq$ [1–15].

However, contrary to ordinary hadrons, multiquark states are not color-irreducible, in the sense that they can be decomposed along a finite number of combinations of ordinary mesonic or baryonic clusters [16, 17]. Schematically, ignoring here flavor and spin indices, a tetraquark interpolating operator, $(\bar{q}\bar{q}qq)$, which is globally color invariant, can be decomposed by means of Fierz rearrangements into a form where clusters of mesonic operators have emerged:

$$(\bar{q}\bar{q}qq) = \sum(\bar{q}q)(\bar{q}q), \tag{1}$$

where the parentheses signify color invariance of the included operator. Similar decompositions can be done with the pentaquark and hexaquark interpolating operators:

$$(\overline{q}qqqq) = \sum(\overline{q}q)(qqq), \tag{2}$$

$$(qqqqqq) = \sum(qqq)(qqq), \qquad (\bar{q}\bar{q}\bar{q}qqq) = \sum(\bar{q}\bar{q}\bar{q})(qqq) + \sum(\bar{q}q)(\bar{q}q)(\bar{q}q). \tag{3}$$

Hadronic clusters, being color-singlets, mutually interact by means of short-range forces, like meson-exchanges or contacts. They would then form loosely bound states, in similarity with atomic molecules. These are called *hadronic molecules* or *molecular states* [18–20].

In contrast, multiquark states, formed directly from confining interactions acting on all quarks, would form *compact bound states*, called *compact multiquark states* [21–24].

Multiquark states can thus be formed by two different mechanisms, each leading to a different structure. The issue is to find, by theoretical justification, also guided by experimental data, the most faithful description.

*e-mail: sazdjian@ijclab.in2p3.fr

 https://doi.org/10.1051/epjconf/202227000011

In the following, we concentrate on tetraquark states, ignoring details coming from quark flavors and spins, which do not play a fundamental role.

2 Energy balance

A first hint is provided by the study of the energy balance of the system [25]. This is most easily done in the *static limit* of the theory, with very heavy quarks, fixed at spatial positions. The system would choose configurations with *minimal energy*.

The problem is analytically solved in the strong coupling limit of lattice theory [26] and confirmed by direct lattice numerical calculations [27–32]. In the strong coupling limit, the potential energy is concentrated on the *Wilson lines*, which are the path-ordered gluon field phase factors, with constant linear energy density.

Figure 1 represents schematic geometric configurations of the two types of structure. In Fig. 1a, one has a compact tetraquark, formed by confining interactions, through diquark–antidiquark global interaction. In Fig. 1b, one has two meson clusters.

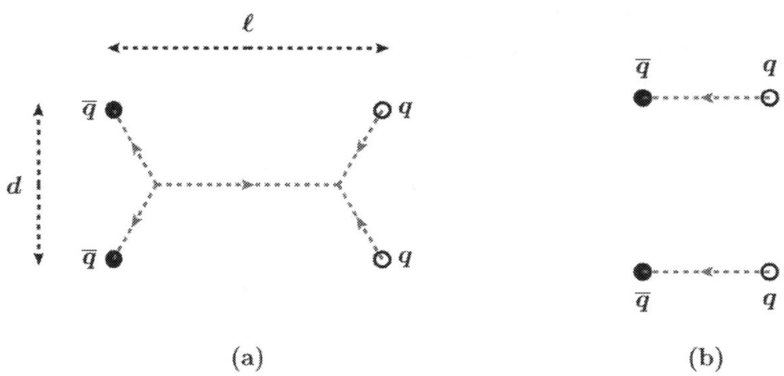

Figure 1. (a) Compact tetraquark, formed by confining interactions, through diquark–antidiquark global interaction. (b) Two meson clusters.

The compact tetraquark formation is energetically favored if the diquark interdistance d is much smaller than the quark-antiquark interdistance ℓ:

$$d \ll \ell. \tag{4}$$

This shows that both structures have nonzero probabilities to be formed. However, quarks have finite masses and move in space. Even if a compact tetraquark has been formed, there is a sizable probability that the quarks, during their motion, reach the two-meson-cluster configuration, in which case the system dislocates or decays.

An interesting and nontrivial case is when the compact tetraquark mass lies below the two-meson threshold, which prevents dislocation. Nevertheless, there will be, through fluctuations, constant transitions to the two-meson virtual states. This will have the tendancy to expand the compact system into a more loosely bound system, close to a molecular-type state. This is a dynamical mechanism, whose precise outcome necessitates the resolution of the four-body bound state problem, in the presence of the confining forces. For the time being, this problem has not yet been satisfactorily solved. One is obliged, to analyze the problem with sufficient accuracy, to have recourse to approximation schemes. We shall consider below the framework provided by the effective field theory approach. A detailed account of this work can be found in [33].

3 Effective field theory approach

According to the energy balance analysis, there is always a probability that a compact tetraquark be formed from the action of the confining forces (actually an infinite tower of such states). We assume that in first approximation, because of the compactness of the state, the latter can be assimilated to a pointlike object and treated as an elementary particle.

We are mainly interested in the case where the mass of the latter particle lies below the lowest possible two-meson threshold. This might be the case when the compact tetraquark is the ground state of the corresponding spectrum.

The two mesons are designated by M_1 and M_2, with masses m_1 and m_2, respectively. The bare mass of the tetraquark is m_{t0}.

Quark flavors and spins will be ignored, as not playing a fundamental role here.

Because of the initial general structure of the tetraquark, containing two meson clusters, the pointlike tetraquark has necessarily a coupling to the mesons M_1 and M_2. The coupling is assumed scalar and is designated by g'; the latter is dimensionless and is factored by the mass term $(m_1 + m_2)$.

The coupling g' generates, through meson loops, radiative corrections inside the tetraquark propagator, modifying the parameters of the bare propagator. In particular, the bare mass m_{t0} gets changed into a physical mass m_t. This is graphically represented in Fig. 2.

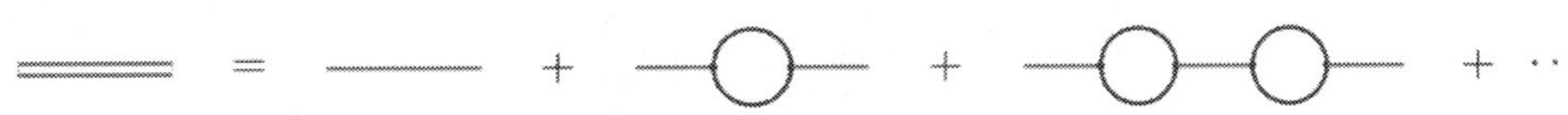

Figure 2. The full tetraquark propagator, including the radiative corrections, coming from the coupling to the two meson clusters.

Mutual interactions of mesons are neglected (in first approximation); they play a less important role than the direct tetraquark-two-meson coupling.

The full tetraquark propagator becomes

$$D_t(s) = \frac{i}{s - m_{t0}^2 + i(m_1 + m_2)^2 g'^2 J(s)}, \qquad (5)$$

where s stands for p^2 and J is the standard loop function. The divergence of J is absorbed by the bare mass term, yielding a renormalized mass m_{t1}:

$$m_{t1}^2 = m_{t0}^2 - i(m_1 + m_2)^2 g'^2 J^{\mathrm{div}}. \qquad (6)$$

The renormalized tetraquark propagator is now

$$D_t(s) = \frac{i}{s - m_{t1}^2 + i(m_1 + m_2)^2 g'^2 J^{\mathrm{r}}(s)}, \qquad (7)$$

where J^{r} is the renormalized finite part of J. Notice that g' does not undergo any renormalization. The mass term m_{t1} does not yet represent the physical mass of the tetraquark. The latter is determined from the pole position of the propagator.

We stick here to the case of heavy quarks and heavy mesons, treating the problem in its nonrelativistic limit, referred to the two-meson threshold. The nonrelativistic energy E is introduced through the standard definition

$$\sqrt{s} = (m_1 + m_2) + E. \qquad (8)$$

The nonrelativistic energy corresponding to the renormalized mass m_{t1} of the tetraquark is defined similarly:

$$E_{t1} = m_{t1} - (m_1 + m_2), \qquad E_{t1} < 0. \tag{9}$$

Since we are considering the case of stable tetraquarks (under the strong interactions), the mass of the tetraquark is expected to lie below the two-meson threshold. The physical nonrelativistic energy of the tetraquark will be designated by E_t and is also expected to be negative.

To simplify notations, we shall use henceforth the reduced dimensionless energy variables e through the definitions

$$e \equiv \frac{E}{2m_r}, \qquad e_{t1} \equiv \frac{E_{t1}}{2m_r}, \qquad e_t \equiv \frac{E_t}{2m_r}, \qquad m_r = \frac{m_1 m_2}{(m_1 + m_2)}. \tag{10}$$

The quantity $-e_t$ represents the nonrelativistic binding energy of the tetraquark by reference to the two-meson threshold, although it is different from the binding energy defined from the confining forces by reference to the quark masses.

One finds for the nonrelativistic energy e_t of the tetraquark the equation

$$-e_t + e_{t1} + \frac{g'^2}{16\pi}\sqrt{-e_t} = 0, \tag{11}$$

whose solution is

$$\sqrt{-e_t} = \frac{1}{2}\Big[-\frac{g'^2}{16\pi} + \sqrt{\Big(\frac{g'^2}{16\pi}\Big)^2 - 4e_{t1}}\Big]. \tag{12}$$

The binding energy $-e_t$ is a decreasing function of $g'^2/(16\pi)$ and comes out smaller than $-e_{t1}$,

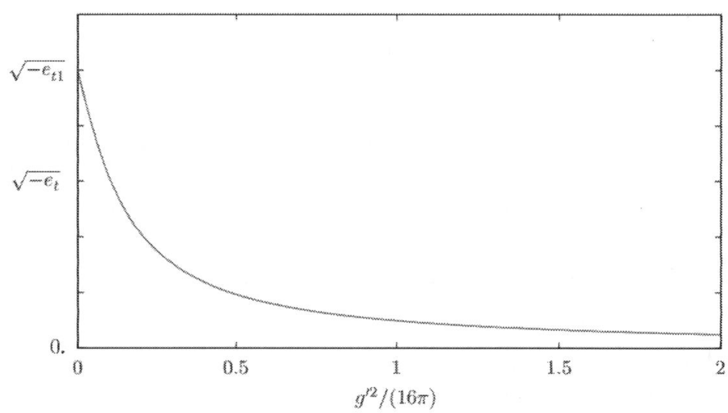

Figure 3. Variation of the square-root of the binding energy as a function of $g'^2/(16\pi)$.

reaching the value 0 when $g' \to \infty$. This is represented in Fig. 3.

We find a very rapid decrease of the binding energy. With ordinary values of $g'^2/(16\pi)$, of the order of 1, the binding energy decreases by a factor of 1/100. The state takes the appearance of a *shallow* bound state.

4 Compositeness

The comparison of the molecular and compact schemes is reminiscent of a general problem, already raised in the past in the case of the deuteron state, denoted under the term of *compositeness* [34].

 https://doi.org/10.1051/epjconf/202227000011

Weinberg has shown that this question can receive, in the nonrelativistic limit, a precise and model-independent answer, by relating the probability of a state as being elementary (or compact) to observable quantities, represented by the scattering length and the effective range of the two constituents of the molecular scheme in the S-wave state of their scattering amplitude. Designating by Z this probability, one has the following relations for the scattering length a and the effective range r_e, adapted to the tetraquark problem:

$$a = \frac{2(1-Z)}{(2-Z)}R, \qquad r_e = -\frac{Z}{(1-Z)}R, \qquad R = (-2m_rE_t)^{-1/2}, \tag{13}$$

where R is the radius of the bound state, m_r the reduced mass of the two-meson system and E_t the tetraquark nonrelativistic energy.

Investigations about the compositeness criterion and its applicability to various tetraquark candidates, as well as to ordinary hadrons, can be found in [35–50].

The contribution of the tetraquark state, in the s-channel, to the two-meson elastic scattering amplitude is obtained by inserting the tetraquark propagator between two tetraquark-two-meson couplings (cf. Fig. 4).

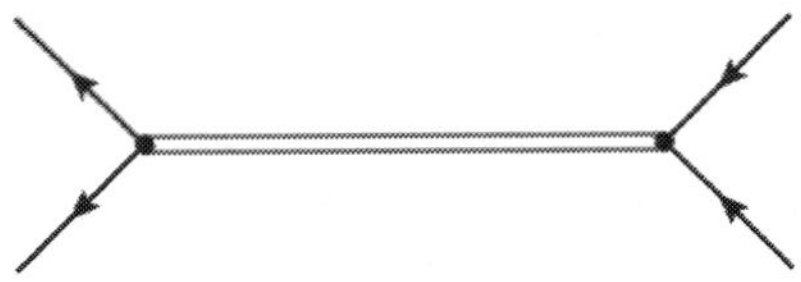

Figure 4. The tetraquark contribution, in the s-channel, to the two-meson elastic scattering amplitude.

From the scattering length and the effective range one obtains Z:

$$Z = \frac{\sqrt{-e_t}}{\sqrt{-e_t} + \frac{1}{2}\frac{g'^2}{16\pi}}. \tag{14}$$

When $Z = 1$, we have the case of a pure compact tetraquark, while for $Z = 0$, we have a pure molecular state. The variation of Z with respect to $g'^2/(16\pi)$ is represented in Fig. 5.

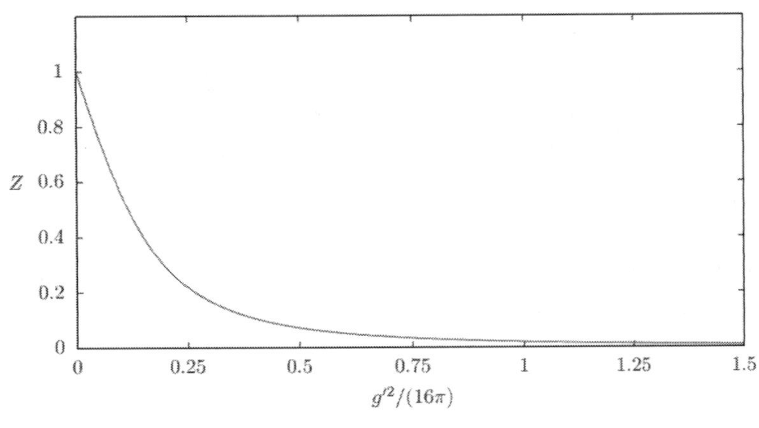

Figure 5. Variation of Z as a function of $g'^2/(16\pi)$.

Like the binding energy, Z is a rapidly decreasing function of $g'^2/(16\pi)$. For values of the latter of the order of 1, Z is very close to zero. This confirms the interpretation that the coupling of the compact tetraquark to its internal mesonic clusters qualitatively deforms its initial structure, bringing it into a form closer to a molecular-type state.

5 Conclusion

A compact tetraquark, formed from the confining forces acting between quarks and gluons, rapidly evolves, under the influence of the clustering phenomenon, towards a molecular-type state. The origin of the state is, however, of compact nature: nowhere, in the present model, did we consider direct interactions between mesons.

The experimental test for this phenomenon is provided by the measure of the elementariness coefficient Z. Pure molecular states are characterized by the value $Z = 0$. A value $Z \neq 0$, beyond uncertainties, reflects the existence of an original compact tetraquark. Many tetraquark candidates fall in this category.

Because of the shrinking of the tetraquark binding energy to values close to zero, shallowness of many bound states may receive a natural explanation from the above mechanism.

The present study can also be extended to the case of resonances and to the case when direct meson-meson interactions are incorporated.

Funding. This research has received financial support from the EU research and innovation programme Horizon 2020, under Grant agreement No. 824093, and from the joint CNRS/RFBR Grant No. PRC Russia/19-52-15022.

References

[1] S. Choi et al. (Belle), Phys. Rev. Lett. **91**, 262001 (2003), hep-ex/0309032
[2] B. Aubert et al. (BaBar), Phys. Rev. Lett. **90**, 242001 (2003), hep-ex/0304021
[3] D. Besson et al. (CLEO), Phys. Rev. D **68**, 032002 (2003), [Erratum: Phys. Rev. D 75 (2007) 119908], hep-ex/0305100
[4] B. Aubert et al. (BaBar), Phys. Rev. Lett. **95**, 142001 (2005), hep-ex/0506081
[5] M. Ablikim et al. (BESIII), Phys. Rev. Lett. **110**, 252001 (2013), 1303.5949
[6] Z. Liu et al. (Belle), Phys. Rev. Lett. **110**, 252002 (2013), [Erratum: Phys. Rev. Lett. 111 (2013) 019901], 1304.0121
[7] M. Ablikim et al. (BESIII), Phys. Rev. Lett. **111**, 242001 (2013), 1309.1896
[8] R. Aaij et al. (LHCb), Phys. Rev. Lett. **112**, 222002 (2014), 1404.1903
[9] R. Aaij et al. (LHCb), Phys. Rev. Lett. **115**, 072001 (2015), 1507.03414
[10] R. Aaij et al. (LHCb), Sci. Bull. **65**, 1983 (2020), 2006.16957
[11] R. Aaij et al. (LHCb), Phys. Rev. D **102**, 112003 (2020), 2009.00026
[12] L. Wu, Nucl. Part. Phys. Proc. **312-317**, 15287 (2021), 2012.15473
[13] E. Spadaro Norella (LHCb), PoS **LHCP2021**, 051 (2021)
[14] R. Aaij et al. (LHCb), Nature Commun. **13**, 3351 (2022), 2109.01056
[15] T. Gershon (LHCb) (2022), 2206.15233
[16] R.L. Jaffe, Nucl. Phys. A **804**, 25 (2008)
[17] W. Lucha, D. Melikhov, H. Sazdjian, Phys. Rev. D **100**, 094017 (2019), 1908.10164
[18] M.B. Voloshin, L.B. Okun, JETP Lett. **23**, 333 (1976), [Pisma Zh. Eksp. Teor. Fiz. 23, 369 (1976)]
[19] A. De Rújula, H. Georgi, S.L. Glashow, Phys. Rev. Lett. **38**, 317 (1977)

[20] N.A. Törnqvist, Z. Phys. C **61**, 525 (1994), hep-ph/9310247
[21] R. Jaffe, F. Wilczek, Phys. Rev. Lett. **91**, 232003 (2003), hep-ph/0307341
[22] E. Shuryak, I. Zahed, Phys. Lett. B **589**, 21 (2004), hep-ph/0310270
[23] L. Maiani, F. Piccinini, A.D. Polosa, V. Riquer, Phys. Rev. D **71**, 014028 (2005), hep-ph/0412098
[24] L. Maiani, V. Riquer, F. Piccinini, A.D. Polosa, Phys. Rev. D **72**, 031502 (2005), hep-ph/0507062
[25] W. Lucha, D. Melikhov, H. Sazdjian, Prog. Part. Nucl. Phys. **120**, 103867 (2021), 2102.02542
[26] H.G. Dosch, Phys. Rev. D **28**, 412 (1983)
[27] C. Alexandrou, G. Koutsou, Phys. Rev. D **71**, 014504 (2005), hep-lat/0407005
[28] F. Okiharu, H. Suganuma, T.T. Takahashi, Phys. Rev. Lett. **94**, 192001 (2005), hep-lat/0407001
[29] F. Okiharu, H. Suganuma, T.T. Takahashi, Phys. Rev. D **72**, 014505 (2005), hep-lat/0412012
[30] H. Suganuma, T. Iritani, F. Okiharu, T.T. Takahashi, A. Yamamoto, AIP Conf. Proc. **1388**, 195 (2011), 1103.4015
[31] N. Cardoso, M. Cardoso, P. Bicudo, Phys. Rev. D **84**, 054508 (2011), 1107.1355
[32] P. Bicudo, M. Cardoso, O. Oliveira, P.J. Silva, Phys. Rev. D **96**, 074508 (2017), 1702.07789
[33] H. Sazdjian, Symmetry **14**, 515 (2022), 2202.01081
[34] S. Weinberg, Phys. Rev. **137**, B672 (1965)
[35] V. Baru, J. Haidenbauer, C. Hanhart, Y. Kalashnikova, A.E. Kudryavtsev, Phys. Lett. B **586**, 53 (2004), hep-ph/0308129
[36] M. Cleven, F.K. Guo, C. Hanhart, U.G. Meissner, Eur. Phys. J. A **47**, 120 (2011), 1107.0254
[37] C. Hanhart, Y.S. Kalashnikova, A.V. Nefediev, Eur. Phys. J. A **47**, 101 (2011), 1106.1185
[38] T. Hyodo, D. Jido, A. Hosaka, Phys. Rev. C **85**, 015201 (2012), 1108.5524
[39] F. Aceti, E. Oset, Phys. Rev. D **86**, 014012 (2012), 1202.4607
[40] T. Sekihara, T. Hyodo, D. Jido, PTEP **2015**, 063D04 (2015), 1411.2308
[41] Z.H. Guo, J.A. Oller, Phys. Rev. D **93**, 096001 (2016), 1508.06400
[42] X.W. Kang, Z.H. Guo, J.A. Oller, Phys. Rev. D **94**, 014012 (2016), 1603.05546
[43] U.G. Meissner, J.A. Oller, Phys. Lett. B **751**, 59 (2015), 1507.07478
[44] J.A. Oller, Annals Phys. **396**, 429 (2018), 1710.00991
[45] Z.H. Guo, J.A. Oller, Phys. Rev. D **103**, 054021 (2021), 2012.11904
[46] A. Esposito, L. Maiani, A. Pilloni, A.D. Polosa, V. Riquer, Phys. Rev. D **105**, L031503 (2022), 2108.11413
[47] Y. Li, F.K. Guo, J.Y. Pang, J.J. Wu, Phys. Rev. D **105**, L071502 (2022), 2110.02766
[48] V. Baru, X.K. Dong, M.L. Du, A. Filin, F.K. Guo, C. Hanhart, A. Nefediev, J. Nieves, Q. Wang, Phys. Lett. B **833**, 137290 (2022), 2110.07484
[49] T. Kinugawa, T. Hyodo, EPJ Web Conf. **262**, 01019 (2022), 2112.00249
[50] J. Song, L.R. Dai, E. Oset, Eur. Phys. J. A **58**, 133 (2022), 2201.04414

Weak decays of the double-beauty tetraquark

Birten Barsbay[1,*]

[1]Department of Physics, Doğuş University, Dudullu-Ümraniye, 34775 Istanbul, Turkey

Abstract. The spectroscopic parameters and decay channels of the axial-vector tetraquark $T_{b:\bar{s}}^{\mathrm{AV}}$ built of heavy diquark bb and light antidiquark $\overline{us}$ are explored using QCD sum rule method. Because this state is stable against strong and electromagnetic decays, we consider its semileptonic and nonleptonic decay modes. Partial widths of weak processes $T_{b:\bar{s}}^{\mathrm{AV}} \to \mathcal{Z}_{b:\bar{s}}^{0} l \bar{\nu}_l$ and $T_{b:\bar{s}}^{\mathrm{AV}} \to \mathcal{Z}_{b:\bar{s}}^{0} M$ are calculated using weak form factors $G_i(q^2)$ computed by QCD three-point sum rules: $\mathcal{Z}_{b:\bar{s}}^{0}$ is the scalar tetraquark with quark content $bc\overline{us}$ and M is a vector meson. The full width and mean lifetime of the tetraquark $T_{b:\bar{s}}^{\mathrm{AV}}$ obtained in this work shed light on features of double-beauty tetraquarks and are useful for their experimental investigations.

1 Introduction

Recently interest in double-charm and double-beauty tetraquarks was renewed after the discovery of the doubly charmed baryon $\Xi_{cc}^{++} = ccu$ by the LHCb Collaboration [1]. Thus, information on mass of this baryon was used in Ref. [2] to estimate the masses of the tetraquarks $T_{bb;\bar{u}\bar{d}}^{-}$ and $T_{cc;\bar{u}\bar{d}}^{+}$ in the context of a phenomenological model. The obtained prediction for $m = (10389 \pm 12)$ MeV confirms that the isoscalar state $T_{bb;\bar{u}\bar{d}}^{-}$ with spin-parity $J^P = 1^+$ is stable against strong and electromagnetic decays, whereas the tetraquark $T_{cc;\bar{u}\bar{d}}^{+}$ lies above the open charm threshold $D^0 D^{*+}$ and can decay to these mesons.

In the framework of the QCD sum rule method the spectroscopic parameters and weak decays of the double-beauty axial-vector tetraquark $T_{bb;\bar{u}\bar{d}}^{-}$ were calculated in Ref. [3]. The mass of the state $T_{bb;\bar{u}\bar{d}}^{-}$ found in this work is $m = (10035 \pm 260)$ MeV, which demonstrates its stability against the strong and electromagnetic decays. The full width and mean lifetime of $T_{bb;\bar{u}\bar{d}}^{-}$ were evaluated in Refs. [3, 4] using its various semileptonic and nonleptonic decay channels. The obtained results, $\Gamma_{\mathrm{full}} = (7.72 \pm 1.23) \times 10^{-8}$ MeV and $\tau = 8.53_{-1.18}^{+1.57}$ fs, are useful for the experimental investigation of the double-beauty tetraquark $T_{bb;\bar{u}\bar{d}}^{-}$.

In this work, we calculate partial widths of the semileptonic $T_{b:\bar{s}}^{\mathrm{AV}} \to \mathcal{Z}_{b:\bar{s}}^{0} l \bar{\nu}_l$ and nonleptonic weak decays $T_{b:\bar{s}}^{\mathrm{AV}} \to \mathcal{Z}_{b:\bar{s}}^{0} M$ of the axial-vector tetraquark $T_{b:\bar{s}}^{\mathrm{AV}}$ with quark content $bb\overline{us}$, where $\mathcal{Z}_{b:\bar{s}}^{0}$ is the scalar tetraquark $[bc][\overline{us}]$ built of a color-antitriplet diquark and triplet antidiquark, and M is one of the vector mesons ρ^-, $K^*(892)$, $D^*(2010)^-$, and D_s^{*-}. To this end, we calculate the weak form factors $G_i(q^2)$ from the three-point sum rules at momentum transfers q^2 where this method is applicable [5–8]. We determine model functions $\mathcal{G}_i(q^2)$ and find the partial widths of aforementioned weak processes. The full width and mean lifetime of $T_{b:\bar{s}}^{\mathrm{AV}}$ are obtained as well.

*e-mail: bbarsbay@dogus.edu.tr

2 Weak Decays

The mass of the tetraquark $T^{\rm AV}_{b:\overline{s}}$ $m_{\rm AV} = 10215$ MeV is 480/477 MeV below the thresholds 10695/10692 MeV for its strong decays to mesons $B^-B^*_s$ and $B^{*-}\overline{B}^0_s$, respectively, and 431 MeV below the threshold 10646 MeV for electromagnetic decay $B^-\overline{B}^0_s\gamma$ [9]. This means that $T^{\rm AV}_{b:\overline{s}}$ is stable against strong and electromagnetic decays and transforms to ordinary mesons only through weak processes. Therefore, to evaluate the full width and lifetime of $T^{\rm AV}_{b:\overline{s}}$, we analyze the semileptonic and nonleptonic weak decays $T^{\rm AV}_{b:\overline{s}} \to \mathcal{Z}^0_{b:\overline{s}}\overline{l}\nu_l$ and $T^{\rm AV}_{b:\overline{s}} \to \mathcal{Z}^0_{b:\overline{s}}M$.

2.1 Semileptonic decays $T^{\rm AV}_{b:\overline{s}} \to \mathcal{Z}^0_{b:\overline{s}}\overline{l}\nu_l$

Weak decays of $T^{\rm AV}_{b:\overline{s}}$ can be generated by subprocesses $b \to W^-c$ and $b \to W^-u$. The processes triggered by the transition $b \to W^-c$ are dominant relative to $b \to W^-u$ ones: the latter decays are suppressed relative to the dominant modes by a factor $|V_{ub}|^2/|V_{cb}|^2 \simeq 0.01$, with $V_{q_1q_2}$ being the Cabibbo-Khobayasi-Maskawa (CKM) matrix elements. In the present work, we restrict ourselves by analysis of the dominant weak decays of $T^{\rm AV}_{b:\overline{s}}$.

The effective Hamiltonian that describes the subprocess $b \to W^-c$ at the tree-level is determined by the expression

$$\mathcal{H}^{\rm eff} = \frac{G_F}{\sqrt{2}} V_{cb}\overline{c}\gamma_\mu(1-\gamma_5)b\overline{l}\gamma^\mu(1-\gamma_5)\nu_l, \tag{1}$$

where G_F and V_{cb} are the Fermi coupling constant and CKM matrix element, respectively. The matrix element of $\mathcal{H}^{\rm eff}$ between the initial and final tetraquarks contains the leptonic and hadronic factors

$$\langle \mathcal{Z}^0_{b:\overline{s}}(p')|\mathcal{H}^{\rm eff}|T^{\rm AV}_{b:\overline{s}}(p)\rangle = L_\mu H^\mu. \tag{2}$$

We are interested in calculation of H^μ, because the leptonic part of the matrix element L_μ is universal for all semileptonic decays and does not contain information on tetraquarks. Then, H^μ is the matrix element of the current

$$J^{\rm tr}_\mu = \overline{c}\gamma_\mu(1-\gamma_5)b \tag{3}$$

between the initial and final particles. It can be modeled in terms of four weak form factors $G_i(q^2)$ which parametrize long-distance dynamics of the weak transformation [10, 11]

$$\begin{aligned}\langle \mathcal{Z}^0_{b:\overline{s}}(p')|J^{\rm tr}_\mu|T^{\rm AV}_{b:\overline{s}}(p)\rangle &= \widetilde{G}_0(q^2)\epsilon_\mu + \widetilde{G}_1(q^2)(\epsilon p')P_\mu \\ &+ \widetilde{G}_2(q^2)(\epsilon p')q_\mu + i\widetilde{G}_3(q^2)\varepsilon_{\mu\nu\alpha\beta}\epsilon^\nu p^\alpha p'^\beta.\end{aligned} \tag{4}$$

The scaled functions $\widetilde{G}_i(q^2)$ are connected with the dimensionless form factors $G_i(q^2)$ by the equalities

$$\widetilde{G}_0(q^2) = \widetilde{m}G_0(q^2),\ \widetilde{G}_j(q^2) = \frac{G_j(q^2)}{\widetilde{m}},\ j = 1, 2, 3. \tag{5}$$

Here, $\widetilde{m} = m_{\rm AV} + m_{\mathcal{Z}}$, p_μ and ϵ_μ are the momentum and polarization vector of the tetraquark $T^{\rm AV}_{b:\overline{s}}$, p' is the momentum of the scalar state $\mathcal{Z}^0_{b:\overline{s}}$, $P_\mu = p'_\mu + p_\mu$, and $q_\mu = p_\mu - p'_\mu$ is the momentum transferred to the leptons. q^2 varies within the limits $m_l^2 \leq q^2 \leq (m_{\rm AV} - m_{\mathcal{Z}})^2$, where m_l is the mass of a lepton l.

The sum rules for the form factors $G_i(q^2)$ can be obtained by analyzing the three-point correlation function

$$\Pi_{\mu\nu}(p, p') = i^2 \int d^4x d^4y e^{i(p'y-px)}\langle 0|\mathcal{T}\{J_{\mathcal{Z}}(y)J^{\rm tr}_\nu(0)J^\dagger_\mu(x)\}|0\rangle, \tag{6}$$

with $J_\mu(x)$ and $J_{\mathcal{Z}}(y)$ being the interpolating currents for the tetraquarks $T^{\rm AV}_{b:\overline{s}}$ and $\mathcal{Z}^0_{b:\overline{s}}$, respectively. In accordance with standard prescriptions, we express $\Pi_{\mu\nu}(p,p')$ using the masses and couplings of the tetraquarks and determine the physical side of the sum rules $\Pi^{\rm Phys}_{\mu\nu}(p,p')$. The function $\Pi^{\rm Phys}_{\mu\nu}(p,p')$ can be written down in the form

$$\Pi^{\rm Phys}_{\mu\nu}(p,p') = \frac{\langle 0|J_{\mathcal{Z}}|\mathcal{Z}^0_{b:\overline{s}}(p')\rangle\langle \mathcal{Z}^0_{b:\overline{s}}(p')|J^{\rm tr}_\nu|T^{\rm AV}_{b:\overline{s}}(p,\epsilon)\rangle}{(p^2-m^2_{\rm AV})(p'^2-m^2_{\mathcal{Z}})} \times\langle T^{\rm AV}_{b:\overline{s}}(p,\epsilon)|J^\dagger_\mu|0\rangle+\cdots, \tag{7}$$

where contribution of the ground-state particles is presented explicitly, whereas effects due to excited and continuum states are denoted by dots.

The phenomenological side of the sum rules can be simplified by substituting into Eq. (7) expressions of the matrix elements. To this aim, we apply Eq. (4) and introduce the matrix elements of the tetraquarks $T^{\rm AV}_{b:\overline{s}}$ and $\mathcal{Z}^0_{b:\overline{s}}$,

$$\langle 0|J_\mu|T^{\rm AV}_{b:\overline{s}}(p)\rangle = m_{\rm AV}f_{\rm AV}\epsilon_\mu, \tag{8}$$

and

$$\langle 0|J_{\mathcal{Z}}|\mathcal{Z}^0_{b:\overline{s}}(p')\rangle = f_{\mathcal{Z}}m_{\mathcal{Z}}, \tag{9}$$

respectively. Then it is not difficult to find that

$$\Pi^{\rm Phys}_{\mu\nu}(p,p') = \frac{f_{\rm AV}m_{\rm AV}f_{\mathcal{Z}}m_{\mathcal{Z}}}{(p^2-m^2_{\rm AV})(p'^2-m^2_{\mathcal{Z}})}\left\{\widetilde{G}_0(q^2)\left(-g_{\mu\nu}+\frac{p_\mu p_\nu}{m^2_{\rm AV}}\right)+\left[\widetilde{G}_1(q^2)P_\mu +\widetilde{G}_2(q^2)q_\mu\right]\left(-p'_\nu+\frac{m^2_{\rm AV}+m^2_{\mathcal{Z}}-q^2}{2m^2_{\rm AV}}p_\nu\right) -i\widetilde{G}_3(q^2)\varepsilon_{\mu\nu\alpha\beta}p^\alpha p'^\beta\right\}+\cdots. \tag{10}$$

The function $\Pi^{\rm OPE}_{\mu\nu}(p,p')$ constitutes the second side of the sum rules and in terms of quark-gluon degrees of freedom has the following form:

$$\begin{aligned}\Pi^{\rm OPE}_{\mu\nu}(p,p') = &\int d^4x d^4y e^{i(p'y-px)}\left\{{\rm Tr}\left[\gamma_5\widetilde{S}^{ba'}_s(x-y)\right.\right.\\ &\left.\times\gamma_5 S^{a'b}_u(x-y)\right]\left({\rm Tr}\left[\gamma_\mu\widetilde{S}^{aa'}_b(y-x)\gamma_5 S^{bi}_c(y)\gamma_\nu(1-\gamma_5)\right.\right.\\ &\left.\times S^{ib'}_b(-x)\right]+{\rm Tr}\left[\gamma_\mu\widetilde{S}^{ia'}_b(-x)(1-\gamma_5)\gamma_\nu\widetilde{S}^{bi}_c(y)\gamma_5\right.\\ &\left.\left.\times S^{ab'}_b(y-x)\right]\right)-{\rm Tr}\left[\gamma_5\widetilde{S}^{b'a}_s(x-y)\gamma_5 S^{a'b}_u(x-y)\right]\\ &\times\left({\rm Tr}\left[\gamma_\mu\widetilde{S}^{aa'}_b(y-x)\gamma_5 S^{bi}_c(y)\gamma_\nu(1-\gamma_5)S^{ib'}_b(-x)\right]\right.\\ &\left.\left.+{\rm Tr}\left[\gamma_\mu\widetilde{S}^{ia'}_b(-x)(1-\gamma_5)\gamma_\nu\widetilde{S}^{bi}_c(y)\gamma_5 S^{ab'}_b(y-x)\right]\right)\right\},\end{aligned} \tag{11}$$

where $S_{Q(q)}(x)$ are quark propagators and

$$\widetilde{S}_{Q(q)}(x) = CS_{Q(q)}(x)C \tag{12}$$

with C being the charge conjugation matrix.

To extract the sum rules for the form factors $G_i(q^2)$, we equate invariant amplitudes corresponding to the same Lorentz structures in $\Pi^{\rm Phys}_{\mu\nu}(p,p')$ and $\Pi^{\rm OPE}_{\mu\nu}(p,p')$, perform a double

Borel transformation over the variables p'^2 and p^2 to suppress contributions of the higher excited and continuum states, and perform continuum subtraction. For instance, to extract the sum rule for $\widetilde{G}_0(q^2)$, we use the structure $g_{\mu\nu}$, whereas for $\widetilde{G}_3(q^2)$, the term $\sim \varepsilon_{\mu\nu\alpha\beta}p^\alpha p'^\beta$ is employed. The sum rules for the scaled form factors $\widetilde{G}_i(q^2)$ can be written in a single formula

$$\widetilde{G}_i(\mathbf{M}^2, \mathbf{s}_0, q^2) = \frac{1}{f_{\rm AV} m_{\rm AV} f_{\mathcal{Z}} m_{\mathcal{Z}}} \int_{\mathcal{M}^2}^{s_0} ds e^{(m_{\rm AV}^2 - s)/M_1^2} \int_{\widetilde{\mathcal{M}}^2}^{s_0'} ds' \rho_i(s, s') e^{(m_{\mathcal{Z}}^2 - s')/M_2^2}, \tag{13}$$

where $\rho_i(s, s')$ are spectral densities calculated as the imaginary parts of the corresponding terms in $\Pi_{\mu\nu}^{\rm OPE}(p, p')$. They include the perturbative and nonperturbative contributions, and are calculated with dimension-six accuracy. In Eq. (13), $\mathbf{M}^2 = (M_1^2, M_2^2)$ are the Borel parameters, and $\mathbf{s}_0 = (s_0,\ s_0')$ are the continuum threshold parameters that separate the main contribution to the sum rules from the continuum effects. The pair of parameters $(M_1^2,\ s_0)$ corresponds to a channel of the initial tetraquark $T_{b:\overline{s}}^{\rm AV}$, whereas $(M_2^2,\ s_0')$ describe the final-state particle $\mathcal{Z}_{b:\overline{s}}^0$.

The form factors $\widetilde{G}_i(q^2)$ determine the differential decay rate $d\Gamma/dq^2$ of the semileptonic decay $T_{b:\overline{s}}^{\rm AV} \to \mathcal{Z}_{b:\overline{s}}^0 l\overline{\nu}_l$, the explicit expression of which can be found in Ref. [3]. The sum rules give reliable results for $G_i(q^2)$ in the region $m_l^2 \leq q^2 \leq 8\ {\rm GeV}^2$, which is not enough to find the partial width of the processes under consideration. To calculate the width of the semileptonic decay, $d\Gamma/dq^2$ must be integrated over q^2 in the boundaries $m_l^2 \leq q^2 \leq (m_{\rm AV} - m_{\mathcal{Z}})^2$, i.e., in the limits $m_l^2 \leq q^2 \leq 11.87\ {\rm GeV}^2$. This region is wider than the one where the sum rules lead to reliable predictions. This problem can be solved by introducing extrapolating (fit) functions $\mathcal{G}_i(q^2)$, which at the momentum transfers q^2 accessible for the sum rule computations coincide with $G_i(q^2)$, but have forms suitable to carry out integrations over q^2.

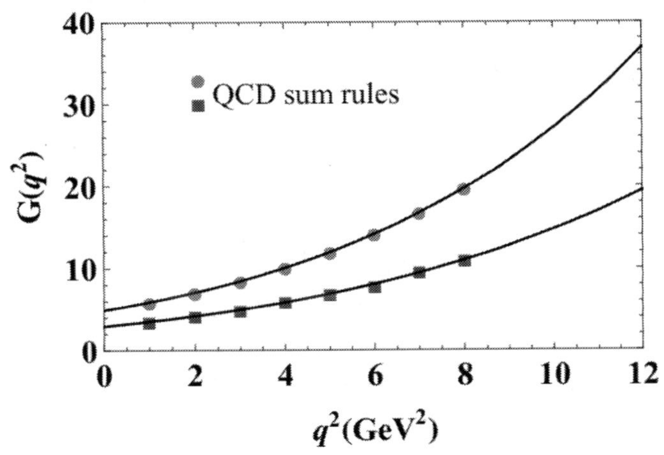

Figure 1. Sum rule results for the form factors $G_0(q^2)$ (red circles) and $G_1(q^2)$ (blue squares). The solid curves are fit functions $\mathcal{G}_0(q^2)$ and $\mathcal{G}_1(q^2)$.

To this end, we use the functions

$$\mathcal{G}_i(q^2) = \mathcal{G}_0^i \exp\left[g_1^i \frac{q^2}{m_{\rm AV}^2} + g_2^i \left(\frac{q^2}{m_{\rm AV}^2} \right)^2 \right], \tag{14}$$

where parameters $\mathcal{G}_0^i$, g_1^i, and g_2^i are fitted to satisfy the sum rules' predictions. The parameters of the functions $\mathcal{G}_i(q^2)$ extracted from numerical analysis are collected in Table 1.

Table 1. Parameters of the extrapolating functions $\mathcal{G}_i(q^2)$.

$\mathcal{G}_i(q^2)$	$\mathcal{G}_0^i$	g_1^i	g_2^i
$\mathcal{G}_0(q^2)$	4.91	19.29	−15.34
$\mathcal{G}_1(q^2)$	2.94	18.73	−20.09
$\mathcal{G}_2(q^2)$	−22.67	20.50	−22.95
$\mathcal{G}_3(q^2)$	−21.14	20.77	−23.62

In Fig. 1, we show fit functions $\mathcal{G}_i(q^2)$ which are in nice agreement with relevant sum rule predictions.

The Fermi constant, CKM matrix elements, and masses of leptons in numerical computations are set equal to

$$G_F = 1.16637 \times 10^{-5}\ \mathrm{GeV}^{-2},\ |V_{cb}| = (42.2 \pm 0.08) \times 10^{-3}, \tag{15}$$

$m_e = 0.511$ MeV, $m_\mu = 105.658$ MeV, and $m_\tau = (1776.82\ \pm 0.16)$ MeV [12].

2.2 Nonleptonic decays $T_{b:\overline{s}}^{\mathrm{AV}} \to \mathcal{Z}_{b:\overline{s}}^{0}\rho^-(K^*(892),\ D^*(2010)^-,\ D_s^{*-})$

In this section, we analyze the nonleptonic weak decays $T_{b:\overline{s}}^{\mathrm{AV}} \to \mathcal{Z}_{b:\overline{s}}^{0}\rho^-(K^*(892),\ D^*(2010)^-,\ D_s^{*-})$ of the tetraquark $T_{b:\overline{s}}^{\mathrm{AV}}$ in the framework of the QCD factorization approach, which allows us to calculate partial widths of these processes. This method was applied to study nonleptonic decays of conventional mesons [13, 14], but can be also used for investigating decays of tetraquarks. Thus, the nonleptonic decays of scalar exotic mesons $T_{b:\overline{s}}^{-}$, $T_{b:\overline{s}}^{-}$, Z_{bc}^{0}, and $T_{bs;\overline{u}\overline{d}}^{-}$ were explored using this approach in Refs. [15–18], respectively.

At the quark level, the effective Hamiltonian for this decay is given by the expression

$$\mathcal{H}_{\mathrm{n.-lep}}^{\mathrm{eff}} = \frac{G_F}{\sqrt{2}} V_{cb} V_{ud}^{*} \left[c_1(\mu)Q_1 + c_2(\mu)Q_2\right], \tag{16}$$

where

$$Q_1 = \left(\overline{d}_i u_i\right)_{\mathrm{V-A}} \left(\overline{c}_j b_j\right)_{\mathrm{V-A}},\ Q_2 = \left(\overline{d}_i u_j\right)_{\mathrm{V-A}} \left(\overline{c}_j b_i\right)_{\mathrm{V-A}}. \tag{17}$$

In Eq. (17) i and j are the color indices, and notation $(\overline{q}_1 q_2)_{\mathrm{V-A}}$ means

$$(\overline{q}_1 q_2)_{\mathrm{V-A}} = \overline{q}_1 \gamma_\mu (1 - \gamma_5) q_2. \tag{18}$$

The short-distance Wilson coefficients $c_1(\mu)$ and $c_2(\mu)$ are given at the factorization scale μ.

The amplitude of the decay $T_{b:\overline{s}}^{\mathrm{AV}} \to \mathcal{Z}_{b:\overline{s}}^{0}\rho^-$ is determined by the expression

$$\mathcal{A} = \frac{G_F}{\sqrt{2}} V_{bc} V_{ud}^{*} a(\mu) \langle \rho^-(q)| \left(\overline{d}_i u_i\right)_{\mathrm{V-A}} |0\rangle \langle \mathcal{Z}_{b:\overline{s}}^{0}(p')| \left(\overline{c}_j b_j\right)_{\mathrm{V-A}} |T_{b:\overline{s}}^{\mathrm{AV}}(p)\rangle, \tag{19}$$

where

$$a(\mu) = c_1(\mu) + \frac{1}{N_c} c_2(\mu), \tag{20}$$

with $N_c = 3$ being the number of quark colors. The matrix element $\langle \rho^-(q)| \left(\overline{d}_i u_i\right)_{\mathrm{V-A}} |0\rangle$ in $\mathcal{A}$ can be given as follows

$$\langle \rho^-(q)| \left(\overline{d}_i u_i\right)_{\mathrm{V-A}} |0\rangle = f_\rho m_\rho \epsilon_\mu^{*}(q). \tag{21}$$

Then, it is evident that

$$\mathcal{A} = i\frac{G_F}{\sqrt{2}} f_\rho V_{bc} V_{ud}^* a(\mu) \left[\widetilde{G}_0(q^2)\epsilon_\mu(p)\epsilon^{*\mu}(q) + 2\widetilde{G}_1(q^2)(p'\epsilon(p))(p'\epsilon^*(q)) + i\widetilde{G}_3(q^2)\varepsilon_{\mu\nu\alpha\beta}\epsilon^{*\mu}(q)\epsilon^\nu(p)p^\alpha p'^\beta \right]. \quad (22)$$

A similar analysis can be carried out for the decay modes $T_{b:\overline{s}}^{\mathrm{AV}} \to \mathcal{Z}_{b:\overline{s}}^0 K^*(892)[D^*(2010)^-, D_s^{*-}]$ as well. We replace in relevant expressions m_ρ and f_ρ by spectroscopic parameters of the $K^*(892)$, $D^*(2010)^-$, and D_s^{*-} mesons, and substitute $V_{ud} \to V_{us}$, V_{cd}, and V_{cs}.

The width of the nonleptonic decay $T_{b:\overline{s}}^{\mathrm{AV}} \to \mathcal{Z}_{b:\overline{s}}^0 \rho^-$ can be evaluated using the expression

$$\Gamma = \frac{|\mathcal{A}|^2}{24\pi m_{\mathrm{AV}}^2} \lambda\left(m_{\mathrm{AV}}, m_{\mathcal{Z}}, m_\rho\right), \quad (23)$$

where

$$\lambda(a,b,c) = \frac{1}{2a}\left[a^4 + b^4 + c^4 - 2\left(a^2b^2 + a^2c^2 + b^2c^2\right)\right]^{1/2}.$$

The key component in Eq. (23), i.e., $|\mathcal{A}|^2$ has a simple form

$$|\mathcal{A}|^2 = \sum_{j=0,1,2} H_j \widetilde{G}_j^2 + H_3 \widetilde{G}_0 \widetilde{G}_1, \quad (24)$$

where the explicit expressions of H_j are given in Ref. [9]

In Eq. (24), the weak form factors $\widetilde{G}_j$ are real functions of q^2, and their values are fixed at $q^2 = m_M^2$ where M is one of the vector mesons ρ^-, $K^*(892)$, $D^*(2010)^-$, and D_s^{*-}.

3 Numerical Results

In this section we present the results of our numerical computations. In general, some of the input parameters used in the form factors $\widetilde{G}_i(\mathbf{M}^2, \mathbf{s}_0, q^2)$ should be defined before numerical analysis. All input information necessary for such analysis, i.e., the vacuum condensates of quark, gluon, and mixed operators, the spectroscopic parameters of the final-state mesons and tetraquarks $T_{b:\overline{s}}^{\mathrm{AV}}$ and $\mathcal{Z}_{b:\overline{s}}^0$, as well as the CKM matrix elements can be found in Ref. [9]. The Borel and continuum threshold parameters $\mathbf{M}^2$ and $\mathbf{s}_0$ are chosen to satisfy all constraints of sum rule analysis. Also we have to take into account that $\widetilde{G}_i(\mathbf{M}^2, \mathbf{s}_0, q^2)$ depends on the masses and couplings of the initial and final tetraquarks, which have been evaluated also in the context of the sum rule approach. We fix the auxiliary parameters (M_1^2, s_0) and (M_2^2, s_0') similar to the mass computations, because they meet standard restrictions of three-point sum rule calculations and do not produce additional uncertainties in the spectroscopic parameters of relevant tetraquarks.

Predictions obtained for the partial widths of the semileptonic $T_{b:\overline{s}}^{\mathrm{AV}} \to \mathcal{Z}_{b:\overline{s}}^0 l\overline{\nu}_l$ and nonleptonic $T_{b:\overline{s}}^{\mathrm{AV}} \to \mathcal{Z}_{b:\overline{s}}^0 \rho^-(K^*(892),\ D^*(2010)^-,\ D_s^{*-})$ decays are presented below

$$\begin{aligned}
\Gamma(T_{b:\overline{s}}^{\mathrm{AV}} &\to \mathcal{Z}_{b:\overline{s}}^0 e^-\overline{\nu}_e) = (5.34 \pm 1.43) \times 10^{-8}\ \mathrm{MeV}, \\
\Gamma(T_{b:\overline{s}}^{\mathrm{AV}} &\to \mathcal{Z}_{b:\overline{s}}^0 \mu^-\overline{\nu}_\mu) = (5.32 \pm 1.41) \times 10^{-8}\ \mathrm{MeV}, \\
\Gamma(T_{b:\overline{s}}^{\mathrm{AV}} &\to \mathcal{Z}_{b:\overline{s}}^0 \tau^-\overline{\nu}_\tau) = (2.15 \pm 0.54) \times 10^{-8}\ \mathrm{MeV}, \\
\Gamma(T_{b:\overline{s}}^{\mathrm{AV}} &\to \mathcal{Z}_{b:\overline{s}}^0 \rho^-) = (3.47 \pm 0.92) \times 10^{-10}\ \mathrm{MeV}, \\
\Gamma(T_{b:\overline{s}}^{\mathrm{AV}} &\to \mathcal{Z}_{b:\overline{s}}^0 K^*(892)) = (1.47 \pm 0.37) \times 10^{-11}\ \mathrm{MeV}, \\
\Gamma(T_{b:\overline{s}}^{\mathrm{AV}} &\to \mathcal{Z}_{b:\overline{s}}^0 D^*(2010)^-) = (1.54 \pm 0.39) \times 10^{-11}\ \mathrm{MeV}, \\
\Gamma(T_{b:\overline{s}}^{\mathrm{AV}} &\to \mathcal{Z}_{b:\overline{s}}^0 D_s^{*-}) = (4.97 \pm 1.32) \times 10^{-10}\ \mathrm{MeV}.
\end{aligned} \quad (25)$$

The partial widths of different weak decays from Eq. (25) can be used to estimate the full width and mean lifetime of the axial-vector tetraquark $T_{b:\overline{s}}^{\rm AV}$

$$\Gamma_{\rm full} = (12.9 \pm 2.1) \times 10^{-8}\ {\rm MeV},\ \tau = 5.1^{+0.99}_{-0.71} \times 10^{-15}\ {\rm s}. \tag{26}$$

These predictions can be employed to explore the double-heavy tetraquarks.

References

[1] R. Aaij *et al.* [LHCb], Phys. Rev. Lett. **119**, no.11, 112001 (2017)
[2] M. Karliner, S. Nussinov and J. L. Rosner, Phys. Rev. D **95**, no.3, 034011 (2017)
[3] S. S. Agaev, K. Azizi, B. Barsbay and H. Sundu, Phys. Rev. D **99**, no.3, 033002 (2019)
[4] S. S. Agaev, K. Azizi, B. Barsbay, and H. Sundu, Eur. Phys. J. A **57**, 106 (2021)
[5] M. A. Shifman, A. I. Vainshtein and V. I. Zakharov, Nucl. Phys. B **147**, 385-447 (1979)
[6] M. A. Shifman, A. I. Vainshtein and V. I. Zakharov, Nucl. Phys. B **147**, 448-518 (1979)
[7] L. J. Reinders, H. Rubinstein and S. Yazaki, Phys. Rept. **127**, 1 (1985)
[8] S. Narison, World Sci. Lect. Notes Phys. **26**, 1-527 (1989)
[9] S. S. Agaev, K. Azizi, B. Barsbay and H. Sundu, Chin. Phys. C **45**, no.1, 013105 (2021)
[10] M. Wirbel, B. Stech and M. Bauer, Z. Phys. C **29**, 637 (1985)
[11] P. Ball, V. M. Braun and H. G. Dosch, Phys. Rev. D **44**, 3567-3581 (1991)
[12] M. Tanabashi *et al.* [Particle Data Group], Phys. Rev. D **98**, no.3, 030001 (2018)
[13] M. Beneke, G. Buchalla, M. Neubert and C. T. Sachrajda, Phys. Rev. Lett. **83**, 1914-1917 (1999)
[14] M. Beneke, G. Buchalla, M. Neubert and C. T. Sachrajda, Nucl. Phys. B **591**, 313-418 (2000)
[15] S. S. Agaev, K. Azizi, B. Barsbay, and H. Sundu, Phys. Rev. D **101**, 094026 (2020)
[16] S. S. Agaev, K. Azizi, B. Barsbay, and H. Sundu, Eur. Phys. J. A **56**, 177 (2020)
[17] H. Sundu, S. S. Agaev and K. Azizi, Eur. Phys. J. C **79**, 753 (2019)
[18] S. S. Agaev, K. Azizi and H. Sundu, Phys. Rev. D **100**, 094020 (2019)

Spectroscopy of the excited D_s mesons

Francesco Debernardis[1,*], on behalf of the LHCb Collaboration

[1]INFN and University of Bari

Abstract. The discoveries of the $D^*_{s0}(2317)^+$ and $D_{s1}(2460)^+$ mesons challenge our understanding of quantum chromodynamics. After almost 20 years, the nature of these hadrons is still subject of debate and many models have been proposed to explain their unexpected masses: standard charmed-strange mesons, DK molecules and tetraquarks. The LHCb experiment has studied the production of excited D^+_s meson in prompt proton-proton collisions and from b-hadron decays. Precise measurement of their properties and observation of new D^+_s states have been reported. The latest results on the spectroscopy of the charmed-strange mesons and the prospect to investigate their nature will be presented.

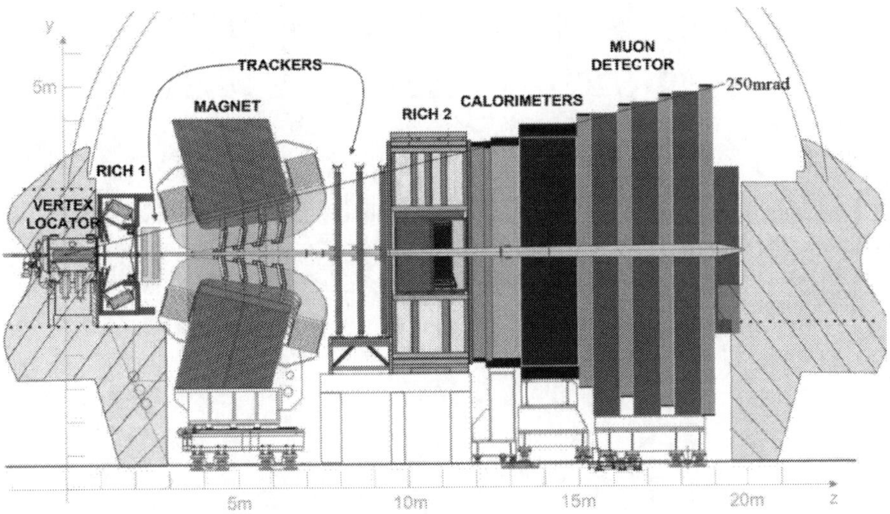

Figure 1. The LHCb apparatus during LHC Run 1 ($\sqrt{s} = 7 \div 8$ TeV) and Run 2 ($\sqrt{s} = 13$ TeV). pp collisions happen in the vertex locator.

1 The LHCb contribution to the spectrum of the excited D^+_s states

The LHCb experiment has largely contributed to the D^+_s spectroscopy in Run 1 and Run 2 with 9/fb integrated luminosity [1] by exploiting its high detection performance [2] and a large D^+_s production cross-section at LHC equal to $\sigma(pp \to D^+_s X) = 353 \pm 9 \pm 76$ μb [3]. As shown in Fig.1 the apparatus is a single-arm forward spectrometer [2]. At upstream, the

*e-mail: francesco.debernardis@ba.infn.it

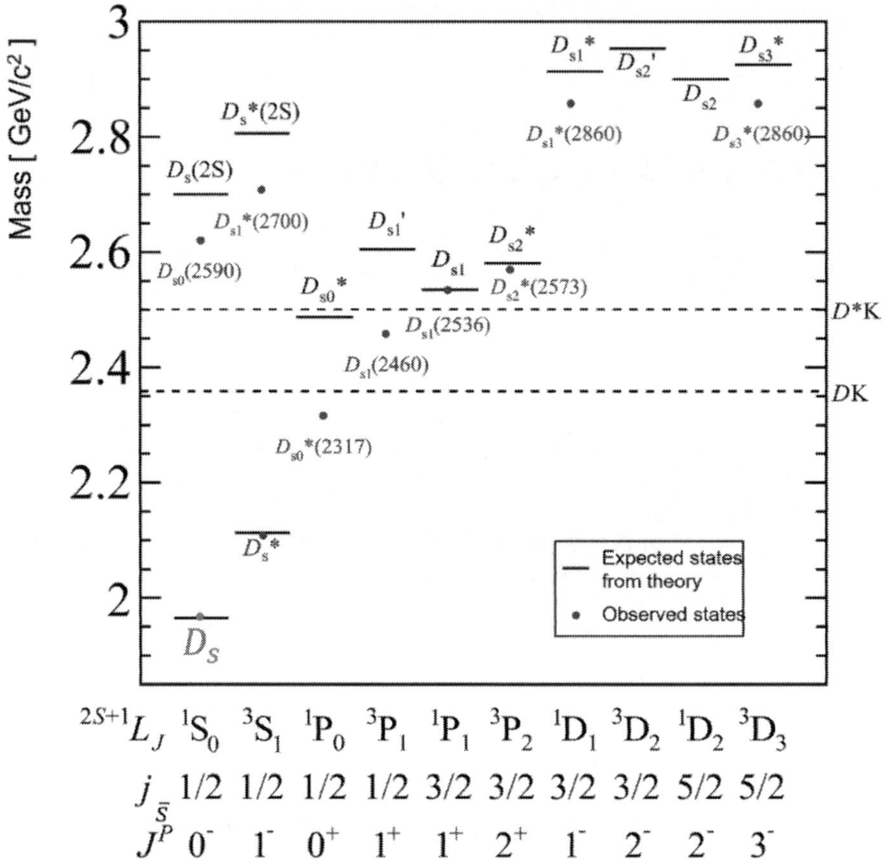

Figure 2. Spectrum of the excited D_s^+ mesons. The D_s ground state is shown in red. The classification is based on masses and quantum numbers. L is the orbital angular momentum between constituent quarks (S, P, D correspond to $L = 0, 1, 2$ respectively); $S = 0, 1$ is the sum of quark spins and J is the total spin of meson; P is the parity.

Vertex Locator detects particles near pp collision point. Momentum measurements are made by tracker stations thanks to a magnet that deflects charged particles. RICH detectors are instead used for the particle identification, distinguishing kaons from pions and calorimeters are used for energy particle measurements. At downstream of the apparatus, the muon detector is devoted to muon identification.

The spectrum of the excited D_s^+ states is shown in Fig.2, where expected and observed excited states are identified according to their mass and quantum numbers. The LHCb collaboration has reported the recent observation of the new state $D_{s0}(2590)^+$ (Fig.2) in the $B^0 \to D^- D^+ K^+ \pi^-$ decay [9], which is interpreted as the radially excitation of the D_s^+ ground state. Another remarkable observation is the $D_{s3}^*(2860)^+$ resonance, the first spin-3 meson ever observed in the B-meson production. In particular, the $D_{s1}^*(2860)^+$ and $D_{s3}^*(2860)^+$ resonances have been distinguished in a broad structure at 2.86 GeV/c^2 by fitting the Dalitz plot of exclusive three-body decays $B_s^0 \to \overline{D}^0 K^- \pi^+$ [4, 5].

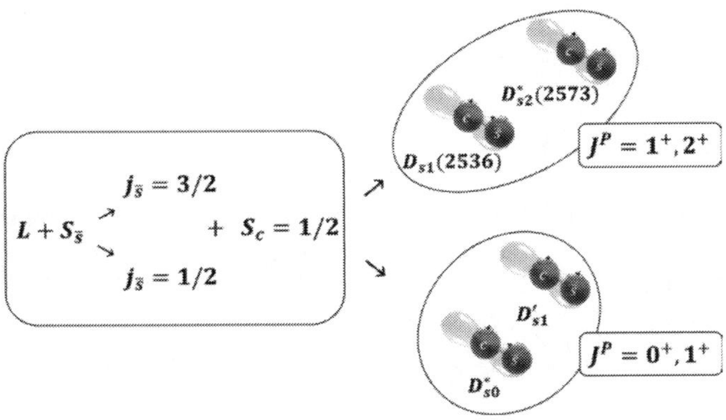

Figure 3. Excited *P*-wave states ($\vec{L} = 1$) in the heavy quark limit model.

2 The heavy quark limit model and unresolved puzzles

A D_s^+ meson state, composed of heavy c and light $\bar{s}$ quarks, is well described by the heavy quark limit model in the QCD theory pattern. Indeed, according to the QCD theory, the strong coupling constant α_S is small at high energy scale of the heavy c quark, causing interactions at short distance scale $<< R_{had}$[1] perturbatively described. At the contrary, α_S is large at small energy scale of the light $\bar{s}$ quark, causing interactions with non perturbative confinement phenomena of quarks and gluons at large length scale $\approx R_{had}$. Considering the two mass scales bound together in the heavy quark limit, the heavy c quark is much smaller than the hadron size and it is surrounded by a strongly interacting cloud of light quarks, antiquarks, and gluons with which the light $\bar{s}$ interacts without seeing the heavy quark spin effect. In this way the heavy quark spin decouples and the bounding energy of the states depends on spin-orbit coupling of the light $\bar{s}$ quark only.

The heavy quark limit model is able to predict the existence of D_s^+ excited states, according to the sum rules of the quarks orbital momentum and spin. In Fig.3 the excited P-wave states ($L = 1$) prediction scheme is shown: after the spin-orbit coupling is made, involving only the spin $\vec{S}_{\bar{s}} = 1/2$ of the light quark, the total angular momentum of that quark is $\vec{j}_{\bar{s}} = \vec{L} + \vec{S}_{\bar{s}}$ that can be equal to $1/2$ or $3/2$. Then the spin of the meson is obtained summing $\vec{j}_{\bar{s}}$ to the spin $\vec{S}_c = 1/2$ of the heavier quark. Four new possible states emerge: the resonances $D_{s1}(2536)$ and $D^*_{s2}(2573)$ with spin parity 1^+ and 2^+, and the resonances D^*_{s0} and D'_{s1} with spin-parity 0^+ and 1^+ respectively. The first state two states have been observed, thus confirmed, by studying the DK and D^*K mass spectra [6, 8].

However a puzzle characterizes $D_{s1}(2536)^+$ particle: since the c quark is not infinitely heavy, the $D_{s1}(2536)^+$ can contain an admixture of another state from $j_{\bar{s}} = 1/2$ and can have a small decay fraction in an S-wave. But the angular analysis of the $D_{s1}(2536)^+ \to D^{*+}K_S^0$ decay [7] shows an unexpected S-wave dominant contribution $\approx 70\%$ against the expectations.

Another more remarkable puzzle concerns the latter two P-wave states, featured by $\vec{j}_{\bar{s}} = 1/2$ quantum number, such that it is not clear that they have been observed. Indeed, the theory predicted them as broad states with masses large enough so that the dominant decays would have been the DK and/or D^*K final states. However, two narrow resonances, named $D^*_{s0}(2317)^+$ and $D_{s1}(2460)^+$, were observed by the *BABAR* [10] and the *CLEO* [11] collaborations, having mass smaller than all theoretical predictions and decaying to the isospin-violating $D_s^{(*)+}\pi^0$ channels. The observation of such surprising states hinted at either the theoretical models are

[1] Considering the hadron size R_{had}.

inadequate or the $D^*_{s0}(2317)^+$ and $D_{s1}(2460)^+$ mesons are not $c\bar{s}$ states but 4-quark systems, as corroborated by several observed anomalies.

2.1 The puzzle of the $D_{s1}(2460)^+$

We focus now on the $D_{s1}(2460)^+$ state, trying to explain well some its anomalies. For instance, in the heavy quark limit model, the following ratio R is expected ≈ 1, contrary to the measured value [12, 13]:

$$R = \frac{BR(B \to DD_{s1}(2460))}{BR(B \to DD^*_s)} \approx \frac{1}{3} \neq 1$$

The discrepancy could be explained by interpreting the $D_{s1}(2460)$ meson as a 4-quark state whose production is suppressed in $B \to DD_{s1}(2460)$ decays, than the D^*_s production, with only two quarks. Another anomaly is evident comparing the D^{*+}_s and $D_{s1}(2460)^+$ decay modes, shown in Tab. 1. For the D^{*+}_s meson, the radiative decay mode is much more likely

Table 1. Known decay modes and measured decay rates of the D^{*+}_s and $D_{s1}(2460)^+$ states [13].

D^{*+}_s		$D_{s1}(2460)^+$	
Decay mode	Branching fraction (%)	Decay mode	Branching fraction (%)
$D^+_s\gamma$	93.5 ± 0.7	$D^+_s\gamma$	18 ± 4
$D^+_s\pi^0$	5.8 ± 0.7	$D^{*+}_s\pi^0$	48 ± 11
$D^+_s e^+e^-$	0.67 ± 0.16	$D^+_s\pi^+\pi^-$	4.3 ± 1.3
		Total	70 ± 12

because it is not isospin-violating, unlike the $D^+_s\pi^0$ mode that violates the isospin conservation. Same remarks should be also valid for the $D_{s1}(2460)^+$ in a $c\bar{s}$ scenario, but its behavior is practically the opposite. Indeed, the isospin-violating $D^{*+}_s\pi^0$ decay mode is much more frequent than radiative $D^+_s\gamma$ mode, suggesting a 4-quark component ($c\bar{s}u\bar{u}$ or $c\bar{s}d\bar{d}$). It can also be noticed that the $D_{s1}(2460)^+$ decay channel spectrum is not complete. New decay channels, not observed so far, could be studied, such as Dalitz decays featured by two leptons in the final state, produced from a virtual photon.

3 LHCb Upgrades and Prospects

During the LHCb Upgrade I, new detectors and new data acquisition electronics systems have been installed, upgrading all the apparatus for the future LHC Run 3 and Run 4 [14]. However, the future of the LHCb experiment is characterized by the LHC luminosity increase up to $1.5\times10^{34}\ cm^{-2}sec^{-1}$ planned for Run 5. It is a great opportunity to improve the precision of all the LHCb physics reach [15], contributing also to disentangle puzzles in the excited D^+_s states described so far. Several analysis prospects could be considered, profiting from the large D^+_s production cross-section and D^+_s signal purity as evident in Fig.4. For instance, some analysis that could be mentioned are:

- Search for muonic Dalitz decay $D_{s1}(2460) \to D_s + \gamma^* \to D_s + \mu^+\mu^-$
  with muons in the final state generated by the virtual photon. The analysis is inspired by the observed $\chi_{c1}, \chi_{c2} \to J/\psi\mu^+\mu^-$ decays [16], justifying the feasibility of the search.

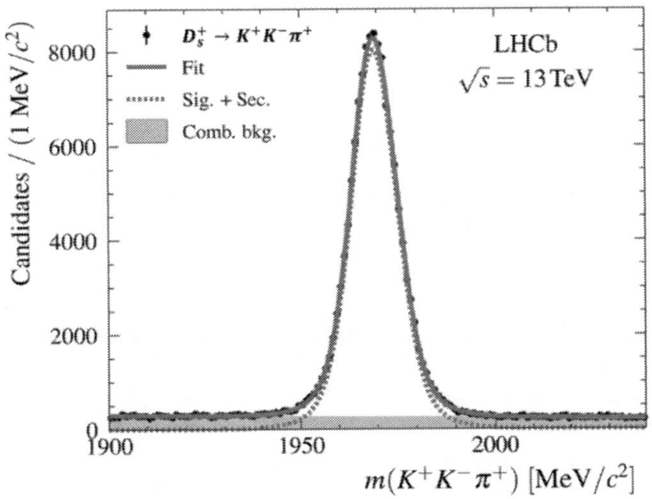

Figure 4. D_s^+ signal in $K^+K^-\pi^+$ decay at LHCb [3].

- Production measurement of the $D_{s1}(2460)^+$ state via $D_s^+\gamma$ decay.

- Quantum numbers measurement of the $D^*_{s0}(2317)^+$ state. In particular spin-parity measurement by studying the decay $B_s^0 \to D^*_{s0}(2317)^-\pi^+ \to D_s^-\pi^0\pi^+$

In conclusion, the debate on the D_s^+ excited states is still open, in particular on the P-wave $D^*_{s0}(2317)^+$ and $D_{s1}(2460)^+$ particles. Their properties disagree from the heavy quark limit model that is suitable instead for the most of the other D_s^+ excited states.
The LHCb experiment has largely contributed to the spectroscopy of the D_s^+ mesons and it will continue more thanks to its detection capability and to the high statistics in the new LHC luminosity era, trying to disentangle the debates on the described puzzles.

References

[1] LHCb Collaboration
[2] Alves Jr. A.A. et al., Journal of instrumentation, **3**, S08005 (2008)
[3] Aaij R. et al. (LHCb Collaboration), JHEP **03** 159 (2016)
[4] Aaij R. et al. (LHCb Collaboration), Physical Review D **90** 072003 (2014)
[5] Aaij R. et al. (LHCb Collaboration), Physical Review Letters **113** 162001 (2014)
[6] Albrecht H. et al. (ARGUS Collaboration), Physics Letters B **230**, 162 (1989)
[7] Balagura V. et al. (BELLE Collaboration), Physical Review D **77**, 032001 (2008)
[8] Kubota Y. et al. (CLEO Collaboration), Physical Review Letters **72**, 1972 (1994)
[9] Aaij R. et al. (LHCb Collaboration), Physical Review Letters **126**, 122002 (2021)
[10] Aubert B. et al. (BABAR Collaboration), Physical Review Letters **90**, 242001 (2003)
[11] Besson D. et al. (CLEO Collaboration), Physical Review D **68**, 032002 (2003)
[12] Datta A. and O'donnell P.J., Physics Letters B **572**, 164 (2003)
[13] Zyla P.A. et al., Particle Data Group **083C01** (2020 and 2021 update)
[14] (LHCb Collaboration), CERN-LHCC-2012-007, CERN-LHCC-2017-003
[15] (LHCb Collaboration), CERN-LHCC-2018-027
[16] Aaij R. et al. (LHCb Collaboration), Physical Review Letters **119**, 221801 (2017)

https://doi.org/10.1051/epjconf/202227000014

Multiquark-Oriented QCD Sum Rules

Wolfgang Lucha[1,*], *Dmitri* Melikhov[2,3,4,**], and *Hagop* Sazdjian[5,***]

[1]Institute for High Energy Physics, Austrian Academy of Sciences, Nikolsdorfergasse 18, A-1050 Vienna, Austria
[2]D. V. Skobeltsyn Institute of Nuclear Physics, M. V. Lomonosov Moscow State University, 119991 Moscow, Russia
[3]Joint Institute for Nuclear Research, 141980 Dubna, Russia
[4]Faculty of Physics, University of Vienna, Boltzmanngasse 5, A-1090 Vienna, Austria
[5]Université Paris-Saclay, CNRS/IN2P3, IJCLab, 91405 Orsay, France

Abstract. We propose to increase the factual reliability of descriptions of exotic multiquark hadrons utilizing the approach to bound states of strongly interacting constituents known as QCD sum rules, by allowing exclusively all contributions that potentially bear some relevance for multiquark states to enter the correlation functions that form the main ingredient of this framework. The route to this goal is illustrated for the (presumably least involved) special case of tetraquark states.

1 Goal: Adaptation of QCD Sum-Rule Approach to Multiquark States

Evaluation of *correlation functions* of operators interpolating one's hadrons of interest at both phenomenological (hadron) and fundamental (QCD) levels by application of operator product expansion [1] and Borel transformations, and assuming a complete cancellation of hadron and perturbative-QCD contributions above optimized [2–4] effective thresholds, provides analytic relationships, denoted as *QCD sum rules* [5], between, on the one hand, observable properties of hadrons and, on the other hand, basic parameters of QCD. In this context QCD accounts for both perturbative contributions, given by integrals over spectral densities, and nonperturbative power corrections (that involve vacuum condensates multiplied by powers of Borel variables).

The colour-singlet bound states governed by QCD encompass not just all quark–antiquark mesons and three-quark baryons, labelled *conventional* hadrons, but also *multiquark* hadrons: tetraquarks, pentaquarks, hexaquarks, etc. For attempts to take hold of such exotic hadrons by QCD sum rules, a systematic scrutiny [6–10] points to the necessity of adapting the traditional wisdom to the peculiarities of multiquarks: Retaining just contributions potentially capable of providing information on multiquark features by getting rid of all contributions not related to multiquarks yields the novel sort of *multiquark-adequate* QCD sum rules; for some tetraquark categories [11], all of the latter involve, at least, two gluon exchanges of appropriate topology.

2 Proof of Assertion by Brief Sketch of Underlying Line of Argument

We illustrate our considerations for the simplest type of multiquark hadrons, the *tetraquark* T, a meson constituted of two antiquarks $\overline{q}_a, \overline{q}_c$ and two quarks q_b, q_d (of masses m_a, m_b, m_c, m_d):

*e-mail: Wolfgang.Lucha@oeaw.ac.at
**e-mail: dmitri_melikhov@gmx.de
***e-mail: sazdjian@ijclab.in2p3.fr

 https://doi.org/10.1051/epjconf/202227000014

$$T = [\overline{q}_a\, q_b\, \overline{q}_c\, q_d]\,, \qquad a,b,c,d \in \{u,d,s,c,b\}\,.$$

Leaving aside the spin degrees of freedom, by application of a Fierz transformation [12] every *tetraquark-interpolating operator* may be proven [13] to be equivalent to a linear combination of just two products θ of colour-singlet quark–antiquark bilinear currents $j_{\overline{a}b}(x) \equiv \overline{q}_a(x)\, q_b(x)$:

$$\theta_{\overline{a}b\overline{c}d}(x) \equiv j_{\overline{a}b}(x)\, j_{\overline{c}d}(x)\,, \qquad \theta_{\overline{a}d\overline{c}b}(x) \equiv j_{\overline{a}d}(x)\, j_{\overline{c}b}(x)\,.$$

Accordingly, we regard it preferable to exploit (potential) manifestations of tetraquarks by means of intermediate-state poles in the scattering amplitudes of two conventional mesons (of momenta p_1, p_2) to two conventional mesons, by scrutiny of vacuum averages of time-ordered products of four quark–antiquark bilinear currents $j^{(\dagger)}$, i.e., of *four-point correlation functions*

$$\left\langle \mathrm{T}\!\left(j(y)\, j(y')\, j^\dagger(x)\, j^\dagger(x')\right)\right\rangle .$$

Configuration-space contractions of the latter yield all required types of correlation functions:

- two-point correlation functions of two operators θ, interpolating the tetraquark investigated,

$$\left\langle \mathrm{T}\!\left(\theta(y)\, \theta^\dagger(x)\right)\right\rangle = \lim_{\substack{x'\to x\\ y'\to y}} \left\langle \mathrm{T}\!\left(j(y)\, j(y')\, j^\dagger(x)\, j^\dagger(x')\right)\right\rangle ;$$

- three-point correlators of one current θ and two conventional-meson interpolating currents j

$$\left\langle \mathrm{T}\!\left(j(y)\, j(y')\, \theta^\dagger(x)\right)\right\rangle = \lim_{x'\to x} \left\langle \mathrm{T}\!\left(j(y)\, j(y')\, j^\dagger(x)\, j^\dagger(x')\right)\right\rangle .$$

Next, we need to formulate an exact and easily applicable *criterion* [14, 15] that allows for unequivocal identification of all contributions impacting any correlation function under study:

Tetraquark-phile [16, 17] contributions depend in a *nonpolynomial* form on the Mandelstam variable $s \equiv (p_1+p_2)^2$ and – as can be established by fulfilment of the Landau equations [18] – enable true intermediate four-quark states, by exhibiting branch cuts that start at *branch points*

$$\hat{s} = (m_a + m_b + m_c + m_d)^2 \,.$$

As an example, we focus to the subset of *flavour-exotic* tetraquarks, which encompass two quarks and two antiquarks carrying four *unequal* quark flavours, generically labelled a, b, c, d.

For such choice of quark-flavour contents, in the course of analysis it proves advantageous to discriminate two categories of four-point correlation functions, differing in the distributions of the four quark flavours among the pairs of interpolating currents $j^{(\dagger)}$ in initial or final states:

- *flavour-retaining* correlation functions [6, 8, 9], identified by identical flavour distributions,

$$\left\langle \mathrm{T}\!\left(j_{\overline{a}b}(y)\, j_{\overline{c}d}(y')\, j^\dagger_{\overline{a}b}(x)\, j^\dagger_{\overline{c}d}(x')\right)\right\rangle , \qquad \left\langle \mathrm{T}\!\left(j_{\overline{a}d}(y)\, j_{\overline{c}b}(y')\, j^\dagger_{\overline{a}d}(x)\, j^\dagger_{\overline{c}b}(x')\right)\right\rangle ;$$

- *flavour-rearranging* correlation functions [6–9], identified by unequal flavour distributions,

$$\left\langle \mathrm{T}\!\left(j_{\overline{a}b}(y)\, j_{\overline{c}d}(y')\, j^\dagger_{\overline{a}d}(x)\, j^\dagger_{\overline{c}b}(x')\right)\right\rangle .$$

Armed with our rigorous criterion [14, 15] above, for every two- or three-point correlation function it is then a straightforward task to identify the tetraquark-phile [16, 17] contributions:

Flavour-retaining ones [6, 8–10] with no or at most a single gluon exchange [Fig. 1(a,b)] are spatially separable and vanish at all or involve s in merely polynomial form, implying that just ones with two or more relevant gluon exchanges do affect tetraquarks [Fig. 1(c) and Fig. 2(a)].

Flavour-reordering ones [6–10] are not spatially separable and will support tetraquarks only if resorting to two or more gluon exchanges, as revealed by the Landau equations [Fig. 2(b,c)].

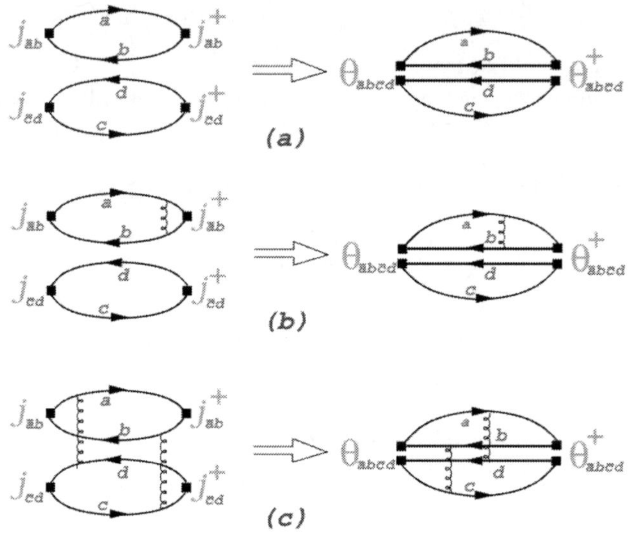

Figure 1. Flavour-preserving correlation functions of two tetraquark-interpolating operators θ, resulting from double configuration-space contractions of a pair of operators j interpolating conventional mesons: generic contributions of lowest (a), next-to-lowest (b) and next-to-next-to-lowest (c) perturbative orders.

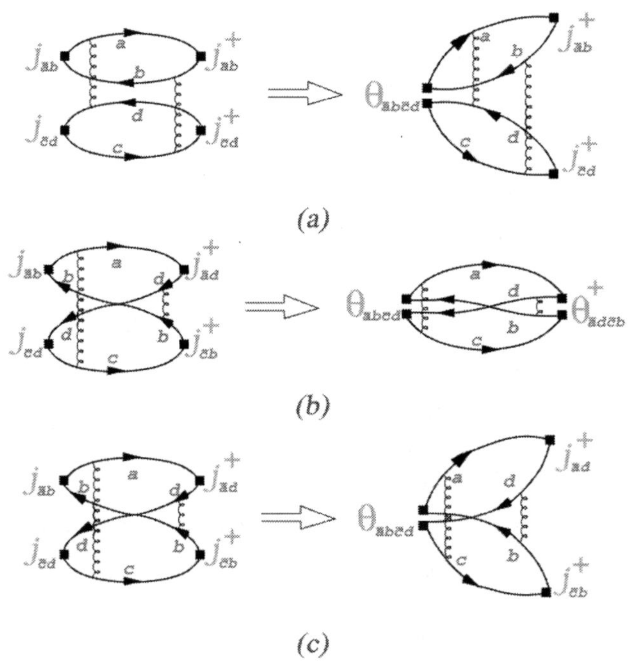

Figure 2. Exemplary *tetraquark-phile* contributions of lowest perturbative order to (a) *flavour-retaining* correlation functions of one tetraquark-interpolating current θ and two quark–antiquark bilinear currents j and to *flavour-reordering* correlation functions of either (b) a pair of tetraquark-interpolating operators θ or (c) one tetraquark-interpolating operator θ and two ordinary-meson interpolating currents j, inferred by single (a,c) or double (b) configuration-space contraction of two meson-interpolating currents j, resp.

3 Outcome: QCD Sum Rules Tailored to Peculiarities of Tetraquarks

Upon feeding the above concepts or insights into the traditional framework of QCD sum rules, it becomes feasible to disentangle, in any deduced relation, the unavoidable intertwinement of expressions genuinely related to tetraquarks and those involving exclusively ordinary mesons:

- Every flavour-preserving relation (Fig. 3) consists of two kinds of QCD sum rules [6, 8–10]; the first of these provides two unconnected QCD sum rules for conventional mesons (Fig. 4) and the second one the required *tetraquark-adequate* QCD sum rule (Fig. 5) [6]. Refraining from one contraction of interpolating currents $j^{(\dagger)}$ disentangles the three-point analogue [6].
- By investing a little bit more efforts, every flavour-rearranging relation may be unscrambled [6–10] and systematically reorganized as two relationships, one that *does not* [Fig. 6(a)] and one that *does* [Fig. 6(b)] involve two-meson s-channel cuts and potentially tetraquark poles.

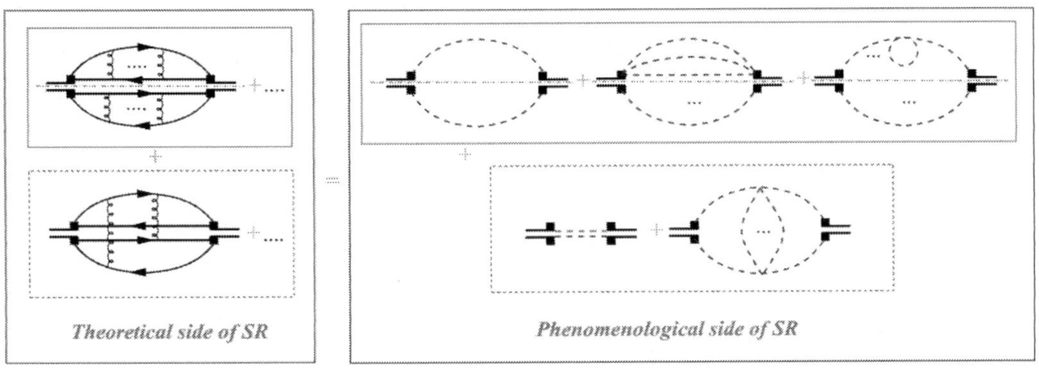

Figure 3. Decomposing a generic *flavour-retaining* relationship into a pair of *conventional*-meson QCD sum rules (top row) and a QCD sum rule that possibly involves, among others, tetraquarks (bottom row).

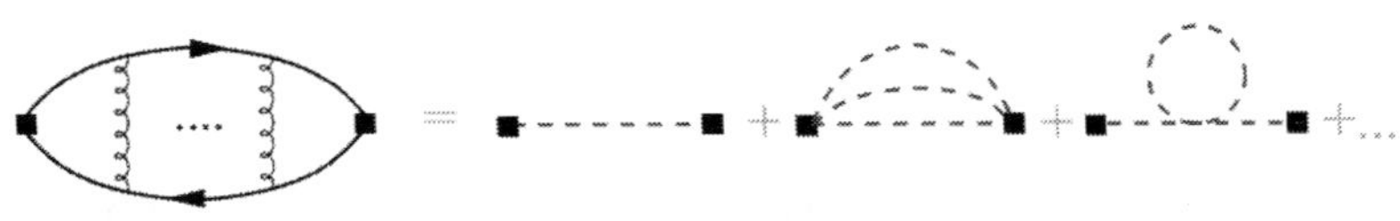

Figure 4. Generic QCD sum rule for *ordinary* mesons (blue dashed lines) following traditional folklore.

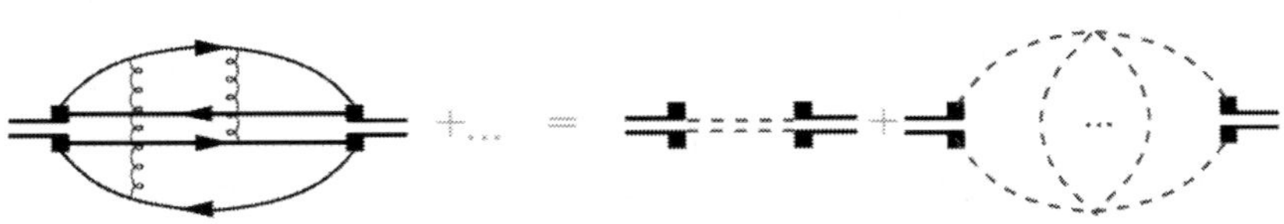

Figure 5. Tetraquark-adequate QCD sum rule, possibly involving tetraquarks (blue dashed double line).

In addition to the perturbative contributions to the QCD sum rules aimed at, various power corrections have to be considered, too. Their inclusion is more tricky but accomplishable [19].

Ultimately, incorporating the aforementioned findings should offer a *reliable* extraction of the fundamental properties of tetraquarks [their masses M, decay constants $f_{\bar{a}b\bar{c}d} \equiv \langle 0|\theta_{\bar{a}b\bar{c}d}|T\rangle$ and $f_{\bar{a}d\bar{c}b} \equiv \langle 0|\theta_{\bar{a}d\bar{c}b}|T\rangle$, and momentum-space amplitudes $A(T \to j_{\bar{a}b}\, j_{\bar{c}d})$ and $A(T \to j_{\bar{a}d}\, j_{\bar{c}b})$] from *tetraquark-adequate QCD sum rules* for all two- and three-point correlation functions of

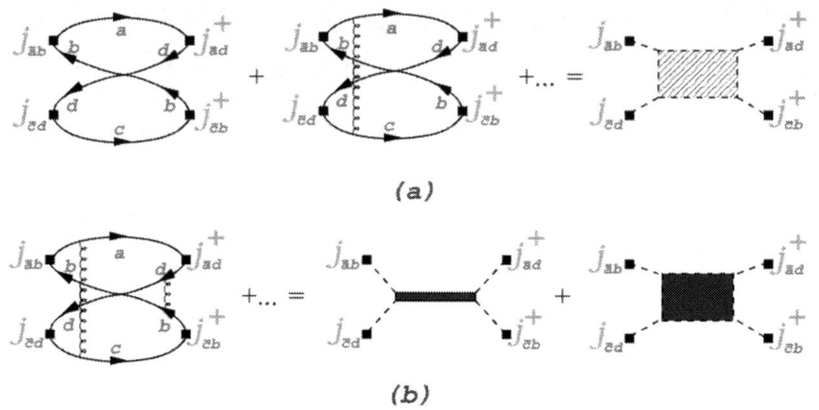

Figure 6. Decomposing each *flavour-rearranging* relation into (a) one without four-quark or two-meson s-channel cut and (b) a *tetraquark-adequate* QCD sum rule supporting tetraquarks (blue horizontal bar).

interest involving flavour-preserving or -reordering tetraquark-phile spectral densities $\rho_{\rm p,r}$ and $\Delta_{\rm p,r}$ and effective thresholds $s_{\rm eff}(\tau)$ depending on Borel parameters τ, symbolically of the form

$$(f_{\bar a b \bar c d})^2 \exp(-M^2\,\tau) = \int_{\hat s}^{s_{\rm eff}(\tau)} {\rm d}s \exp(-s\,\tau)\,\rho_{\rm p}(s) + \text{Borel-transformed power corrections}\,,$$

$$f_{\bar a b \bar c d}\,A(T \to j_{\bar a b}\,j_{\bar c d}) \exp(-M^2\,\tau) = \int_{\hat s}^{s_{\rm eff}(\tau)} {\rm d}s \exp(-s\,\tau)\,\Delta_{\rm p}(s) + \text{Borel-transformed power corrections}\,,$$

$$f_{\bar a b \bar c d}\,f_{\bar a d \bar c b} \exp(-M^2\,\tau) = \int_{\hat s}^{s_{\rm eff}(\tau)} {\rm d}s \exp(-s\,\tau)\,\rho_{\rm r}(s) + \text{Borel-transformed power corrections}\,,$$

$$f_{\bar a d \bar c b}\,A(T \to j_{\bar a b}\,j_{\bar c d}) \exp(-M^2\,\tau) = \int_{\hat s}^{s_{\rm eff}(\tau)} {\rm d}s \exp(-s\,\tau)\,\Delta_{\rm r}(s) + \text{Borel-transformed power corrections}\,.$$

Last but not least, the *cluster reducibility* [20] of any multiquark (its group-theory-implied potential decomposition into a cluster of ordinary hadrons) requires to regard a multiquark, on an equal footing, as tightly bound compact and loosely bound molecular-type hadron [21, 22].

Acknowledgements. D. M. and H. S. express sincere gratitude for support by joint CNRS/RFBR Grant PRC Russia/19-52-15022, D. M. for support by the Austrian Science Fund (FWF), Project P29028-N27, H. S. for support by EU research and innovation program Horizon 2020 under Grant Agreement 824093.

References

[1] K. G. Wilson, *Non-Lagrangian models of current algebra*, Phys. Rev. **179** (1969) 1499

[2] W. Lucha, D. Melikhov and S. Simula, *Effective continuum threshold in dispersive sum rules*, Phys. Rev. D **79** (2009) 096011, arXiv:0902.4202 [hep-ph]
[3] W. Lucha, D. Melikhov and S. Simula, *Bound-state parameters from dispersive sum rules for vacuum-to-vacuum correlators*, J. Phys. G **37** (2010) 035003, arXiv:0905.0963 [hep-ph]
[4] W. Lucha, D. Melikhov and S. Simula, *Extraction of ground-state decay constant from dispersive sum rules: QCD vs potential models*, Phys. Lett. B **687** (2010) 48, arXiv: 0912.5017 [hep-ph]
[5] M. A. Shifman, A. I. Vainshtein and V. I. Zakharov, *QCD and resonance physics. Theoretical foundations*, Nucl. Phys. B **147** (1979) 385
[6] W. Lucha, D. Melikhov and H. Sazdjian, *Tetraquark-adequate formulation of QCD sum rules*, Phys. Rev. D **100** (2019) 014010, arXiv:1901.03881 [hep-ph]
[7] W. Lucha, D. Melikhov and H. Sazdjian, *Tetraquark-adequate QCD sum rules for quark-exchange processes*, Phys. Rev. D **100** (2019) 074029, arXiv:1909.06324 [hep-ph]
[8] W. Lucha, D. Melikhov and H. Sazdjian, *Tetraquark-adequate QCD sum rules*, PoS (EPS-HEP2019) 536, arXiv:1908.08803 [hep-ph]
[9] W. Lucha, D. Melikhov and H. Sazdjian, *Multiquark-adequate QCD sum rules: the case of flavour-exotic tetraquarks*, EPJ Web Conf. **222** (2019) 03016, arXiv:1910.12064 [hep-ph]
[10] W. Lucha, D. Melikhov and H. Sazdjian, *Cluster reducibility of multiquark operators and tetraquark-adequate QCD sum rules*, Supl. Rev. Mex. Fís. **3** (2022) 0308035, arXiv: 2111.10239 [hep-ph]
[11] W. Lucha, D. Melikhov and H. Sazdjian, *Exotic tetraquark mesons in large-N_c limit: an unexpected great surprise*, EPJ Web Conf. **192** (2018) 00044, arXiv:1808.05519 [hep-ph]
[12] M. Fierz, *Zur Fermischen Theorie des β-Zerfalls*, Z. Phys. **104** (1937) 553
[13] R. L. Jaffe, *Two types of hadrons*, Nucl. Phys. A **804** (2008) 25
[14] W. Lucha, D. Melikhov and H. Sazdjian, *Narrow exotic tetraquark mesons in large-N_c QCD*, Phys. Rev. D **96** (2017) 014022, arXiv:1706.06003 [hep-ph]
[15] W. Lucha, D. Melikhov and H. Sazdjian, *Tetraquark and two-meson states at large N_c*, Eur. Phys. J. C **77** (2017) 866, arXiv:1710.08316 [hep-ph]
[16] W. Lucha, D. Melikhov and H. Sazdjian, *Exotic states and their properties from large-N_c QCD*, PoS (EPS-HEP 2017) 390, arXiv:1709.02132 [hep-ph]
[17] W. Lucha, D. Melikhov and H. Sazdjian, *Are there narrow flavour-exotic tetraquarks in large-N_c QCD?*, Phys. Rev. D **98** (2018) 094011, arXiv:1810.09986 [hep-ph]
[18] L. D. Landau, *On analytic properties of vertex parts in quantum field theory*, Nucl. Phys. **13** (1959) 181
[19] W. Lucha, D. Melikhov and H. Sazdjian, *OPE and quark–hadron duality for two-point functions of tetraquark currents in the $1/N_c$ expansion*, Phys. Rev. D **103** (2021) 014012, arXiv:2012.03894 [hep-ph]
[20] W. Lucha, D. Melikhov and H. Sazdjian, *Cluster reducibility of multiquark operators*, Phys. Rev. D **100** (2019) 094017, arXiv:1908.10164 [hep-ph]
[21] W. Lucha, D. Melikhov and H. Sazdjian, *Tetraquarks in large-N_c QCD*, Prog. Part. Nucl. Phys. **120** (2021) 103867, arXiv:2102.02542 [hep-ph]
[22] H. Sazdjian, *The interplay between compact and molecular structures in tetraquarks*, Symmetry **14** (2022) 515, arXiv:2202.01081 [hep-ph]

Exclusive semileptonic B-meson decays using lattice QCD and unitarity

Guido Martinelli[1,*], *Manuel* Naviglio[2,3,**], *Silvano* Simula[4,***], and *Ludovico* Vittorio[3,5,****]

[1]Physics Department, University of Rome "La Sapienza" and INFN, Sezione di Roma, Piazzale Aldo Moro 5, I-00185 Rome, Italy
[2]Physics Department, University of Pisa, Largo Bruno Pontecorvo 3, I-56127 Pisa, Italy
[3]INFN, Sezione di Pisa, Largo Bruno Pontecorvo 3, I-56127 Pisa, Italy
[4]INFN, Sezione di Roma Tre, Via della Vasca Navale 84, I-00146 Rome, Italy
[5]Scuola Normale Superiore, Piazza dei Cavalieri 7, I-56126 Pisa, Italy

Abstract. We present the results of the application of the Dispersion Matrix approach to exclusive semileptonic B-meson decays. This method allows to determine the hadronic form factors in a non-perturbative and completely model-independent way. Starting from lattice results available at large values of the momentum transfer, the behaviour of the form factors in their whole kinematical range is obtained without introducing any parameterization of their momentum dependence. We will focus on the determination of the Cabibbo-Kobayashi-Maskawa matrix elements $|V_{cb}|$ and $|V_{ub}|$ through the analysis of $B_{(s)} \to D^{(*)}_{(s)}\ell\nu$ and $B_{(s)} \to \pi(K)\ell\nu$ decays. New theoretical determinations of the Lepton Flavour Universality ratios relevant for these transitions will be also presented.

1 State-of-the-art of exclusive semileptonic B-meson decays

Exclusive semileptonic B-meson decays are one of the most challenging processes in the phenomenology of flavour physics, since they are affected by two unsolved problems.

On the one hand, we have the so-called $|V_{cb}|$ *puzzle*, *i.e.* the discrepancy between the inclusive and the exclusive determinations of the CKM matrix element $|V_{cb}|$. According to the FLAG Review 2021 [1], there is a $\sim 2.8\sigma$ tension between the exclusive estimate (which depends on the form factors parametrization) and the inclusive one, namely

$$|V_{cb}|_{\rm excl} \times 10^3 = 39.36(68), \qquad |V_{cb}|_{\rm incl} \times 10^3 = 42.00(65). \tag{1}$$

Two new estimates of the inclusive value have also recently appeared, namely $|V_{cb}|_{\rm incl} \times 10^3 = 42.16(50)$ [2] and $|V_{cb}|_{\rm incl} \times 10^3 = 41.69(63)$ [3], which are compatible with the inclusive FLAG value in Eq. (1) and corroborate its truthfulness.

On the other hand, a strong tension exists between the theoretical values and the measurements of $R(D^{(*)})$, which are a fundamental test of Lepton Flavour Universality (LFU) and are defined as

$$R(D^{(*)}) \equiv \frac{\Gamma(B \to D^{(*)}\tau\nu_\tau)}{\Gamma(B \to D^{(*)}\ell\nu_\ell)}, \tag{2}$$

*e-mail: guido.martinelli@roma1.infn.it
**e-mail: manuel.naviglio@phd.unipi.it
***e-mail: silvano.simula@roma3.infn.it
****e-mail: ludovico.vittorio@sns.it

where $\ell = e, \mu$ denotes a light lepton. The HFLAV Collaboration [4] has recently computed the world averages of the available measurements of the $R(D^{(*)})$ ratios and of their SM theoretical expectations, obtaining

$$R(D)_{\rm SM} = 0.299 \pm 0.003, \qquad R(D)_{\rm exp} = 0.339 \pm 0.026 \pm 0.014 \tag{3}$$

for the $B \to D$ case and

$$R(D^*)_{\rm SM} = 0.254 \pm 0.005, \qquad R(D^*)_{\rm exp} = 0.295 \pm 0.010 \pm 0.010 \tag{4}$$

for the $B \to D^*$ one. As clearly stated by HFLAV Collaboration, the averages of the measurements of $R(D)$ and $R(D^*)$ exceed the corresponding SM predictions by 1.4σ and 2.8σ, respectively. If the experimental correlation between these two quantities, namely $\rho = -0.38$, is taken into account, the resulting difference with the SM is increased to the 3.3σ level [4].

2 The Dispersion Matrix approach to hadronic Form Factors

Let us now focus on semileptonic $B \to D^{(*)}$ decays for massless leptons (namely $\ell = e, \mu$ in what follows). In case of production of a *pseudoscalar* (PS) meson, *i.e.* the $B \to D\ell\nu$ case, the differential decay width reads

$$\frac{d\Gamma}{dq^2} = \frac{G_F^2}{24\pi^3}\eta_{EW}^2|V_{cb}|^2|\vec{p}_D|^3|f^+(q^2)|^2, \tag{5}$$

where $\vec{p}_D$ represents the 3-momentum of the produced D meson. In this case, there is only one Form Factor (FF) appearing in the theoretical expression of the decay width, namely $f^+(q^2)$. In case of production of a *vector* meson, *i.e.* the $B \to D^*\ell\nu$ case, the expression of the differential decay width is instead

$$\begin{aligned}\frac{d\Gamma(B \to D^*(\to D\pi)\ell\nu)}{dw\, d\cos\theta_\ell\, d\cos\theta_v\, d\chi} = \frac{3}{4}\frac{G_F^2}{(4\pi)^4}\eta_{EW}^2|V_{cb}|^2 m_B^3 r^2\sqrt{w^2-1}(1+r^2-2rw)\\ \times\{(1-\cos\theta_\ell)^2\sin^2\theta_v|H_+|^2 + (1+\cos\theta_\ell)^2\sin^2\theta_v|H_-|^2\\ + 4\sin^2\theta_\ell\cos^2\theta_v|H_0|^2 - 2\sin^2\theta_\ell\sin^2\theta_v\cos 2\chi H_+H_-\\ - 4\sin\theta_\ell(1-\cos\theta_\ell)\sin\theta_v\cos\theta_v\cos\chi H_+H_0\\ + 4\sin\theta_\ell(1+\cos\theta_\ell)\sin\theta_v\cos\theta_v\cos\chi H_-H_0\},\end{aligned} \tag{6}$$

where we have defined $r \equiv m_{D^*}/m_B$ and we have introduced the helicity amplitudes

$$H_0(w) = \frac{\mathcal{F}_1(w)}{\sqrt{m_B^2 + m_D^2 - 2m_Bm_Dw}}, \qquad H_\pm(w) = f(w) \mp m_Bm_{D^*}\sqrt{w^2-1}\,g(w). \tag{7}$$

In Eq.(6) θ_l, θ_v, χ represent the helicity angles and the FFs are $f(w)$, $g(w)$, $\mathcal{F}_1(w)$. Note that in Eqs. (5) and (6) we refer equivalently to the momentum transfer q^2 or to the recoil w, since they are related by $q^2(w) = m_B^2 + m_{D^{(*)}}^2 - 2m_Bm_{D^{(*)}}w = m_B^2(1 + r^2 - 2rw)$.

Let us finally stress again that in the previous expressions we are assuming a massless produced lepton. For massive leptons ($\ell = \tau$), instead, one should add the FFs $f_0(q^2)$ for semileptonic $B \to D$ decays and $P_1(w)$ for semileptonic $B \to D^*$ one.

Now, our goal is to describe the FFs entering in exclusive semileptonic B-meson decays by using the novel Dispersion Matrix (DM) method [5], which was originally proposed in [6]. The DM method allows us to study the FFs in a non-perturbative and model-independent way.

Starting from the available LQCD computations of the FFs at high momentum transfer (or, equivalently, at low recoil), we can extrapolate their behaviour in the opposite kinematical region without assuming any functional dependence of the FFs on q^2 (or, equivalently, on w) and using only non-perturbative inputs. Moreover, the resulting bands of the FFs will be independent of the experimental determinations of the differential decay widths.

From the mathematical point of view, the starting point is to focus on one FF, for instance f, and then consider the matrix

$$\mathbf{M} = \begin{pmatrix} \chi & \phi f & \phi_1 f_1 & \dots & \phi_N f_N \\ \phi f & \frac{1}{1-z^2} & \frac{1}{1-z z_1} & \dots & \frac{1}{1-z z_N} \\ \phi_1 f_1 & \frac{1}{1-z_1 z} & \frac{1}{1-z_1^2} & \dots & \frac{1}{1-z_1 z_N} \\ \dots & \dots & \dots & \dots & \dots \\ \phi_N f_N & \frac{1}{1-z_N z} & \frac{1}{1-z_N z_1} & \dots & \frac{1}{1-z_N^2} \end{pmatrix}, \tag{8}$$

where we have introduced the conformal variable z defined as

$$z(t) = \frac{\sqrt{t_+ - t} - \sqrt{t_+ - t_-}}{\sqrt{t_+ - t} + \sqrt{t_+ - t_-}}, \qquad t_\pm = (m_B \pm m_{D^{(*)}})^2 \tag{9}$$

with $t \equiv q^2$, or, equivalently, as

$$z(w) = \frac{\sqrt{w+1} - \sqrt{2}}{\sqrt{w+1} + \sqrt{2}}. \tag{10}$$

In this expression, $\phi_i f_i \equiv \phi(z_i) f(z_i)$ (with $i = 1, 2, \dots N$) are the known values of the quantity $\phi(z) f(z)$ corresponding to the values z_i at which the FFs have been computed on the lattice. The kinematical function $\phi(z)$ has a specific expression for each FF [7]. Finally, the susceptibility $\chi(q_0^2)$ is related to the derivative with respect to q_0^2 of the Fourier transform of suitable Green functions of bilinear quark operators and follows from the dispersion relation associated to a particular spin-parity quantum channel. Note that they have been computed for the first time on the lattice in [8] for $b \to c$ quark transitions by choosing $q_0^2 = 0$. At this point, one can demonstrate from first principles that $\det \mathbf{M} \geq 0$. Then, the positivity of the determinant, which we will refer to as *unitarity filter* hereafter, allows to compute the lower and the upper bounds of the FF of interest for each generic value of z, *i.e.*

$$f_{\rm lo}(z) \leq f(z) \leq f_{\rm up}(z). \tag{11}$$

To be more quantitative, we have that [5]

$$\beta - \sqrt{\gamma} \leq f \leq \beta + \sqrt{\gamma}, \tag{12}$$

where (after some algebraic manipulations)

$$\beta = \frac{1}{\phi(z) d(z)} \sum_{j=1}^{N} f_j \phi_j d_j \frac{1 - z_j^2}{z_0 - z_j}, \qquad \gamma = \frac{1}{1 - z_0^2} \frac{1}{\phi^2(z) d^2(z)} (\chi - \overline{\chi}),$$

$$\overline{\chi} = \sum_{i,j=1}^{N} f_i f_j \phi_i d_i \phi_j d_j \frac{(1 - z_i^2)(1 - z_j^2)}{1 - z_i z_j}$$

where $d(z)$, d_i are simply kinematical functions [5]. Unitarity is satisfied only when $\gamma \geq 0$, which implies $\chi \geq \overline{\chi}$. One can show that the values of β and γ depend on z, while the value of $\overline{\chi}$ does not depend on z and it depends only on the set of input data. Consequently, the unitarity condition $\chi \geq \overline{\chi}$ does not depend on z.

In what follows, we will describe the results of the phenomenological applications of the DM method to several exclusive semileptonic B-meson decays. We will analyze in detail two instructive cases, *i.e.* $B \to D^* \ell \nu$ and $B \to \pi \ell \nu$ transitions, and then we will give an overview of all the results relevant for phenomenology obtained so far.

3 The DM application to semileptonic $B \to D^* \ell \nu$ decays

Let us firstly discuss in detail semileptonic $B \to D^*$ decay, which is very challenging due to the high number of FFs involved. In [9] we have computed the unitarity bands of the FFs, starting from the final results of the computations on the lattice performed by the FNAL/MILC Collaborations [10]. There, in the ancillary files, the authors give the synthetic values of the FFs $g(w), f(w), \mathcal{F}_1(w)$ and $\mathcal{F}_2(w)$ at three non-zero values of the recoil variable w, namely $w = \{1.03, 1.10, 1.17\}$, together with their correlations. Note that the FF $\mathcal{F}_2(w)$ is directly related to the $P_1(w)$ one, in fact $P_1(w) = \mathcal{F}_2(w)\sqrt{r}/(1+r)$, where $r \equiv m_{D^*}/m_B \simeq 0.38$. In Fig. 1 we show the results of our DM analysis as red bands. The DM unitarity bands are built up through bootstrap events that satisfy exactly both the unitarity filter discussed in the previous Section and the Kinematical Constraints (KCs)

$$\mathcal{F}_1(1) = m_B(1-r)f(1), \qquad P_1(w_{max}) = \frac{\mathcal{F}_1(w_{max})}{m_B^2(1+w_{max})(1-r)\sqrt{r}}, \tag{13}$$

where $w_{\max}$ is the maximum value of the recoil, namely $w_{\max} \simeq 1.5$. Starting from the unitary bands shown in Fig. 1, we are then able to compute new estimates of both the CKM matrix element $|V_{cb}|$ and the ratio $R(D^*)$.

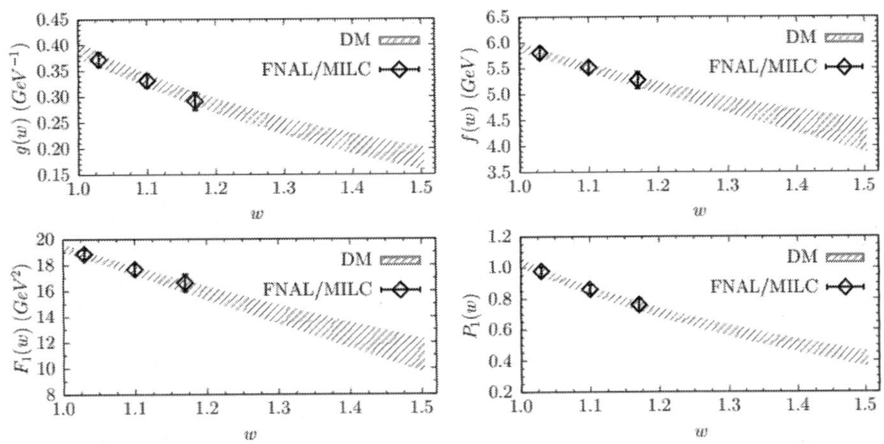

Figure 1. The bands of the FFs $g(w)$, $f(w)$, $\mathcal{F}_1(w)$ and $P_1(w)$ computed by the DM method after imposing both the unitarity filter and the KCs. The FNAL/MILC values [10] used as inputs for the DM method are represented by the black diamonds.

For what concerns $|V_{cb}|$, the Belle Collaboration [11, 12] has published two different sets of measurements of the differential decay widths $d\Gamma/dx$ ($x = w, \cos\theta_l, \cos\theta_v, \chi$) in the form of ten bins for each kinematical variable. Thus, we have computed the theoretical estimate of each one-dimensional differential decay width (starting from the complete expression in the Eq. (6) and using the unitarity bands of the FFs) and we have compared them to the experimental data in order to obtain bin-per-bin estimates of $|V_{cb}|$. They are shown in Fig. 2, where the blue squares come from the first set of Belle measurements [11] while the red circles from the second one [12]. For each kinematical variable and for each experiment, we compute the average of these bin-per-bin estimates as

$$|V_{cb}| = \frac{\sum_{i,j=1}^{10} (\mathbf{C}^{-1})_{ij} |V_{cb}|_j}{\sum_{i,j=1}^{10} (\mathbf{C}^{-1})_{ij}}, \qquad \sigma^2_{|V_{cb}|} = \frac{1}{\sum_{i,j=1}^{10} (\mathbf{C}^{-1})_{ij}}, \tag{14}$$

where $\mathbf{C}$ is the covariance matrix and $|V_{cb}|_i$ is the $|V_{cb}|$ estimate for the i-th bin. This procedure generates the dashed blue (red) bands in Fig. 2 for the first (second) set of Belle measurements. We finally combine the resulting eight mean values as

$$\mu_x = \frac{1}{N} \sum_{k=1}^{N} x_k, \qquad \sigma^2_x = \frac{1}{N} \sum_{k=1}^{N} \sigma^2_k + \frac{1}{N} \sum_{k=1}^{N} (x_k - \mu_x)^2, \tag{15}$$

obtaining

$$|V_{cb}| \times 10^3 = 41.3 \pm 1.7.$$

The DM method allows us also to obtain fully-theoretical expectation values of other quantities relevant for phenomenology, *i.e.* the anomaly $R(D^*)$, the τ-polarization P_τ and finally the D^* longitudinal polarization F_L, namely

$$R(D^*) = 0.275 \pm 0.008, \qquad P_\tau = -0.529 \pm 0.007, \qquad F_L = 0.414 \pm 0.012.$$

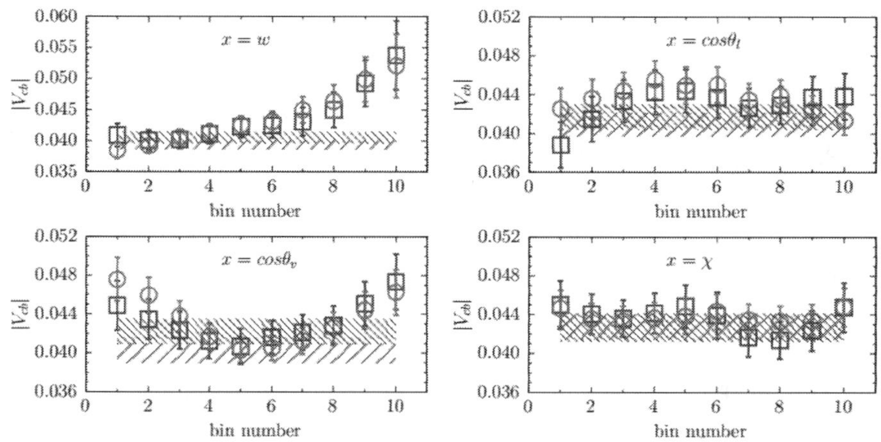

Figure 2. The bin-per-bin estimates of $|V_{cb}|$ and their weighted means for each kinematical variable ($x = w, \cos\theta_l, \cos\theta_v, \chi$). The blue squares and the red circles correspond respectively to the first [11] and to the second [12] set of the Belle measurements. The dashed blue (red) bands are the results of the average (14) in the case of the blue squares (red circles) for each variable.

4 The DM application to semileptonic $B \to \pi \ell \nu$ decays

The DM method can be applied to any semileptonic charged-current decays of hadrons. In this Section we investigate its potential in the analysis of the $b \to u$ quark transitions by discussing the case of the $B \to \pi\ell\nu$ decays [14]. In this case, since the pion is a PS meson, the formalism is analogous to the one characterizing $B \to D\ell\nu$ decays described in Sec. 2.

Thus, $B \to \pi\ell\nu$ transitions are characterized by two FFs: $f_+^\pi(q^2)$ and $f_0^\pi(q^2)$. These FFs have been studied by the RBC/UKQCD [15] and the FNAL/MILC [16] Collaborations. For both channels the lattice computations of the FFs are available in the large-q^2 region[1].

In Fig. 3 we show the red (blue) DM bands obtained by using as inputs the RBC/UKQCD (FNAL/MILC) data. Note that using the BCL fits of [16] the mean value and the uncertainty of the FFs extrapolated at zero momentum transfer are not stable against variation of the truncation order of the series expansion of the FFs. On the contrary, the DM approach is completely independent of this issue, since no approximation due to the truncation of a series expansion is present. The DM method is equivalent to the results of all possible fits which satisfy unitarity and, at the same time, reproduce exactly the input data. This property is particularly useful in $B \to \pi\ell\nu$ decays, where the extrapolation to $q^2 = 0$ is quite long.

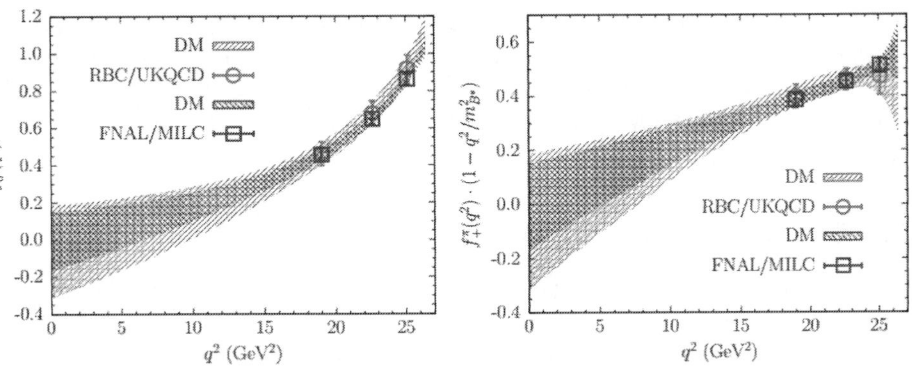

Figure 3. The scalar $f_0^\pi(q^2)$ (left panel) and vector $f_+^\pi(q^2)$ (right panel) FFs entering the semileptonic $B \to \pi\ell\nu_\ell$ decays computed by the DM method as a function of the 4-momentum transfer q^2 using the LQCD inputs from RBC/UKQCD (red points) and FNAL/MILC (blue squares) Collaborations. In the right panel, the vector FF is multiplied by the factor $(1 - q^2/m_{B^*}^2)$ with $m_{B^*} = 5.325$ GeV.

For the extraction of the CKM matrix element we compute bin-per-bin values of $|V_{ub}|$ for each q^2-bin of each available experiment. Recall that the branching fractions of this transition have been measured by several experiments [18–21]. Thus, we evaluate the CKM matrix element for the n-th experiment ($n = 1, \ldots, 6$ for the semileptonic $B \to \pi$ decays) through expressions analogous to Eq. (14). Our results are shown in Fig. 4 when one uses the combination of the RBC/UKQCD and FNAL/MILC data as inputs of the DM method.

Our final result for $|V_{ub}|$ is evaluated making use of the averaging procedure given by Eq. (15) and reads

$$|V_{ub}|^{B\pi} \cdot 10^3 = 3.62 \pm 0.47. \tag{16}$$

Let us mention here that we are currently investigating strategies to improve the precision of the determination of $|V_{ub}|$ within our DM approach. Some results can be found in [22], where our improved determination of the CKM matrix element reads

$$|V_{ub}|^{B\pi}_{\text{impr}} \cdot 10^3 = 3.88 \pm 0.32. \tag{17}$$

[1]To be more specific, the authors of Ref. [15] provide synthetic LQCD values of the FFs (together with their statistical and systematic correlations) at $q^2 = \{19.0, 22.6, 25.1\}$ GeV^2. In [16], instead, only the results of BCL fits [17] of the FFs extrapolated to the continuum limit and to the physical pion point are available. Thus, from the marginalized BCL coefficients we evaluate the mean values, uncertainties and correlations of the FFs at the same three values of q^2 given in Ref. [15].

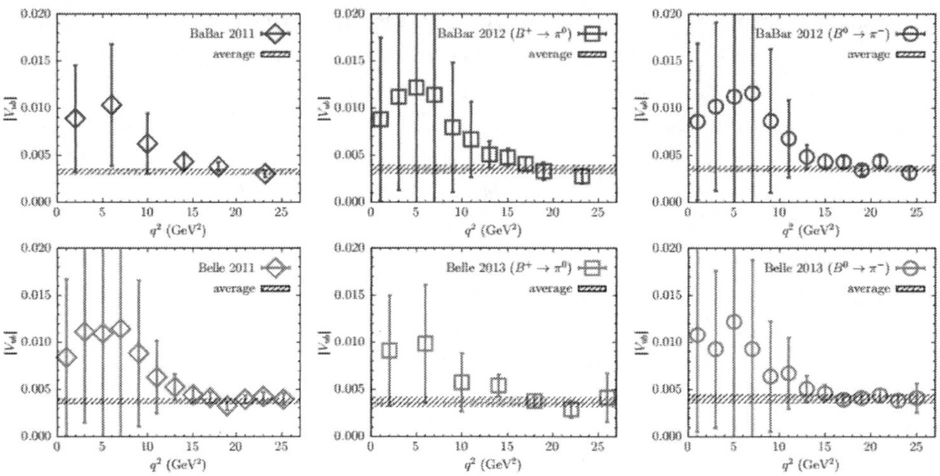

Figure 4. *Bin-per-bin estimates of $|V_{ub}|$ for each of the six experiments of Refs. [18–21] specified in the insets of the panels as a function of q^2. The theoretical DM bands of the FFs correspond to the use of the combination of the available LQCD data as inputs [14]. The (black) dashed bands represent the correlated weighted averages for each experiment.*

5 Conclusions

In this contibution we have reviewed the main features of the DM approach, which is an interesting tool to implement unitarity and LQCD calculations in the analysis of exclusive charged-current semileptonic decays of mesons and baryons. In Fig. 5 we have condensed the results obtained so far from the application of the DM method to the semileptonic $B \to D^{(*)}$ [7, 9], $B_s \to D_s^{(*)}$ [23], $B \to \pi$ and $B_s \to K$ [14] decays. The DM values of the CKM matrix elements in the left panel represent the averages of all the DM determination of $|V_{cb}|$ and $|V_{ub}|$ from the various decay channels. For both the CKM matrix elements, the DM determinations are compatible with the corresponding inclusive values within the 1σ level. Furthermore, the DM values are practically identical to the indirect determinations coming for the latest analysis by the UTfit Collaboration [24]. The values of the LFU observables (for both the $B \to D^{(*)}$ and the $B_s \to D_s^{(*)}$ decays) are, instead, shown in the right panel, together with the experimental average and the SM one by HFLAV. By using the FNAL/MILC computations of the FFs for the $B \to D^*$ channel, we observe that the tension between theoretical expectations and measurements of $R(D^{(*)})$ is reduced.

Table 1. Numerical values of the CKM matrix elements $|V_{cb}|$ and $|V_{ub}|$ plotted in Fig. 5.

	Decay channel	DM values	FLAG '21	Inclusive	UTfit '22
$\lvert V_{cb}\rvert \times 10^3$	$B_{(s)} \to D_{(s)}^{(*)}$	41.2 (8)	39.48 (68)	42.16 (50)	41.27 (89)
$\lvert V_{ub}\rvert \times 10^3$	$B_{(s)} \to \pi(K)$	3.85 (27)	3.63 (14)	4.13 (26)	3.77 (22)

References

[1] FLAG Collaboration, [arXiv:2111.09849]
[2] Bordone M., Capdevila B. and Gambino P., Phys. Lett. B **822**, 136679 (2021)
[3] Bernlochner F. *et al.*, [arXiv:2205.10274]
[4] HFLAV Collaboration, [arXiv:2206.07501]

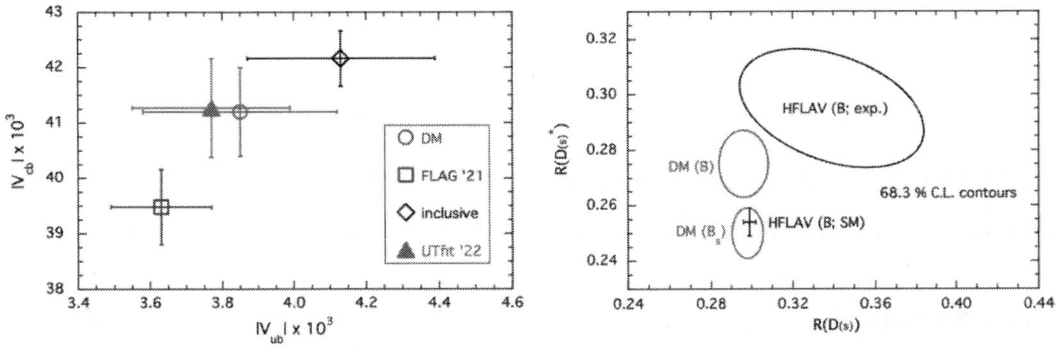

Figure 5. ***Left panel:*** *$|V_{cb}|$ vs $|V_{ub}|$ plot. The numerical values corresponding to the various entries (DM estimates, FLAG results [1], inclusive values [2, 25], UTfit indirect determinations [24]) can be found in Tab. 1.* ***Right panel:*** *$R(D^*_{(s)})$ vs $R(D_{(s)})$ plot. The numerical values corresponding to the various entries (DM estimates, HFLAV experimental average, HFLAV SM average) can be found in Tab. 2.*

Table 2. Numerical values of the LFU observables plotted in Fig. 5.

	DM values	HFLAV '21 (exp)	HFLAV '21 (SM)
$R(D)$	0.296 (8)	0.339 (26) (14)	0.299 (3)
$R(D^*)$	0.275(8)	0.295 (10) (10)	0.254 (5)
$R(D_s)$	0.298 (5)		
$R(D_s^*)$	0.250 (6)		

[5] Di Carlo M. *et al.*, Phys. Rev. D **104**, 054502 (2021)
[6] Lellouch L., Nucl. Phys. B **479**, 353 (1996)
[7] Martinelli G., Simula S. and Vittorio L., Phys. Rev. D **105**, 034503 (2022)
[8] Martinelli G., Simula S. and Vittorio L., Phys. Rev. D **104**, 094512 (2022)
[9] Martinelli G., Simula S. and Vittorio L., [arXiv:2109.15248]
[10] Fermilab/MILC collaboration, [arXiv:2105.14019]
[11] BELLE Collaboration, [arXiv:1702.01521]
[12] BELLE Collaboration, Phys. Rev. D **100**, 052007 (2019)
[13] Fermilab/MILC collaboration, Phys. Rev. D **92**, 034506 (2015)
[14] Martinelli G., Simula S. and Vittorio L., JHEP **08**, 022 (2022)
[15] RBC/UKQCD Collaboration, Phys. Rev. D **91**, 074510 (2015)
[16] Fermilab/MILC collaboration, Phys. Rev. D **92**, 014204 (2015)
[17] Bourrely C., Caprini I. and Lellouch L., Phys. Rev. D **79**, 013008 (2009)
[18] BaBar Collaboration, Phys. Rev. D **83**, 032007 (2011)
[19] BELLE Collaboration, Phys. Rev. D **83**, 071101 (2011)
[20] BaBar Collaboration, Phys. Rev. D **86**, 092004 (2012)
[21] BELLE Collaboration, Phys. Rev. D **88**, 032005 (2013)
[22] Martinelli G., Naviglio M., Simula S. and Vittorio L., [arXiv:2203.16213]
[23] Martinelli G., Naviglio M., Simula S. and Vittorio L., [arXiv:2204.05925]
[24] Bona M. et al, PoS **EPS-HEP2021**, 512 (2022)
[25] P. A. Zyla et al., PTEP **2020**, 083C01 (2020)

Implications of Heavy Quark Spin Symmetry and NRQCD on $B_c \to J/\psi, \eta_c$ form factors

Martín Novoa-Brunet[1,*]

[1]Istituto Nazionale di Fisica Nucleare, Sezione di Bari, Via Orabona 4, 70126 Bari, Italy

Abstract. We discuss the analysis of $B_c \to J/\psi, \eta_c$ form factors, based on the heavy quark spin symmetry and nonrelativistic QCD (NRQCD) performed in Ref. [1]. We perform an expansion on the inverse of the heavy quarks' masses using power counting rules of NRQCD up to the next-to-leading order. This allows us to classify the heavy quark spin symmetry-breaking terms and express the form factors in terms of universal functions in a selected kinematical range. Using as input the lattice QCD results for the $B_c \to J/\psi$ matrix element of the SM operator, we obtain information on other form factors.

1 Introduction

Over the last years several anomalies have been detected in the heavy flavour sector, most notably in the semileptonic decays involving the $b \to c\ell\bar{\nu}$ and $b \to s\ell^+\ell^-$ transitions [2, 3]. If these deviations originated in New Physics (NP) effects, one should expect to systematically see this effect on all of the modes induced by these quark transitions. In the case of $b \to c\ell\bar{\nu}$, deviations have been measured in $B \to D^{(*)}\ell\bar{\nu}$ decays [2] and other semileptonic decays involving the same transition are being extensively studied right now. In particular, $B_c \to J/\psi, \eta_c$ decays are important to study this transition. One of the main problems for the interpretation of the measurements is the hadronic uncertainties that affect these modes. While, in the case of $B \to D^{(*)}\ell\bar{\nu}$ several studies have been performed to understand better hadronic contributions, including Lattice QCD, Sum Rules and Heavy Quark Effective Theory (HQET) methods, the situation is less apparent on the side $B_c \to J/\psi, \eta_c$ decays, where only a limited amount of information is currently available [4–7].

It has been observed [4] that the semileptonic B_c form factors can be expressed in terms of universal functions in selected kinematical regions, based on the heavy quark spin symmetry for large heavy quark masses leading to several analyses of B_c decays [8–10]. The relations to the universal functions of $B_c \to J/\psi, \eta_c$ decays have been provided at the leading order [4] in the heavy quark mass expansion. The relations obtained in such a systematic expansion have two applications. They can be used to test the form factors obtained by different methods, for a quantitative assessment of their theoretical uncertainty. Moreover, information can be gained on form factors that have not been computed yet, which are needed for analyses based on the generalized low-energy Hamiltonian. When comparing these relations to recent lattice QCD results [7] violations of the order of 50% are found for the vector form factors, suggesting the need for a better understanding of next-to-leading order effects.

*e-mail: martin.novoa@ba.infn.it

In this contribution, we describe the analysis at the next-to-leading order in the expansion performed in Ref. [1], where we established further relations among the form factors and a set of universal functions based on the heavy quark spin symmetry and the power counting rules of nonrelativistic QCD (NRQCD). In Section 2 we introduce the relevant form factors, to then we give an outline of the calculation in Section 3. The details can be found in Ref. [1]. In Section 4 we present the main result and discuss its implications.

2 Form Factors for $B_c \to J/\psi, \eta_c$

We focus on the B_c and J/ψ, η_c matrix elements relevant for semileptonic $B_c \to J/\psi(\eta_c)\ell\bar{\nu}$ decays, which involve the vector (axial) $\bar{Q}'\gamma_\mu(\gamma_5)Q$, (pseudo)scalar $\bar{Q}'(\gamma_5)Q$, and (pseudo)tensor $\bar{Q}'\sigma_{\mu\nu}(\gamma_5)Q$ currents. Due to the nature of QCD these matrix elements cannot be computed perturbatively and thus need to be parametrized in terms of form factors. We choose the following parametrization due to its ability to simplify the expressions that are obtained in the following section.

$$\begin{aligned}
\langle P(v')|\bar{Q}'\gamma_\mu Q|B_c(v)\rangle &= \sqrt{m_P\, m_{B_c}}\,[h_+(w)\,(v+v')_\mu + h_-(w)\,(v-v')_\mu] \\
\langle P(v')|\bar{Q}' Q|B_c(v)\rangle &= \sqrt{m_P\, m_{B_c}}\,h_S(w)(1+w) \\
\langle P(v')|\bar{Q}'\sigma_{\mu\nu} Q|B_c(v)\rangle &= -i\,\sqrt{m_P\, m_{B_c}}\,h_T(w)\,(v_\mu\, v'_\nu - v_\nu\, v'_\mu)
\end{aligned} \tag{1}$$

$$\begin{aligned}
\langle V(v',\epsilon)|\bar{Q}'\gamma_\mu Q|B_c(v)\rangle &= i\,\sqrt{m_V\, m_{B_c}}\,h_V(w)\,\epsilon_{\mu\nu\alpha\beta}\,\epsilon^{*\nu}\,v'^\alpha\,v^\beta \\
\langle V(v',\epsilon)|\bar{Q}'\gamma_\mu\gamma_5\, Q|B_c(v)\rangle &= \sqrt{m_V\, m_{B_c}}\,[h_{A_1}(w)\,(1+w)\,\epsilon^*_\mu - h_{A_2}(w)\,(\epsilon^*\cdot v)\,v_\mu - h_{A_3}(w)\,(\epsilon^*\cdot v)\,v'_\mu] \\
\langle V(v',\epsilon)|\bar{Q}'\gamma_5 Q|B_c(v)\rangle &= -\sqrt{m_V\, m_{B_c}}\,h_P(w)\,(\epsilon^*\cdot v) \\
\langle V(v',\epsilon)|\bar{Q}'\sigma_{\mu\nu} Q|B_c(v)\rangle &= -\sqrt{m_V\, m_{B_c}}\,\epsilon^{\mu\nu\alpha\beta}[h_{T_1}(w)\epsilon^*_\alpha(v+v')_\beta + h_{T_2}(w)\epsilon^*_\alpha(v-v')_\beta \\
&\quad + h_{T_3}(w)(\epsilon^*\cdot v)v_\alpha v'_\beta]
\end{aligned} \tag{2}$$

where $P = \eta_c$, $V = J/\psi$, $v = \frac{p}{m_{B_c}}$, $v' = \frac{p'}{m_{P(V)}}$, $w = v\cdot v'$ and ϵ is the J/ψ polarization vector. We use $\epsilon^{0123} = +1$ and the relation $\sigma_{\mu\nu}\gamma_5 = \frac{i}{2}\epsilon_{\mu\nu\alpha\beta}\,\sigma^{\alpha\beta}$.

The form factors h_i can be related to a set of universal functions in a kinematical range close to $w = 1$ through the Heavy quark expansion. For hadrons comprising a single heavy quark, this has been done in Refs. [11–13]. The modifications for heavy quarkonium are discussed in the next section. The results for the vector and axial form factors computed in [7] will be exploited in our numerical analysis.

3 Expansion of the Heavy Quark Field and QCD Lagrangian

To construct the heavy quark expansion, the heavy quark QCD field $Q(x)$ with mass m_Q is written factorizing a fast oscillation mass term:

$$Q(x) = e^{-i m_Q v\cdot x}\psi(x) = e^{-i m_Q v\cdot x}\Big(\psi_+(x) + \psi_-(x)\Big) \tag{3}$$

with $\psi_\pm = P_\pm\psi(x)$ and $P_\pm = \frac{1\pm\not{v}}{2}$. ψ_+ is the positive energy component of the field (we use the notation adopted in Ref. [14]). v is identified with the heavy meson (quarkonium) 4-velocity with $v^2 = 1$. The equation of motion allows us to relate ψ_- to ψ_+ and obtain an expansion on $1/m_Q$ for both $Q(x)$ and the QCD Lagrangian.

$$Q(x) = e^{-im_Q v\cdot x}\left(1 + \frac{i\not{D}_\perp}{2m_Q} + \frac{(-iv\cdot D)}{2m_Q}\frac{i\not{D}_\perp}{2m_Q} + \ldots\right)\psi_+(x)\,, \tag{4}$$

$$\mathcal{L}_{QCD} = \bar{\psi}_+(x)\left(iv\cdot D + \frac{(iD_\perp)^2}{2m_Q} + \frac{g}{4m_Q}\sigma\cdot G_\perp + \frac{i\not{D}_\perp}{2m_Q}\frac{(-iv\cdot D)}{2m_Q}(i\not{D}_\perp) + \ldots\right)\psi_+(x)\,. \tag{5}$$

In this expression $G_{\perp\mu\nu} = (g_{\mu\alpha} - v_\mu v_\alpha)(g_{\nu\beta} - v_\nu v_\beta)G^{\alpha\beta}$ and $D_{\perp\mu} = (g_{\mu\alpha} - v_\mu v_\alpha)D^\alpha$. In the rest frame $G_{\perp\mu\nu} = G_{ij}$ for $\mu = i = 1,2,3$ and $\nu = j = 1,2,3$, while the other components vanish.

The power counting is delicate since there are two small parameters. On one side, the inverse of the heavy quark mass (HQET), and on another side the 3-velocity in the hadron rest frame (NRQCD). The power counting within NRQCD takes the following form:

$$D_t \sim \tilde{v}^2, \qquad D_\perp \sim \tilde{v}, \qquad \psi_+ \sim \tilde{v}^{3/2}, \qquad E_i = G_{0i} \sim \tilde{v}^3 \quad \text{and} \quad B_i = \frac{1}{2}\epsilon_{ijk}G^{jk} \sim \tilde{v}^4$$

where $\tilde{v} = |\vec{v}| \ll 1$ is the relative heavy quark 3-velocity in the hadron rest frame. In this work, we consider a mixed approach, where we perform the expansion considering terms up NLO in $\tilde{v}$ and NNLO in $1/m_Q$ in our expansion. NLO terms in $\tilde{v}$ involving three derivatives ($O(1/m_Q^3)$) are excluded, we assume that they provide numerically suppressed effects.

The first and second terms in Eq. (5) are $O(\tilde{v}^2)$ and provide the leading order Lagrangian $\mathcal{L}_0$ while the third and fourth terms are $O(\tilde{v}^4)$ and give the NLO Lagrangian $\mathcal{L}_1$.

3.1 Trace formalism

The B_c and $J/\psi, \eta_c$ matrix elements of the various terms can be expressed using the trace formalism by exploiting the spin-symmetry present in the NRQCD expansion. In this formalism, the lowest-lying S-wave $\bar{b}c$ and $\bar{c}c$ bound states are described by 4×4 matrices

$$H^{c\bar{b}}(v) = \frac{1+\not{v}}{2}\left[B_c^{*\mu}\gamma_\mu - B_c\gamma_5\right]\frac{1-\not{v}}{2} \qquad H^{c\bar{c}}(v') = \frac{1+\not{v}'}{2}\left[\Psi^{*\mu}\gamma_\mu - \eta_c\gamma_5\right]\frac{1-\not{v}'}{2} \tag{6}$$

satisfying the relations $\not{v}H(v) = H(v) = -H(v)\not{v}$ and $H(v')\not{v}' = H(v') = -\not{v}'H(v')$. $B_c^{*\mu}$, B_c and $\Psi^{*\mu}$, η_c annihilate vector and pseudoscalar $\bar{b}c$ and $\bar{c}c$ mesons of velocity v and v', respectively. This formalism can be extended to other charmonia states as discussed in Ref. [15].

3.2 Expanding the weak current (Local corrections)

Keeping terms up to $O(\tilde{v}^3)$ and $O(1/m_Q^2)$, the current expansion can be written as

$$\bar{Q}'(x)\Gamma Q(x) = J_0 + \left(\frac{J_{1,0}}{2m_Q} + \frac{J_{0,1}}{2m_{Q'}}\right) + \left(-\frac{J_{2,0}}{4m_Q^2} - \frac{J_{0,2}}{4m_{Q'}^2} + \frac{J_{1,1}}{4m_Q m_{Q'}}\right) \tag{7}$$

$$\begin{aligned} J_0 &= \bar{\psi}'_+\Gamma\psi_+ & J_{0,1} &= \bar{\psi}'_+\left(-i\overleftarrow{\not{D}}'_\perp\right)\Gamma\psi_+ & J_{0,2} &= \bar{\psi}'_+ i\overleftarrow{\not{D}}'_\perp\left(iv'\cdot\overleftarrow{D}\right)\Gamma\psi_+ \\ J_{1,0} &= \bar{\psi}'_+\Gamma i\overrightarrow{\not{D}}_\perp\psi_+ & J_{2,0} &= \bar{\psi}'_+\Gamma\left(iv\cdot\overrightarrow{D}\right)i\overrightarrow{\not{D}}_\perp\psi_+ & J_{1,1} &= \bar{\psi}'_+\left(-i\overleftarrow{\not{D}}'_\perp\right)\Gamma\left(i\overrightarrow{\not{D}}_\perp\right)\psi_+ \end{aligned} \tag{8}$$

Using the trace formalism, we parametrize the matrix elements of the various terms in the expansion. For instance, the matrix element of J_0 and $J_{1,0}$ are parametrized as

$$\langle M'(v')|J_0|M(v)\rangle = -\Delta(w)\mathrm{Tr}\left[\bar{H}'(v')\Gamma H(v)\right] \tag{9}$$

$$\langle M'(v')|\bar{\psi}'_+\Gamma^\alpha iD_\alpha\psi_+ M(v)\rangle = -\mathrm{Tr}\left[\Delta_\alpha(v,v')\bar{H}'(v')\Gamma^\alpha H(v)\right] \tag{10}$$

and involve the form factor $\Delta(w)$ and the function $\Delta_\alpha(v, v')$ respectively

$$\Delta_\alpha(v, v') = \Delta_+(w)(v + v')_\alpha + \Delta_-(w)(v - v')_\alpha - \Delta_3(w)\gamma_\alpha \tag{11}$$

The remaining terms, including the higher derivative terms, are parametrized analogously [1].

3.3 Expanding the states (Non-local corrections)

In addition to the corrections obtained by the expansion of the weak currents, we must consider the corrections to the states given by the corrections to the Lagrangian in Eq. (5). They can be written as

$$\begin{aligned}\langle M'(v')|i\int d^4x\, T[J_0(0)\mathcal{L}_1(x)]|M(v)\rangle = &-\frac{1}{2m_Q^2}\chi_1(w)\mathrm{Tr}\left[\bar{H}'(v')\Gamma H(v)\right]\\ &-\frac{1}{4m_Q}\mathrm{Tr}\left[\chi_{2\mu\nu}(w)\bar{H}'(v')\Gamma P_+\left(-\frac{i}{2}\right)\sigma^{\mu\nu}H(v)\right]\end{aligned} \tag{12}$$

for the quark field and analogously for the antiquark field. The $\chi_i(w)$ functions are arbitrary functions, in the case of $\chi_{2\mu\nu}(w)$ we parametrize it as

$$\chi_{2\mu\nu} = \chi_2^A i\sigma_{\mu\nu} + \chi_2^B(v_\mu\gamma_\nu - v_\nu\gamma_\mu) + \chi_2^C(v'_\mu\gamma_\nu - v'_\nu\gamma_\mu) \tag{13}$$

3.4 Relations between universal functions

The x-dependence of the matrix element $\mathcal{M}_0(x) = \langle M'(v')|\bar{\psi}'_+(x)\Gamma\psi_+(x)|M(v)\rangle$ can be obtained exploiting the dependence in the effective theory [1, 16, 17] obtaining

$$\mathcal{M}_0(x) = e^{-i\phi\cdot x}\mathcal{M}_0(0) \tag{14}$$

with $\phi = \tilde{\Lambda}v - \tilde{\Lambda}'v'$. For $B_c \to J/\psi(\eta_c)$, the binding energies $\tilde{\Lambda}^{(\prime)}$ are given by $\tilde{\Lambda} = m_{B_c} - m_b - m_c$ and $\tilde{\Lambda}' = m_{J/\psi(\eta_c)} - 2m_c$.

Exploiting the relation

$$i\partial_\alpha(\bar{\psi}'_+\Gamma\psi_+) = \bar{\psi}'_+(i\overleftarrow{D}_\alpha)\Gamma\psi_+ + \bar{\psi}'_+\Gamma(i\overrightarrow{D}_\alpha)\psi_+ \tag{15}$$

the equations of motion and the x-dependence in Eq. (14) we can reduce the number of independent universal functions.

3.5 Form factors in terms of universal functions

Using the expansions above and relating some of the universal functions, one can obtain expressions for the form factors h_i in terms of universal functions. We show here an example for the $h_{A_2}(w)$ form factor.

$$\begin{aligned}h_{A_2} = &\frac{1}{m_c}\left(\frac{1}{1+w}[\phi_K(w) - \Delta(w)\tilde{\Lambda} - \Delta_3(w)] + \frac{1}{2}\bar{\chi}_2^B(w)\right) + \frac{1}{m_c^2}\frac{1}{2(w+1)}\left(\tilde{\Lambda}w - \tilde{\Lambda}'\right)[\phi_K(w) - \Delta(w)\tilde{\Lambda} - \Delta_3(w)]\\ &+ \frac{1}{4m_c^2}\Big[-(1+w)\psi_2^S(w) - (w-1)\psi_3^S(w) + \psi_4^S(w) + \psi_5^S(w) - 2w\psi_6^S(w) - \psi_1^A(w) + \psi_2^A(w) + \psi_4^A(w)\Big]\\ &+ \frac{1}{4m_bm_c}\Big[-\psi_1^S(w) - (1+w)\psi_2^S(w) + (w+1)\psi_3^S(w)\\ &\qquad + 3\psi_4^S(w) + 3\psi_5^S(w) - (1+w)\psi_1^A(w) + 3\psi_2^A(w) + \psi_3^A(w) + 3\psi_4^A(w)\Big]\end{aligned} \tag{16}$$

The collection of these relations can be found in App. B of Ref.[1].

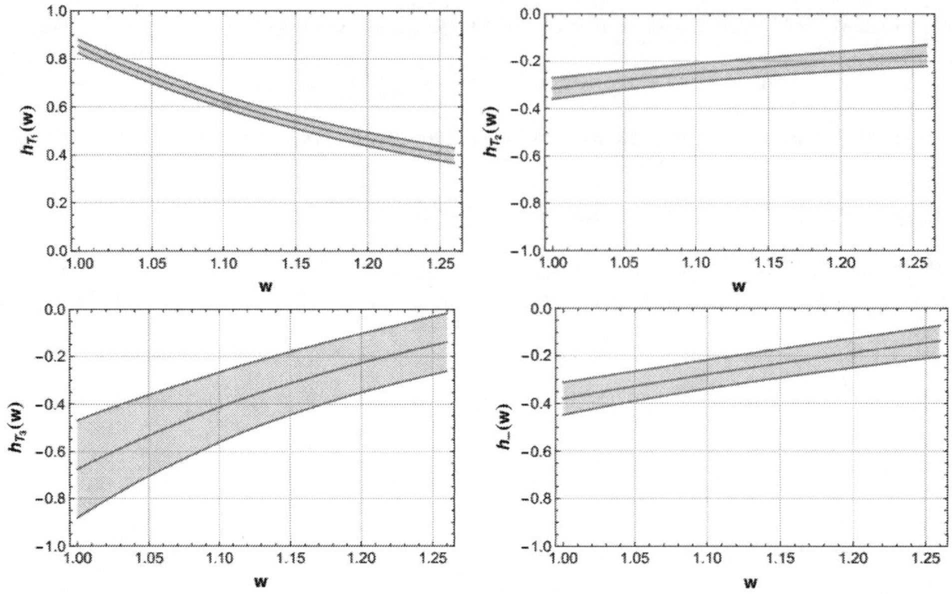

Figure 1. $B_c \to J/\psi(\eta_c)$ form factors obtained applying Eqs. (17) to (20) in the full kinematical range and using lattice QCD results from [7].

4 Relations between Form Factors at $O(1/m_Q)$

We can take advantage of the relations discussed in the previous section to relate unknown form factors, with the ones that have already been computed on the lattice. Using the above expression at $O(1/m_Q)$ one can derive relations between the form factors for $B_c \to J/\psi$ (and similarly for $B_c \to \eta_c$) and the known vector and axial ones.

$$h_{T_1}(w) = \frac{1}{2}\Big((1+w)h_{A_1}(w) - (w-1)h_V(w)\Big) \tag{17}$$

$$h_{T_2}(w) = \frac{1+w}{2(m_b+3m_c)}\Big((m_b-3m_c)h_{A_1}(w) + 2m_c(h_{A_2}(w)+h_{A_3}(w)) - (m_b-m_c)h_V(w)\Big) \tag{18}$$

$$h_{T_3}(w) = h_{A_3}(w) - h_V(w) \tag{19}$$

$$h_-(w) = \frac{m_b-m_c}{2(m_b+3m_c)}(1+w)\Big(3h_{A_1}(w) - h_{A_2}(w) - h_{A_3}(w) - 2h_V(w)\Big) \tag{20}$$

$$h_T(w) = h_+(w) - \frac{m_b+m_c}{2(m_b+3m_c)}(1+w)\Big(3h_{A_1}(w) - h_{A_2}(w) - h_{A_3}(w) - 2h_V(w)\Big) \tag{21}$$

$$h_T(w) = h_S(w) - \frac{m_b+m_c}{(m_b+3m_c)}\Big(3h_{A_1}(w) - h_{A_2}(w) - h_{A_3}(w) - 2h_V(w)\Big). \tag{22}$$

Their extrapolations are displayed in Fig. 1. The value of the universal functions at $w = 1$ is not predicted, however from Eqs. (17), (21) and (22) the relations $h_{T_1}(w=1) = h_{A_1}(w=1)$ and $h_S(w=1) = h_+(w=1)$ are obtained.

As mentioned before, lattice QCD results [7] violate the results in [4] at leading order, however, due to the limited amount of form factors computed on the lattice, consistency checks including the NLO corrections cannot be performed as of now. Future results from lattice QCD for other $B_c \to J\psi$ currents or the $B_c \to \eta_c$ mode are fundamental to check the consistency between the formalism discussed here and lattice computations, leading to a better understanding of all form factors, including the already computed $B_c \to J/\psi$ SM form factors.

5 Conclusions

Using the heavy quark expansion, the heavy quark spin symmetry and NRQCD power counting we have expressed the form factors parametrizing the matrix elements $\langle J/\psi(\eta_c)|\bar{c}\Gamma_i b|B_c\rangle$in terms of universal functions near the zero-recoil point and we established relations among form factors in this kinematical range. Lattice QCD results for the matrix element of the Standard Model operator between B_c and J/ψ allow us to predict the tensor form factors and the h_- form factor for $B_c \to \eta_c$. Further information from lattice QCD is fundamental to further check the relations worked out here and the general coherence of both lattice results and the heavy quark expansion predictions. In the future, a better understanding can be obtained through the use of unitarity constraints and the dispersion matrix [18, 19] in order to constrain the form factor extrapolations out of the zero-recoil point $w \sim 1$.

Acknowledgments

I would like to thank my collaborators P. Colangelo, F. de Fazio, F. Loparco and N. Losacco with whom this work was developed. This study has been carried out within the INFN project (Iniziativa Specifica) QFT-HEP.

References

[1] P. Colangelo, F. De Fazio, F. Loparco, N. Losacco, M. Novoa-Brunet, JHEP **09**, 028 (2022), 2205.08933
[2] Y.S. Amhis et al. (HFLAV), Eur. Phys. J. C **81**, 226 (2021), 1909.12524
[3] D. London, J. Matias (2021), 2110.13270
[4] E.E. Jenkins, M.E. Luke, A.V. Manohar, M.J. Savage, Nucl. Phys. B **390**, 463 (1993), hep-ph/9204238
[5] P. Colangelo, G. Nardulli, N. Paver, Z. Phys. C **57**, 43 (1993)
[6] V. Kiselev, A. Likhoded, A. Onishchenko, Nucl. Phys. B **569**, 473 (2000), hep-ph/9905359
[7] J. Harrison, C.T.H. Davies, A. Lytle (HPQCD), Phys. Rev. D **102**, 094518 (2020), 2007.06957
[8] P. Colangelo, F. De Fazio, Phys. Rev. D **61**, 034012 (2000), hep-ph/9909423
[9] P. Colangelo, F. De Fazio, F. Loparco, Phys. Rev. D **103**, 075019 (2021), 2102.05365
[10] P. Colangelo, F. De Fazio, F. Loparco, Phys. Rev. D **104**, 115024 (2021), 2107.07291
[11] A.F. Falk, H. Georgi, B. Grinstein, M.B. Wise, Nucl. Phys. B **343**, 1 (1990)
[12] A.F. Falk, M. Neubert, Phys. Rev. D **47**, 2965 (1993), hep-ph/9209268
[13] A.K. Leibovich, Z. Ligeti, I.W. Stewart, M.B. Wise, Phys. Rev. D **57**, 308 (1998), hep-ph/9705467
[14] J. Aebischer, B. Grinstein, JHEP **07**, 130 (2021), 2105.02988
[15] P. Colangelo, F. De Fazio, N. Losacco, F. Loparco, M. Novoa-Brunet (2022), 2208.13398
[16] T. Mannel, G.A. Schuler, Z. Phys. C **67**, 159 (1995), hep-ph/9410333
[17] T. Mannel, G.A. Schuler, Phys. Lett. B **349**, 181 (1995), hep-ph/9412337
[18] I. Caprini, L. Lellouch, M. Neubert, Nucl. Phys. B **530**, 153 (1998), hep-ph/9712417
[19] G. Martinelli, S. Simula, L. Vittorio, Phys. Rev. D **104**, 094512 (2021), 2105.07851

Inclusive $b \to \{u, c\}\, \ell^- \bar{\nu}_\ell$ modes: Polarized/unpolarized Λ_b

Francesco Loparco[1,*]

[1]Istituto Nazionale di Fisica Nucleare, Sezione di Bari, Via Orabona 4, I-70126 Bari, Italy

Abstract. The increasing number of flavour anomalies motivates the investigation of new processes where tensions similar to the observed ones may emerge. It is necessary to identify observables sensitive to physics beyond the Standard Model. The analysis which follows concerns the inclusive semileptonic decays of polarized beauty baryons, computed through the Heavy Quark Expansion at $O(1/m_b^3)$ and at the leading order in α_s. New Physics interactions have been taken into account, extending the Standard Model $b \to U\, \ell\, \bar{\nu}_\ell$ low-energy Hamiltonian, where $U = \{c, u\}$ and $\ell = \{e, \mu, \tau\}$, including the full set of $D = 6$ operators with left-handed neutrinos. Among the possible observables one can consider, the ones depending on the spin of the decaying baryon are very appealing and can be considered for physics programmes of future facilities, such as FCC-ee.

1 Introduction

Deviations in a number of observables w.r.t. the Standard Model (SM) predictions, the *flavour anomalies*, have been recently detected. They represent a motivation for searching signals of New Physics (NP). Anomalies have been observed in tree-level and loop-induced decays of B, B_s and B_c mesons [1, 2]. Particularly interesting, hints of lepton flavour universality (LFU) violations have been collected. It is necessary to investigate other heavy hadron decay processes to get a full comprehension of LFU, for this reason inclusive semileptonic beauty baryon decays have been analyzed [3] treating the nonperturbative effects of strong interactions in a systematic way, exploiting an expansion in the inverse heavy quark mass [4, 5]. The calculation of the inclusive semileptonic decay width $H_b \to X_{c,u}\, \ell^- \bar{\nu}_\ell$ of a polarized heavy hadron H_b will be described below, expanding the hadronic matrix elements at $O(1/m_b^3)$ in the heavy quark expansion (HQE), at leading order in α_s. Observables have been computed for the transition $\Lambda_b \to X_{c,u}\, \ell^- \bar{\nu}_\ell$ for non vanishing charged lepton mass and considering the NP generalization of the SM in the effective Hamiltonian, including all the $D = 6$ semileptonic operators with left-handed neutrinos [6–12]. A recent improvement has been achieved in [13] where α_s corrections has been computed for the inclusive $B \to X_c\, \tau\, \bar{\nu}_\tau$ decay width and leptonic invariant mass spectrum up to $O(1/m_b^3)$ in the HQE.

At LHC the Λ_b is produced unpolarized [14–17] because the b quark is produced through strong interactions, a sizeable longitudinal Λ_b polarization is expected for b quarks produced in Z and t quark decays [18–20].

*e-mail: francesco.loparco1@ba.infn.it

2 Generalized effective Hamiltonian

Let us consider a beauty hadron H_b with spin s. The inclusive semileptonic decays $H_b(p,s) \to X_U(p_X)\,\ell^-(p_\ell)\,\bar\nu_\ell(p_\nu)$, induced by the quark transition $b \to U$, with $U = \{c, u\}$, are described by the general low-energy Hamiltonian which extends the SM one:

$$\mathcal{H}_{\rm eff}^{b\to U\ell\bar\nu} = \frac{G_F\,V_{Ub}}{\sqrt{2}}\left[(1+\epsilon_V^\ell)\,O_{\rm SM} + \epsilon_S^\ell\,O_S + \epsilon_P^\ell\,O_P + \epsilon_T^\ell\,O_T + \epsilon_R^\ell\,O_R\right] + \text{h.c.}\,, \tag{1}$$

that comprises the Fermi constant G_F, the CKM matrix element V_{Ub} and the full set of $D = 6$ semileptonic $b \to U$ operators with left-handed neutrinos: $O_{{\rm SM},R} = [\bar u\,\gamma_\mu\,(1\mp\gamma_5)\,b]\,[\bar\ell\,\gamma^\mu\,(1-\gamma_5)\,\nu_\ell]$, $O_S = [\bar u\,b]\,[\bar\ell\,(1-\gamma_5)\nu_\ell]$, $O_P = [\bar u\,\gamma_5\,b]\,[\bar\ell\,(1-\gamma_5)\nu_\ell]$ and $O_T = [\bar u\,\sigma_{\mu\nu}\,(1-\gamma_5)\,b]\,[\bar\ell\,\sigma^{\mu\nu}\,(1-\gamma_5)\,\nu_\ell]$. $\epsilon^\ell_{V,S,P,T,R}$ are complex lepton-flavour dependent couplings. In the case of $\epsilon_i^\ell = 0$ one recovers the SM case. We keep $m_\ell \neq 0$ for all leptons $\ell = \{e, \mu, \tau\}$.

3 Inclusive decay width

Any $D = 6$ operator in Eq. (1) can be written as the product of two currents, $O = J_M^{(i)}\,L^{M(i)}$, where $J_M^{(i)}$ and $L^{M(i)}$ are respectively the hadronic and the leptonic currents. M generically denotes the Lorentz indices of the currents. The H_b inclusive semileptonic differential decay width can be written as

$$d\Gamma = \underbrace{d\Sigma}_{\text{phace space}}\;\frac{G_F^2\,|V_{Ub}|^2}{4\,M_H}\sum_{i,j} g_i^*\,g_j\;\underbrace{(W^{ij})_{MN}}_{\text{hadronic tensor}}\;\underbrace{(L^{ij})^{MN}}_{\text{leptonic tensor}}\;, \tag{2}$$

with the phase-space element $d\Sigma = (2\,\pi)^4\,d^4q\,\delta^4(q-p_\ell-p_\nu)\,[dp_\ell]\,[dp_\nu]$. The notation $[dp] = \frac{d^3p}{(2\,\pi)^3\,2\,p^0}$ has been used. $g_V = 1 + \epsilon_V^\ell$ and $g_{S,P,T,R} = \epsilon^\ell_{S,P,T,R}$ are the Wilson coefficients.
Using the optical theorem, the hadronic tensor $(W^{ij})_{MN} = \frac{1}{\pi}\,{\rm Im}\left[(T^{ij})_{MN}\right]$ is given in terms of the forward amplitude depicted in figure 1

$$(T^{ij})_{MN} = i\int d^4x\,e^{-i\,q\cdot x}\,\langle H_b(p,s)|\mathbb{T}\left\{J_M^{(i)\dagger}(x)\,J_N^{(j)}(0)\right\}|H_b(p,s)\rangle\;. \tag{3}$$

The hadron momentum $p = m_b\,v + k$ is written in terms of the 4-velocity v, the heavy quark

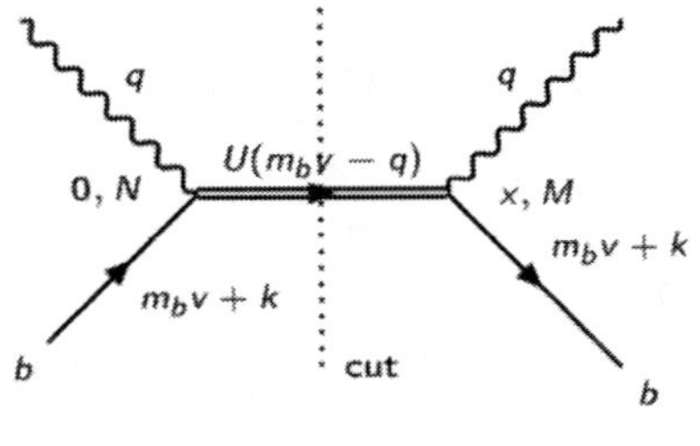

Figure 1. Discontinuity of the forward amplitude across the cut of the semileptonic process.

mass m_b and the residual momentum k, with $|k| \sim O(\Lambda_{\rm QCD})$. The QCD quark field $b(x)$ is redefined as $b(x) = e^{-i\,m_b\,v\cdot x}\,b_v(x)$, with $b_v(x)$ still QCD quark field satisfying the EoM $b_v(x) = \left(P_+ + \frac{i\,\not{D}}{2\,m_b}\right)b_v(x)$, where $P_\pm = \frac{1\pm\not{v}}{2}$ are the velocity projectors.

Introducing $p_X = m_b\,v + k - q$, the Eq. (3) becomes

$$(T^{ij})_{MN} = \langle H_b(v,s)|\bar b_v(0)\,\overline{\Gamma}_M^{(i)}\,\frac{1}{\not{p}_X - m_U}\,\Gamma_N^{(j)}\,b_v(0)|H_b(v,s)\rangle\;, \tag{4}$$

and considering $|k| \sim O(\Lambda_{\rm QCD})$ we have

$$(T^{ij})_{MN} = \sum_{n=0}^{+\infty} \langle H_b(v,s)|\bar{b}_v(0)\,\overline{\Gamma}^{(i)}_M\,(\not{p}_U + m_U)\,(i\not{D}\,(\not{p}_U + m_U))^n\,\Gamma^{(j)}_N\,b_v(0)|H_b(v,s)\rangle\;\frac{(-1)^n}{\Delta_0^{n+1}}\,, \tag{5}$$

where $\overline{\Gamma}^{(i)}_M = \gamma^0\,\Gamma^{(i)\dagger}_M\,\gamma^0$, $p_U = p_X - k$ and $\Delta_0 = p_U^2 - m_U^2$.

Using the trace formalism [21], we can write the n-th term in the series as

$$\langle H_b(v,s)|\bar{b}_v(0)\,\overline{\Gamma}^{(i)}_M\,(\not{p}_U + m_U)\,\underbrace{i\not{D}\,(\not{p}_U + m_U)\ldots i\not{D}\,(\not{p}_U + m_U)}_{\text{n times}}\,\Gamma^{(j)}_N\,b_v(0)|H_b(v,s)\rangle =$$

$$= \left[\overline{\Gamma}^{(i)}_M\,(\not{p}_U + m_U)\prod_{k=1}^{n}\left[\gamma_{\mu_k}\,(\not{p}_U + m_U)\right]\Gamma^{(j)}_N\right]_{ab}\underbrace{\langle H_b(v,s)|\bar{b}_v(0)\,iD^{\mu_1}\ldots iD^{\mu_n}\,b_v(0)|H_b(v,s)\rangle_{ba}}_{(\mathcal{M}^{\mu_1\ldots\mu_n})_{ba}}\,. \tag{6}$$

The expression in Eq. (6) involves the hadronic matrix elements $(\mathcal{M}^{\mu_1\cdots\mu_n})_{ba}$ where a and b are Dirac indices. These matrix elements can be written in terms of nonperturbative parameters, the number of which increases with the order of the expansion:

$$O(1/m_b^n)\ldots\begin{cases} O(1/m_b^3)\begin{cases} O(1/m_b^2)\begin{cases} -2\,M_H\,\hat{\mu}_\pi^2 = \langle H_b|\bar{b}_v\,iD^\mu\,iD_\mu\,b_v|H_b\rangle \\ 2\,M_H\,\hat{\mu}_G^2 = \langle H_b|\bar{b}_v\,(-i\sigma_{\mu\nu})\,iD^\mu\,iD^\nu\,b_v|H_b\rangle \end{cases} \\ 2\,M_H\,\hat{\rho}_D^3 = \langle H_b|\bar{b}_v\,iD^\mu\,(iv\cdot D)\,iD_\mu\,b_v|H_b\rangle \\ 2\,M_H\,\hat{\rho}_{LS}^3 = \langle H_b|\bar{b}_v\,(-i\sigma_{\mu\nu})\,iD^\mu\,(iv\cdot D)\,iD^\nu\,b_v|H_b\rangle \end{cases} \\ \ldots \end{cases}. \tag{7}$$

To compute $\mathcal{M}^{\mu_1\cdots\mu_n}$ one can follow the methods proposed in [21, 22].
In the case of baryons, the dependence on the spin s_μ in Eq. (6) must be kept into account as done up to $O(1/m_b^{-2})$ in [23]. In [3] $\mathcal{M}^{\mu_1\cdots\mu_n}$ have been derived at $O(1/m_b^{-3})$ for a polarized baryon, extending the previous results obtained in [10, 12, 23–25].

From the expressions of the matrix elements $\mathcal{M}^{\mu_1\cdots\mu_n}$ the hadronic tensor can be computed and expanded in Lorentz structures which depend on v, q and s. The results for the effective Hamiltonian Eq. (1) are collected in [3]. The four-fold differential decay rate for the $H_b(p,s) \to X_U(p_X)\,\ell^-(p_\ell)\,\bar{\nu}_\ell(p_\nu)$ transition reads

$$\frac{d^4\Gamma}{dE_\ell\,dq^2\,dq_0\,d\cos\theta_P} = \frac{G_F^2\,|V_{Ub}|^2}{32\,(2\,\pi)^3\,M_H}\sum_{i,j} g_i^*\,g_j\,\frac{1}{\pi}\,\mathrm{Im}\left[(T^{ij})_{MN}\right](L^{ij})^{MN}\,, \tag{8}$$

where $p_\ell = (E_\ell, \vec{p}_\ell)$, $q_0 = q\cdot v$ and θ_P is the angle between the two 3-vectors $\vec{p}_\ell$ and $\vec{s}$ in the H_b rest frame. Double and single distributions are obtained integrating Eq. (8) over the phase-space [26]. The full decay width can be obtained by performing all integrations and can be written as:

$$\Gamma(H_b \to X_U\,\ell^-\,\bar{\nu}_\ell) = \Gamma_b\sum_{i,j} g_i^*\,g_j\left[C_0^{(i,j)} + \frac{\hat{\mu}_\pi^2}{m_b^2}\,C^{(i,j)}_{\hat{\mu}_\pi^2} + \frac{\hat{\mu}_G^2}{m_b^2}\,C^{(i,j)}_{\hat{\mu}_G^2} + \frac{\hat{\rho}_D^3}{m_b^3}\,C^{(i,j)}_{\hat{\rho}_D^3} + \frac{\hat{\rho}_{LS}^3}{m_b^3}\,C^{(i,j)}_{\hat{\rho}_{LS}^3}\right], \tag{9}$$

where $\Gamma_b = \frac{G_F^2\,|V_{Ub}|^2\,m_b^5}{192\,\pi^3}$ is the *partonic term* corresponding to a free quark decay. The indices i and j runs over the contributions of the various operators and of their interferences. The coefficients $C^{(i,j)}$ can be found in [3]. The OPE breaks down in the endpoint region of the

Table 1. Inclusive semileptonic Λ_b branching fractions in SM, obtained for the central values of the parameters.

$b \to c$		$b \to u$	
$\mathcal{B}(\Lambda_b \to X_c\,\mu\,\bar{\nu}_\mu)$	11.0×10^{-2}	$\mathcal{B}(\Lambda_b \to X_u\,\mu\,\bar{\nu}_\mu)$	11.65×10^{-4}
$\mathcal{B}(\Lambda_b \to X_c\,\tau\,\bar{\nu}_\tau)$	2.4×10^{-2}	$\mathcal{B}(\Lambda_b \to X_u\,\tau\,\bar{\nu}_\tau)$	2.75×10^{-4}

spectra, where singularities appear and which should be resummed in a H_b shape function, the convolution with such a function smears the spectra at the endpoint. The effects of the baryon shape function has not been considered here. Perturbative QCD corrections have also been not included: in the SM case they can be found in [27–32]. NLO QCD corrections have also been recently computed in [13, 33].

4 Results for $\Lambda_b \to X_{c,u}\,\ell\,\bar{\nu}_\ell$

In [3] several observables for the modes $\Lambda_b \to X_{c,u}\,\ell\,\bar{\nu}_\ell$ have been studied. Here we consider a few of them. Further ones, as well as the input parameters, can be found in [3]. For the couplings ϵ_i^ℓ in Eq. (1) we fix three NP benchmark points, set in [7, 34] for $U = c$, while for $U = u$ we use the ranges fixed in [35]. Using $G_F = 1.16637(1) \times 10^{-5}\,\text{GeV}^{-2}$, $|V_{cb}| = 0.042$ and $|V_{ub}| = 0.0037$, together with $\tau_{\Lambda_b} = 1.471(9)\,\text{ps}$ [36], we compute the inclusive Λ_b branching fraction for the two quark transitions and for final τ and μ lepton. The results in the SM, for the central values of the parameters and neglecting QCD corrections, are in table 1. For comparison, the available measurements are $\mathcal{B}(\Lambda_b \to \Lambda_c\,\ell^-\,\bar{\nu}_\ell + \text{anything}) = (10.9 \pm 2.2) \times 10^{-2}$, with $\ell = e, \mu$ and $\mathcal{B}(\Lambda_b \to p\,\mu^-\,\bar{\nu}_\mu) = (4.1 \pm 1.0) \times 10^{-4}$[36].

For inclusive semileptonic Λ_b decays it is interesting to consider a ratio analogous to $R(D^*)$ for B meson, to compare the τ and the muon mode using a quantity in which several theoretical uncertainties cancel:

$$R_{\Lambda_b}(X_U) = \frac{\Gamma(\Lambda_b \to X_U\,\tau\,\bar{\nu}_\tau)}{\Gamma(\Lambda_b \to X_U\,\mu\,\bar{\nu}_\mu)} \qquad \text{with} \qquad U = u, c\,. \tag{10}$$

Another ratio sensitive to LFU violating NP effects can be defined. It can be constructed from the distribution $\frac{d\Gamma(\Lambda_b \to X_U\,\ell\,\bar{\nu}_\ell)}{d\cos\theta_P} = A_\ell^U + B_\ell^U \cos\theta_P$. The dependence of $\frac{d\Gamma(\Lambda_b \to X_U\,\ell\,\bar{\nu}_\ell)}{d\cos\theta_P}$ on $\cos\theta_P$ is linear, and NP contributions modify both the slope and the intercept of the curve as displayed in figure 2. Similarly to the ratio defined in Eq. (10), we define the ratio of the slopes $R_S^U = B_\tau^U / B_\mu^U$ which has a definite value in SM, and can deviate due to NP as specified in table 2. A correlation between $R_{\Lambda_b}(X_U)$ and R_S^U can be constructed. As an example, for

Table 2. Ratios $R_{\Lambda_b}(X_u)$ and $R_{\Lambda_b}(X_c)$ (left) and ratios R_S^u and R_S^c (right) for SM and NP at the BP.

	$R_{\Lambda_b}(X_u)$	$R_{\Lambda_b}(X_c)$	R_S^u	R_S^c
SM	0.234	0.214	0.081	0.100
NP	0.238	0.240	0.091	0.074

$\Lambda_b \to X_c\,\tau\,\bar{\nu}_\tau$ with the effective Hamiltonian extended including a tensor operator, we vary the couplings $(\text{Re}[\epsilon_T^\mu], \text{Im}[\epsilon_T^\mu])$ and $(\text{Re}[\epsilon_T^\tau], \text{Im}[\epsilon_T^\tau])$ in the regions determined in [7]. The correlation plot in figure 3 shows that the (challenging) measurement of the two ratios would discriminate SM and NP.

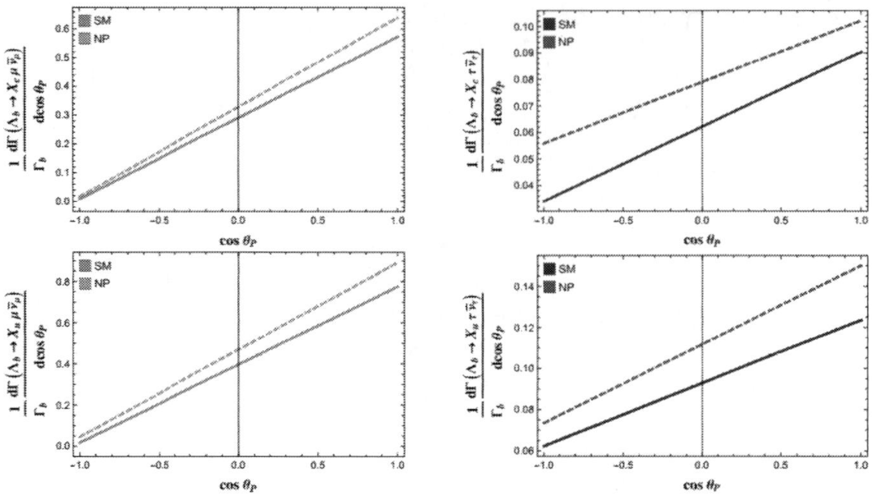

Figure 2. $\frac{1}{\Gamma_b}\frac{d\Gamma}{d\cos\theta_P}$ distribution for $\Lambda_b \to X_U\,\ell\,\bar{\nu}_\ell$, with $U = c$ (top row), $U = u$ (bottom row), $\ell = \mu$ (left column) and $\ell = \tau$ (right column). The solid line is the SM result, the dashed line the NP result at the benchmark point.

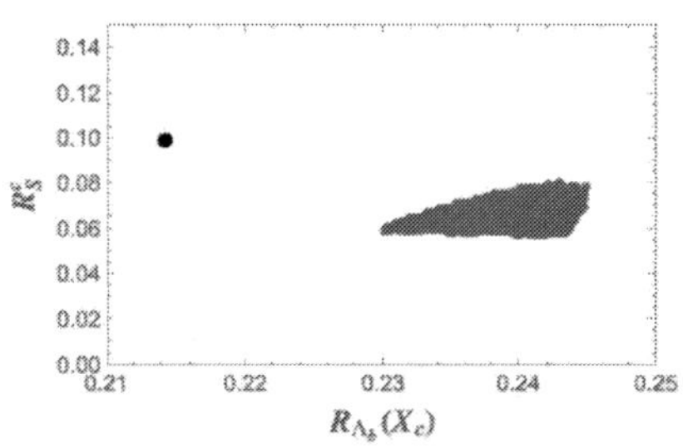

Figure 3. Correlation between $R_{\Lambda_b}(X_c)$ and the ratio R^c_S of the slopes of the $\frac{d\Gamma}{d\cos\theta_P}$ distribution. The dot corresponds to SM, the broad region to NP with the effective couplings varied as specified in the text.

5 Conclusions

The calculation of the fully differential inclusive semileptonic decay width of a polarized heavy hadron at $O(1/m_b^3)$ in the HQE, at leading order in α_s and for non vanishing charged lepton mass has been described. NP generalization of the SM has been considered in the effective Hamiltonian in Eq.(1) including all the $D = 6$ semileptonic operators with left-handed neutrinos. The correlation plot in figure 3 shows that the ratios $R_{\Lambda_b}(X_c)$ and R^c_S can assume different values in the SM and when NP is included. Although this measurement is experimentally challenging, it represents a suitable observable to test violation of LFU.

Acknowledgments

I thank P. Colangelo and F. De Fazio for collaboration. This study has been carried out within the INFN project (Iniziativa Specifica) QFT-HEP.

References

[1] S. Fajfer, J.F. Kamenik, I. Nisandzic, Phys. Rev. **D85**, 094025 (2012), 1203.2654

[2] M. Algueró, B. Capdevila, S. Descotes-Genon, J. Matias, M. Novoa-Brunet, ***b* → *sℓℓ*** *global fits after Moriond 2021 results*, in *55th Rencontres de Moriond on QCD and High Energy Interactions* (2021), 2104.08921

[3] P. Colangelo, F. De Fazio, F. Loparco, JHEP **11**, 032 (2020), 2006.13759
[4] I.I.Y. Bigi, M.A. Shifman, N.G. Uraltsev, A.I. Vainshtein, Phys. Rev. Lett. **71**, 496 (1993), hep-ph/9304225
[5] J. Chay, H. Georgi, B. Grinstein, Phys. Lett. B **247**, 399 (1990)
[6] P. Biancofiore, P. Colangelo, F. De Fazio, Phys. Rev. **D87**, 074010 (2013), 1302.1042
[7] P. Colangelo, F. De Fazio, JHEP **06**, 082 (2018), 1801.10468
[8] S. Bhattacharya, S. Nandi, S. Kumar Patra, Eur. Phys. J. C **79**, 268 (2019), 1805.08222
[9] P. Colangelo, F. De Fazio, Phys. Rev. **D95**, 011701 (2017), 1611.07387
[10] T. Mannel, A.V. Rusov, F. Shahriaran, Nucl. Phys. B **921**, 211 (2017), 1702.01089
[11] S. Kamali, A. Rashed, A. Datta, Phys. Rev. D **97**, 095034 (2018), 1801.08259
[12] S. Kamali, Int. J. Mod. Phys. A **34**, 1950036 (2019), 1811.07393
[13] D. Moreno (2022), 2207.14245
[14] R. Aaij et al. (LHCb), Phys. Lett. B **724**, 27 (2013), 1302.5578
[15] G. Aad et al. (ATLAS), Phys. Rev. D **89**, 092009 (2014), 1404.1071
[16] A.M. Sirunyan et al. (CMS), Phys. Rev. D **97**, 072010 (2018), 1802.04867
[17] R. Aaij et al. (LHCb), JHEP **06**, 110 (2020), 2004.10563
[18] D. Buskulic et al. (ALEPH), Phys. Lett. B **365**, 437 (1996)
[19] G. Abbiendi et al. (OPAL), Phys. Lett. B **444**, 539 (1998), hep-ex/9808006
[20] P. Abreu et al. (DELPHI), Phys. Lett. B **474**, 205 (2000)
[21] B.M. Dassinger, T. Mannel, S. Turczyk, JHEP **03**, 087 (2007), hep-ph/0611168
[22] T. Mannel, S. Turczyk, N. Uraltsev, JHEP **11**, 109 (2010), 1009.4622
[23] A.V. Manohar, M.B. Wise, Phys. Rev. D **49**, 1310 (1994), hep-ph/9308246
[24] Y. Grossman, Z. Ligeti, Phys. Lett. B **332**, 373 (1994), hep-ph/9403376
[25] S. Balk, J.G. Korner, D. Pirjol, Eur. Phys. J. C **1**, 221 (1998), hep-ph/9703344
[26] M. Jezabek, L. Motyka, Acta Phys. Polon. B **27**, 3603 (1996), hep-ph/9609352
[27] A. Czarnecki, M. Jezabek, J.H. Kuhn, Phys. Lett. B **346**, 335 (1995), hep-ph/9411282
[28] M. Jezabek, L. Motyka, Nucl. Phys. B **501**, 207 (1997), hep-ph/9701358
[29] F. De Fazio, M. Neubert, JHEP **06**, 017 (1999), hep-ph/9905351
[30] M. Trott, Phys. Rev. D **70**, 073003 (2004), hep-ph/0402120
[31] V. Aquila, P. Gambino, G. Ridolfi, N. Uraltsev, Nucl. Phys. B **719**, 77 (2005), hep-ph/0503083
[32] A. Alberti, P. Gambino, K.J. Healey, S. Nandi, Phys. Rev. Lett. **114**, 061802 (2015), 1411.6560
[33] T. Mannel, D. Moreno, A.A. Pivovarov, Phys. Rev. D **105**, 054033 (2022), 2112.03875
[34] R.X. Shi, L.S. Geng, B. Grinstein, S. Jäger, J. Martin Camalich, JHEP **12**, 065 (2019), 1905.08498
[35] P. Colangelo, F. De Fazio, F. Loparco, Phys. Rev. D **100**, 075037 (2019), 1906.07068
[36] R.L. Workman (Particle Data Group), PTEP **2022**, 083C01 (2022)

EPJ Web of Conferences **270**, 00018 (2022) https://doi.org/10.1051/epjconf/202227000018
QCD@Work 2022

Semirelativistic Potential Modelling of Bound States: Advocating Due Rigour

Wolfgang Lucha[1,*]

[1]Institute for High Energy Physics, Austrian Academy of Sciences, Nikolsdorfergasse 18, A-1050 Vienna, Austria

Abstract. The Poincaré-covariant quantum-field-theoretic description of bound states by the homogeneous Bethe–Salpeter equation usually exhibits an intrinsic complexity that can be attenuated by allowing this formalism to undergo various simplifications. The resulting approximate outcome's reliability can be assessed by applying several rigorous constraints on the nature of the bound-state spectra; most prominent here are existence, number and location of discrete eigenvalues.

1 Our Battlefield: Semirelativistic Bound States of Least Complexity

A popular (possibly even analytic) approach to semirelativistic bound states of scalar particles is known under the name spinless Salpeter equation, which is just the eigenvalue equation of a Hamiltonian H encompassing both the relativistic kinetic term of the bound-state constituents and a static potential V encoding all interactions. For the case of two bound-state constituents, of masses m_1 and m_2, and relative coordinates and momenta $\boldsymbol{x}$ and $\boldsymbol{p}$, this Hamiltonian H reads

$$H = \sqrt{\boldsymbol{p}^2 + m_1^2} + \sqrt{\boldsymbol{p}^2 + m_2^2} + V(\boldsymbol{x}) , \qquad (1)$$

which, for equal masses $m_1 = m_2 = m$ of its bound-state constituents, simplifies (a little bit) to

$$\widetilde{H} = 2 \sqrt{\boldsymbol{p}^2 + m^2} + V(\boldsymbol{x}) .$$

The spinless Salpeter equation may be regarded as some approximation defined by a sequence of self-evident assumptions that serves to simplify the homogeneous Bethe–Salpeter equation [1], which constitutes within the framework of relativistic quantum field theories the adequate approach to bound states of the fundamental degrees of freedom of such quantum field theory:

1. The most decisive simplification is to disregard any reference to *timelike* coordinate and momentum variables, thereby defining an *instantaneous* Bethe–Salpeter formalism [2].
2. Enabling any bound-state constituent to propagate freely entails Salpeter's equation [3].
3. Skipping impacts of negative-energy contributions gives the *reduced* Salpeter equation.
4. Ignoring the spins of all bound-state constituents leads to the *spinless* Salpeter equation.

2 General Insights: Elementary Constraints on Bound-State Spectra

Even if being not accessible analytically, the spectrum of the operator $\widetilde{H}$ is tightly constrained.

*e-mail: Wolfgang.Lucha@oeaw.ac.at

1. Boundedness from below of Hamiltonian operators is, as kinetic operators are positive, obvious if $V(\boldsymbol{x})$ itself is bounded from below. If not, comparison with the relativistic Coulomb problem may prove advantageous. The Hamiltonian operator $\widetilde{H}$ with the Coulomb interaction

$$V(\boldsymbol{x}) \equiv -\frac{\alpha}{|\boldsymbol{x}|}\ , \qquad \alpha > 0\ ,$$

will be bounded from below [4, 5] if and only if the coupling strength α satisfies the constraint

$$\alpha \leq \frac{4}{\pi} = 1.273239\ldots\ .$$

Within this interval, the spectrum, $\sigma(\widetilde{H})$, of this operator $\widetilde{H}$ is subject to the lower bound [4, 5]

$$\sigma(\widetilde{H}) \geq 2\,m\,\sqrt{1-\left(\frac{\pi\,\alpha}{4}\right)^2}\ . \tag{2}$$

For Coulomb couplings $\alpha \leq 1$, the lower bound (2) may, straightforwardly, be improved [6] to

$$\sigma(\widetilde{H}) \geq 2\,m\,\sqrt{\frac{1+\sqrt{1-\alpha^2}}{2}}\ .$$

2. Number of discrete eigenstates in the spectrum of $\widetilde{H}$: For all potentials $V(\boldsymbol{x})$ that satisfy

$$V(\boldsymbol{x}) = V(r) \in L^{3/2}(\mathbb{R}^3) \cap L^3(\mathbb{R}^3)\ , \qquad V(\boldsymbol{x}) \leq 0\ , \qquad r \equiv |\boldsymbol{x}|\ , \tag{3}$$

the number of bound states of such operator $\widetilde{H}$ may be shown to be bounded from above by [7]

$$N \leq \frac{C}{12\,\pi}\int\limits_0^\infty \mathrm{d}r\,r^2\,[|V(r)|\,(|V(r)|+4\,m)]^{3/2}\ ,$$

with the constant C given by $C = 14.107590867$ for $m > 0$ or $C = 6.074898097$ for $m = 0$ [8].

3. Upper bounds on the eigenvalues of $\widetilde{H}$ arise as rather trivial consequence of the so-called minimum–maximum theorem [9]: for every self-adjoint operator H bounded from below with (ordered) eigenvalues $E_0 \leq E_1 \leq E_2 \leq \cdots$, the d (ordered) eigenvalues $\widehat{E}_0 \leq \widehat{E}_1 \leq \cdots \leq \widehat{E}_{d-1}$ of the operator $\widehat{H}$ defined by a restriction of H to a d-dimensional subspace of the domain of H provide upper bounds to the first d eigenvalues of H below the onset of the *essential* spectrum,

$$E_k \leq \widehat{E}_k \qquad \forall \quad k = 0, 1, 2, \ldots, d-1\ .$$

Its eigenstates can be expanded in terms of some convenient basis [10, 11], e.g., in products of generalized-Laguerre polynomials, $L_k^{(\gamma)}$, and spherical harmonics $\mathcal{Y}$ for angular momentum ℓ:

$$\psi_k(\boldsymbol{x}) \propto r^{\ell+\beta-1}\exp(-\mu\,r)\,L_k^{(2\ell+2\beta)}(2\,\mu\,r)\,\mathcal{Y}(\Omega_x)\ , \qquad k \in \mathbb{N}_0\ , \tag{4}$$

$$L_k^{(\gamma)}(x) \equiv \sum_{t=0}^{k}\binom{k+\gamma}{k-t}\frac{(-x)^t}{t!}\ , \qquad \mu > 0\ , \qquad \beta > -\frac{1}{2}\ .$$

4. Quality of eigenstates [12, 13] tentatively localized by whatever technique forms a crucial aspect that may be quantified by the *relativistic* generalization [14, 15] of the virial theorem of nonrelativistic quantum theory, which relates, for any of the eigenstates $|\chi\rangle$ of the operator (1), the expectation values of the radial derivatives of both the kinetic terms and the potential $V(\boldsymbol{x})$:

$$\left\langle\chi\left|\frac{\boldsymbol{p}^2}{\sqrt{\boldsymbol{p}^2+m_1^2}}+\frac{\boldsymbol{p}^2}{\sqrt{\boldsymbol{p}^2+m_2^2}}\right|\chi\right\rangle = \left\langle\chi\left|\boldsymbol{x}\cdot\frac{\partial V}{\partial \boldsymbol{x}}(\boldsymbol{x})\right|\chi\right\rangle .$$

All of this proved useful in locating the discrete spectra of problems defined by spherically symmetric (central) potentials $V(\boldsymbol{x}) = V(r)$ (where $r \equiv |\boldsymbol{x}|$) of Woods–Saxon [16, 17], Hulthén [17, 18], Yukawa [19], kink-like [8] or generalized-Hellmann [20, 21] type. As an illustration, an application of all these tools is sketched for the totality of generalized-Hellmann potentials.

3 Application: Class of Potentials Generalizing Hellmann's Proposal

Let us take the liberty to generalize the form proposed by Hellmann [22, 23] to a superposition

$$V_{\mathrm{H}}(r) \equiv V_{\mathrm{C}}(r) + V_{\mathrm{Y}}(r) = -\frac{\kappa}{r} - \upsilon\,\frac{\exp(-b\,r)}{r}$$

of a definitely attractive Coulomb potential $V_{\mathrm{C}}(r)$ and a Yukawa-inspired potential $V_{\mathrm{Y}}(r)$, with Coulomb coupling κ, and Yukawa coupling υ and slope parameter b, subject to the constraints

$$\kappa \ge 0\,, \qquad \upsilon \gtreqless 0\,, \qquad b > 0\,.$$

Within the obtained class of *generalized* Hellmann potentials $V_{\mathrm{H}}(r)$, the *qualitative* behaviour of any of its members depends on the relative sign and size of the two coupling strengths κ and υ that govern the Coulomb and Yukawa contributions to such generalized Hellmann potential. These potentials show a significant diversity: they may be unbounded [Fig. 1(a,b)] or bounded [Fig. 1(c,d)] from below, and be singular [Fig. 1(a,b,d)] or finite [Fig. 1(c)] at the origin, $r = 0$.

Needless to stress, using the tools sketched in Sect. 2, it is a breeze to rough out the spectra of the entirety of relativistic Hellmann problems (for some relevant details, consult Ref. [20]).

- Evidently, boundedness from below is the question to be answered first and foremost: Since *any* of our generalized Hellmann potentials is trivially bounded from below by conveniently selected Coulomb potentials, suboptimal bounds on the spectrum $\sigma(\widetilde{H})$ necessarily exist for

$$\kappa + \upsilon \le \frac{4}{\pi}\,.$$

For the sum of couplings non-positive, i.e., $\kappa+\upsilon \le 0$, this lower bound of Coulomb origin is easily improvable: $V_{\mathrm{H}}(r)$ is bounded from below and hence $\sigma(\widetilde{H})$ by the minimum of $V_{\mathrm{H}}(r)$.

Table 1. Binding energies $E_k - 2\,m$ $[m]$ ($k \in \mathbb{N}_0$) of the spinless relativistic Hellmann problem [20, 21], for lowest-lying states of radial excitation n_{r} and orbital angular momentum ℓ, assuming $b = \mu = m$ and $\beta = 1$ for the potential-slope and Laguerre-basis parameters, and trial-space dimensions around $d \gtrsim 29$.

Bound state		Upper bounds		
		$\kappa = \upsilon = \frac{1}{2}$	$\kappa = 1, \upsilon = -1$	$\kappa = 1, \upsilon = -2$
n_{r}	ℓ	[case Fig. 1(b)]	[case Fig. 1(c)]	[case Fig. 1(d)]
0	0	−0.11673	−0.17951	−0.14410
0	1	−0.01579	−0.06294	−0.06157
0	2	−0.00616	−0.02813	−0.02812
1	0	−0.02107	−0.05464	−0.04786
1	1	−0.00509	−0.02810	−0.02762
2	0	−0.00688	−0.02566	−0.02338
Lower bound		−0.58578...	−1	−0.37336...

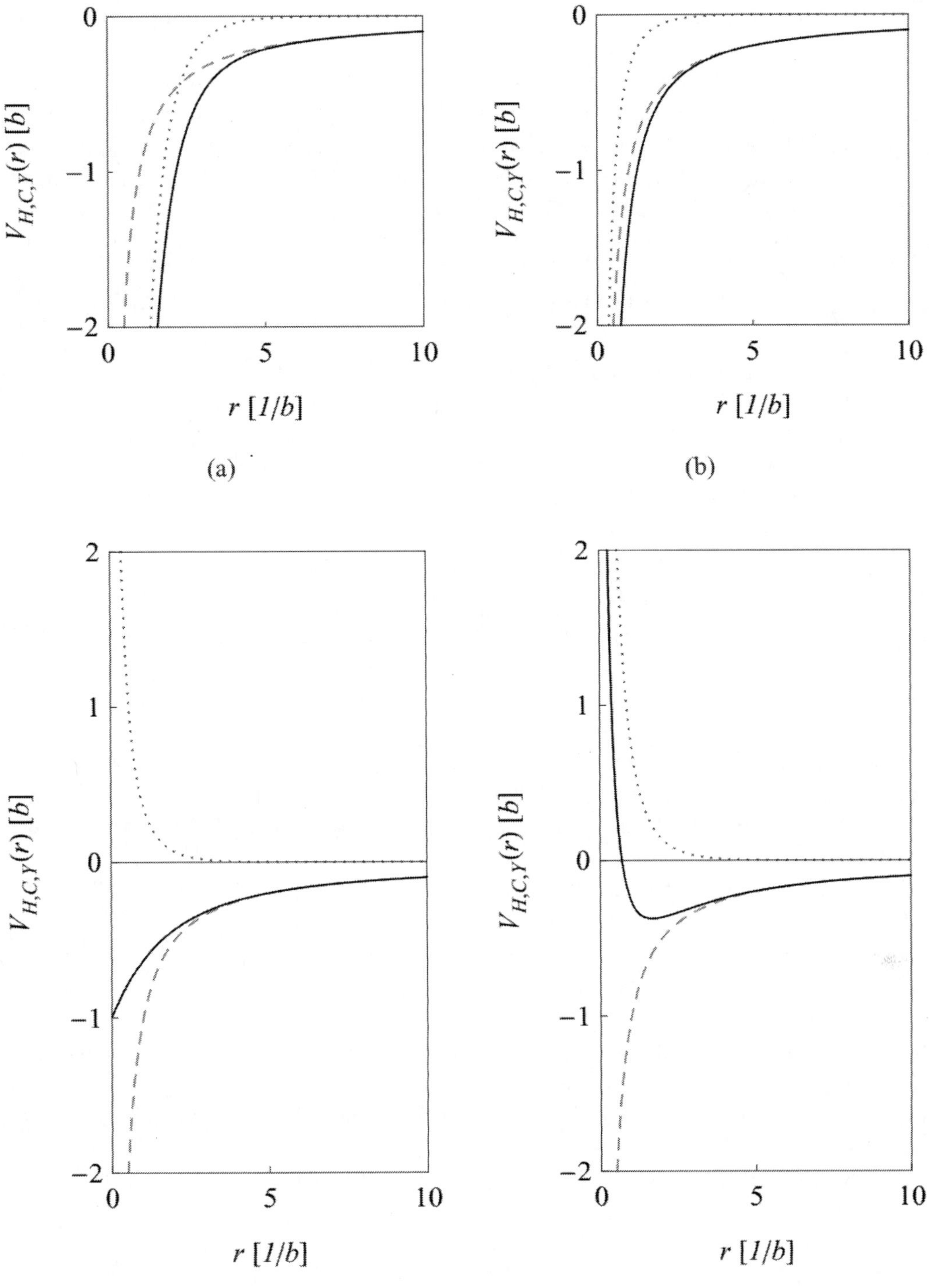

Figure 1. Representatives of four (out of the in total seven) different categories of generalized Hellmann potentials $V_{\rm H}(r)$, specified by the (in)equalities of their couplings (a) $\upsilon > \kappa$, (b) $\upsilon = \kappa$, (c) $\upsilon = -\kappa$, and (d) $\upsilon < -\kappa$, obtained as sums (solid black) of Coulombic (dashed magenta) and Yukawa (dotted blue) terms. The systematic classification of all *generalized* Hellmann potentials can be found in Table 1 of Ref. [21].

- The number of discrete eigenstates of *any* generalized-Hellmann Hamiltonian may never be bounded from above because any generalized Hellmann potential becomes Coulomb-like at large distances, in the limit $r \to \infty$. More technically, any generalized relativistic Hellmann problem fails [20] to satisfy the prerequisite (3) that guarantees finiteness [7] of the number.
- Upper bounds on the eigenvalues corresponding to lowest-lying discrete eigenstates may be conveniently calculated numerically upon spanning that d-dimensional subspace referred to by the minimum–maximum theorem by a Laguerre basis (4), as exemplified in Table 1; they can be optimized by increasing the dimension d and adapting the two parameters μ and/or β.

References

[1] E. E. Salpeter and H. A. Bethe, Phys. Rev. **84** (1951) 1232
[2] W. Lucha and F. F. Schöberl, J. Phys. G: Nucl. Part. Phys. **31** (2005) 1133, arXiv:hep-th/0507281
[3] E. E. Salpeter, Phys. Rev. **87** (1952) 328
[4] I. W. Herbst, Commun. Math. Phys. **53** (1977) 285
[5] I. W. Herbst, Commun. Math. Phys. **55** (1977) 316 (addendum)
[6] A. Martin and S. M. Roy, Phys. Lett. B **233** (1989) 407
[7] I. Daubechies, Commun. Math. Phys. **90** (1983) 511
[8] W. Lucha and F. F. Schöberl, Int. J. Mod. Phys. A **30** (2015) 1550062, arXiv:1412.4950 [hep-ph]
[9] M. Reed and B. Simon, *Methods of Modern Mathematical Physics IV: Analysis of Operators* (Academic Press, New York, 1978) Sect. XIII.1
[10] W. Lucha and F. F. Schöberl, Int. J. Mod. Phys. A **14** (1999) 2309, arXiv:hep-ph/9812368
[11] W. Lucha and F. F. Schöberl, Recent Res. Dev. Phys. **5** (2004) 1423, arXiv:hep-ph/0408184
[12] W. Lucha and F. F. Schöberl, Phys. Rev. A **60** (1999) 5091, arXiv:hep-ph/9904391
[13] W. Lucha and F. F. Schöberl, Int. J. Mod. Phys. A **15** (2000) 3221, arXiv:hep-ph/9909451
[14] W. Lucha and F. F. Schöberl, Phys. Rev. Lett. **64** (1990) 2733
[15] W. Lucha, Mod. Phys. Lett. A **5** (1990) 2473
[16] W. Lucha and F. F. Schöberl, Int. J. Mod. Phys. A **29** (2014) 1450057, arXiv:1401.5970 [hep-ph]
[17] W. Lucha and F. F. Schöberl, EPJ Web Conf. **80** (2014) 00049, arXiv:1407.4624 [hep-ph]
[18] W. Lucha and F. F. Schöberl, Int. J. Mod. Phys. A **29** (2014) 1450181, arXiv:1408.4957 [hep-ph]
[19] W. Lucha and F. F. Schöberl, Int. J. Mod. Phys. A **29** (2014) 1450195, arXiv:1410.5241 [hep-ph]
[20] W. Lucha and F. F. Schöberl, Int. J. Mod. Phys. A **34** (2019) 1950028, arXiv:1812.10756 [hep-ph]
[21] W. Lucha, PoS (EPS-HEP2019) (2020) 537, arXiv:1908.08801 [hep-ph]
[22] H. Hellmann, J. Chem. Phys. **3** (1935) 61
[23] H. Hellmann and W. Kassatotschkin, J. Chem. Phys. **4** (1936) 324

EPJ Web of Conferences **270**, 00019 (2022) https://doi.org/10.1051/epjconf/202227000019
QCD@Work 2022

A journey through the experimental highlights on heavy-ion physics

Alberto Calivà[1,2,*]

[1]University of Salerno, Physics Department "E.R. Caianiello", via Giovanni Paolo II, 132 84084 Fisciano, Italy
[2]National Institute for Nuclear Physics (INFN), Section of Napoli, group of Salerno, via Giovanni Paolo II, 132 84084 Fisciano, Italy

Abstract. Heavy-ion collisions are a unique tool to create in the laboratory the quark-gluon plasma (QGP), a state of strongly-interacting matter where quarks and gluons are deconfined. Significant progress was made over the last years in the understanding of the QGP properties and in the characterization of the phase diagram of QCD matter. In these proceedings, a review of recent experimental highlights on heavy-ion physics from different experiments is presented.

1 Introduction

One of the main goals of studying heavy-ion collisions is to characterize the properties of the quark-gluon plasma (QGP), a deconfined phase of quarks and gluons, and explore the phase diagram of QCD matter. The latter is a two-dimensional space where a state of QCD matter is defined by its temperature and baryochemical potential μ_B. Different regions of the phase diagram can be accessed with heavy-ion collisions at different center-of-mass energies. The space-time evolution of heavy-ion collisions comprises several stages that can be studied using different experimental probes and methods. The overall picture, containing the current knowledge of the QPG properties and of the phase diagram, emerges from a complex and diverse experimental program involving many experiments at different facilities running at center-of-mass energies per nucleon pair covering the range from a few GeV to TeV. In the following, a review of selected highlights on heavy-ion physics from different experiments is presented and the interpretation of the results is discussed.

1.1 Heavy Quarkonia

Quarkonia are flavorless mesons made of heavy quark-antiquark pairs. Charm and bottom quarks are created in the early stages of the collision in hard parton-parton scattering processes. These particles then travel through the hot and dense QCD medium losing a fraction of their energy before hadronization in collisions with other partons and (mostly) by gluon radiation. The energy loss effects can be quantified by the nuclear modification factor R_{AA}, defined as the ratio of the measured quarkonium yield in heavy-ion collisions and the $\langle N_{coll} \rangle$-scaled yield measured in pp collisions at the same energy, where $\langle N_{coll} \rangle$ is the average number

*e-mail: alberto.caliva@cern.ch

of binary nucleon-nucleon collisions. The R_{AA} of inclusive J/ψ at mid-rapidity was measured both at RHIC and LHC energies, as shown in Fig. 1. The R_{AA} continuously decreases with increasing average number of participant nucleons ($\langle N_{part} \rangle$) at RHIC energy [1, 2] while and enhancement is observed at LHC energy starting from $\langle N_{part} \rangle \approx 100$. This effect is due to the recombination of charm-anticharm pairs in the late stages of the QGP evolution close to the phase boundary, as predicted by different theoretical works several years ago [3–6]. Recombination effects, which become stronger going to more central collisions and lower transverse momentum (p_T), are observed also for the $\psi(2S)$.

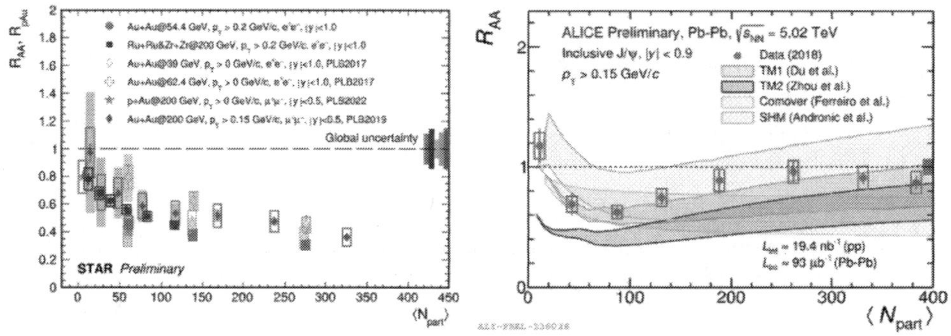

Figure 1. R_{AA} of inclusive J/ψ at mid-rapidity measured at RHIC (left) and LHC (right) energies as a function of the average number of participant nucleons ($\langle N_{part} \rangle$).

The elliptic flow (v_2), i.e. the second coefficient in the Fourier expansion of azimuthal particle distribution with respect to the symmetry plane of the collision, is also sensitive to recombination effects and can be used to further characterize the J/ψ production mechanism. While the v_2 of inclusive J/ψ measured in Au–Au collisions at $\sqrt{s_{NN}} = 200$ GeV at RHIC is consistent with zero, both prompt and non-prompt J/ψ show a significant v_2 up to high p_T, as measured by CMS. These results confirm the absence of significant $c\bar{c}$ recombination effects and early production of charmonia at RHIC energies. On the contrary, the large v_2 of prompt and non-prompt J/ψ measured at the LHC at low p_T implies a high degree of thermalization of charm quarks and a significant contribution from regeneration.

Since the $c\bar{c}$ recombination occurs in the late stages of the collision, the v_2 of different quarkonium states is sensitive to the time when their regeneration happens. Indeed, the v_2 of the prompt $\psi(2S)$ measured by CMS is larger than that of the prompt J/ψ, which could be explained by a sequential regeneration of charmonia [7]. Another interesting effect predicted many years ago and considered a decisive signature of the QGP formation [8] is the sequential suppression of higher-mass quarkonium states. Evidence of this effect is found not only for charmonium but also for bottomonium states, as recently shown by both ATLAS and CMS Collaborations.

1.2 Open Heavy Flavor

Open heavy-flavor hadrons are complementary probes to charmonia to study the in-medium energy loss of heavy quarks and characterize the QGP properties. Since the bottom quark is heavier than charm, it is expected to lose a lower fraction of its energy in collisions with other partons and by gluon radiation (dead cone effect). This is confirmed by the measurement of the R_{AA} of electrons from charm and bottom decays in central Au–Au collisions at $\sqrt{s_{NN}} =$

200 GeV by PHENIX and also by the measurement of prompt and non-prompt D mesons by ALICE [9]. The R_{AA} and v_2 of prompt D mesons are used by ALICE to constrain the spatial diffusion coefficients of charm. Interestingly, charm and bottom quarks are also sensitive to initial state fluctuations in the spatial distribution of nucleons participating in the collision for a given impact parameter. This is demonstrated by the significant v_3 of prompt and non-prompt D^0 mesons measured by CMS in mid-central Pb–Pb collisions at $\sqrt{s_{NN}} = 5.02$ TeV as shown in Fig 2. The bottom quark is less sensitive than charm due to its larger mass.

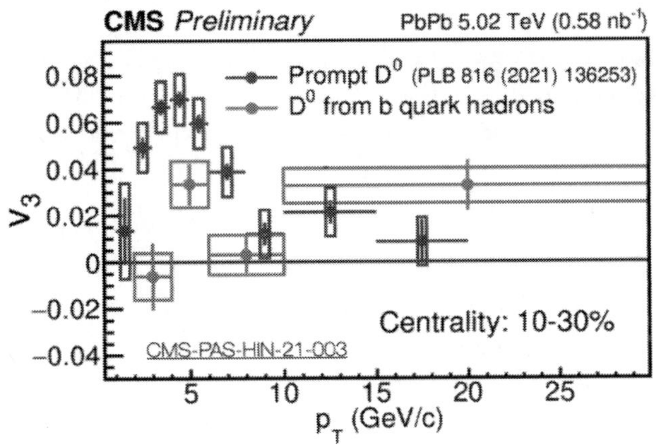

Figure 2. v_3 of prompt and non-prompt D^0 mesons measured by CMS in Pb–Pb collisions at $\sqrt{s_{NN}} =$ 5.02 TeV.

1.3 Light Flavor

In the light-flavor sector, one of the most debated topics is the production mechanism of light (anti)nuclei in high-energy hadronic collisions. The statistical hadronization model [10–16] and nucleon coalescence [17–23] are the main competing models that are based on different microscopic processes. Both models provide similar predictions of (anti)nuclei yields, therefore novel techniques and observables have recently been introduced to distinguish these (anti)nucleosynthesis models. One of them is the (anti)nuclei number fluctuation. The STAR Collaboration has recently reported the first measurement of the deuteron-deuteron femtoscopic correlation function in Au–Au collisions at $\sqrt{s_{NN}} = 3$ GeV and the center-of-mass energy dependence of the proton-deuteron Pearson coefficient, as illustrated in Fig 3. Both measurements are consistent with model calculations incorporating nucleon coalescence, while the Pearson coefficient measurement is also compatible with the canonical statistical hadronization model [16, 24].

The (anti)deuteron number fluctuation has been recently measured also by the ALICE Collaboration [25]. The cumulant ratio $\kappa_2/\kappa_1 = \langle(n_d - \langle n_d\rangle)^2\rangle/\langle n_d\rangle$, where n_d is the measured number of deuterons per event, and the proton-deuteron Pearson correlation coefficient favor the canonical statistical model over the coalescence approach. Interestingly, this measurement indicates a smaller correlation volume, i.e. the volume in which the baryon number conservation is implemented, with respect to the net-proton fluctuation measurement [26].

The above-mentioned (anti)nucleosynthesis models describe also the production of exotic hypernuclei, i.e. nuclei with hyperons. The hypertriton (${}^3_\Lambda$H) yields measured by STAR in Au–Au collisions at $\sqrt{s_{NN}} = 3$, 19.6 and 27 GeV and the first measurement of

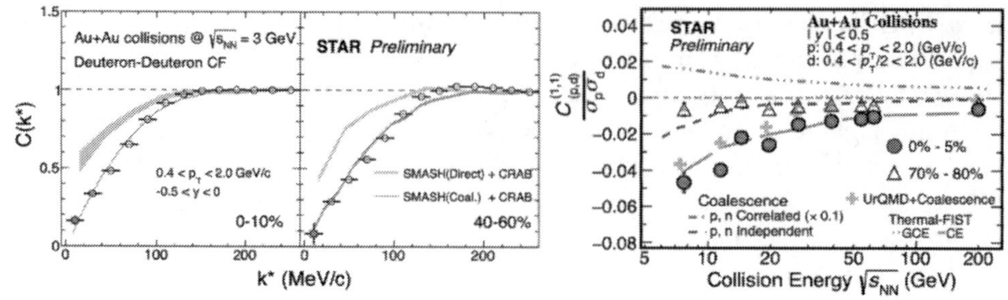

Figure 3. Deuteron-deuteron correlation function measured in Au–Au collisions at $\sqrt{s_{NN}} = 3$ GeV (left) and the center-of-mass energy dependence of the proton-deuteron Pearson coefficient (right) measured by STAR.

the hyperhydrogen-4 ($^4_\Lambda$H) in Au–Au collisions at $\sqrt{s_{NN}} = 3$ GeV, shown in Fig 4, are essential to constrain the production models at high μ_B. Also in this case, the coalescence approach provides a good description of the data while some configurations of SHM and hybrid models fail.

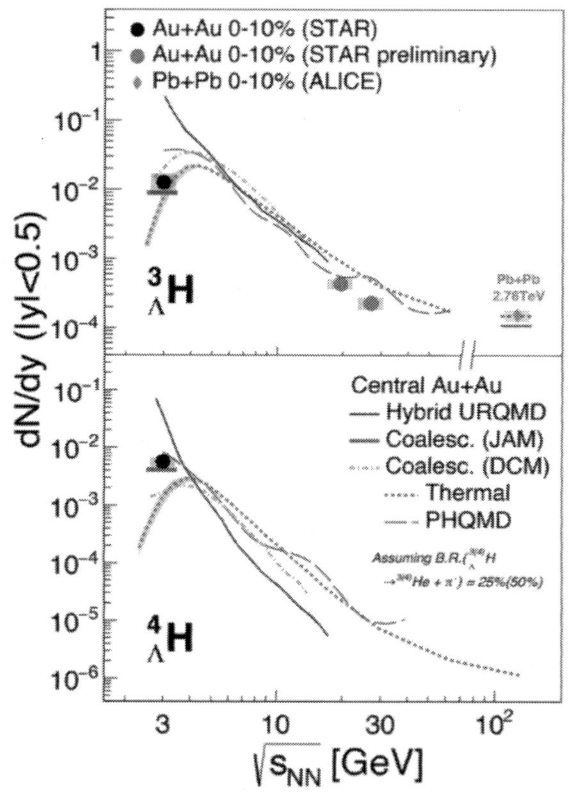

Figure 4. Hypernuclei yields measured by STAR in Au–Au collisions at $\sqrt{s_{NN}} = 3$, 19.6 and 27 GeV in comparison with hypernuclei production models.

In the coalescence approach, the production yield of the hypertriton depends on the radial extension of its wave function. Recent measurements of the hypertriton lifetime, performed by different experiments, consistently yield a value that is very close to that of the free Λ. These results confirm that the hypertriton is a loosely bound object, thus closing the long-standing hypertriton lifetime puzzle [27].

The properties of the hadron gas phase, produced after hadronization, can be studied by measuring the yields of short-lived resonances. Due to their short lifetimes, which are comparable to that of the hadron gas phase, resonance yields are affected by re-scattering effects of their decay daughters with other hadrons in the gas and also by re-generation effects, i.e. the back reaction to their decay. The STAR Collaboration has reported measurements of the yield ratios K^{*0}/K and Φ/K as a function of multiplicity. A significant decrease is observed for K^{*0}/K with increasing multiplicity, while the ratio Φ/K is almost flat. This is due to the competing strangeness enhancement effect i.e. the increase in the ratio of strange to non-strange hadron yields with increasing multiplicity [28]. To further characterize the hadron gas phase, the ALICE Collaboration has measured the yield ratios of resonances with different lifetimes to those of the corresponding stable hadrons. Figure 5 shows the $K^{*\pm}/K^0_S$ and $\Lambda(1520)/\Lambda$ measured as a function of multiplicity in different collision systems and center-of-mass energies. Although the $\Lambda(1520)$ has a longer lifetime than K^*, it shows a stronger suppression when going from peripheral to central Pb–Pb collisions. Obviously, the lifetime is not a good predictor for resonance suppression, which depends on the mean free path of the resonance in the hadron gas phase and on the (partial) chemical equilibrium with the system [29].

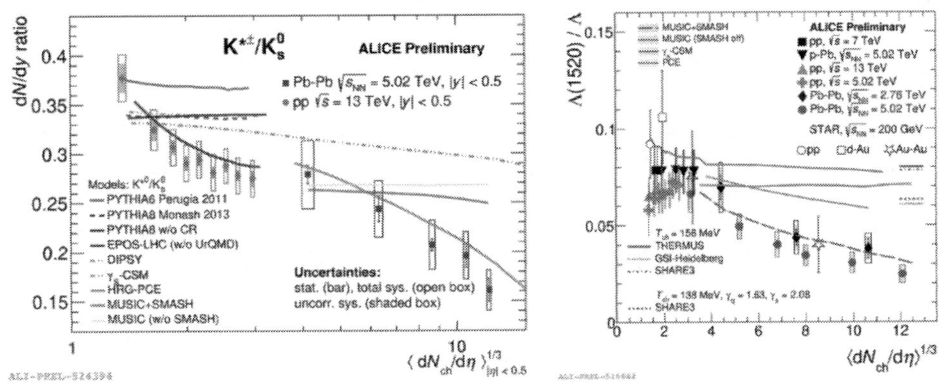

Figure 5. $K^{*\pm}/K^0_S$ and $\Lambda(1520)/\Lambda$ measured by ALICE as a function of multiplicity in different collision systems and center-of-mass energies.

1.4 Direct Photons and Dileptons

Direct photons and dileptons, i.e. lepton-antilepton pairs from virtual photons, are electromagnetic probes that are emitted continuously throughout the entire history of a heavy-ion collision. These particles interact with the medium only through the electroweak force, hence they carry undisturbed information on the production process. The PHENIX Collaboration has recently reported the measurement of non-prompt direct-photon production in Au–Au collisions at $\sqrt{s_{NN}} = 200$ GeV [30]. The non-prompt direct-photon p_T spectrum is obtained by subtracting from the inclusive direct-photon spectrum measured in Au–Au collisions the prompt component, estimated using the $\langle N_{coll} \rangle$-scaled direct-photon spectrum measured in pp collisions at the same energy. The extracted non-prompt direct-photon spectrum is consistent with theoretical calculations including pre-equilibrium and thermal photons at high p_T, while at low p_T, there is a hint of an excess of non-prompt direct-photon production with respect to the expectations both in central and peripheral collisions. The inverse slope parameter $\langle T_{\text{eff}} \rangle$, extracted from an exponential fit to the non-prompt direct-photon spectrum, is studied as a function of p_T and multiplicity. A higher effective average temperature is observed for higher p_T, with no significant multiplicity dependence. This implies that different p_T regions of the spectrum are sensitive to different stages of the collision and that $\langle N_{\text{coll}} \rangle$-scaling of direct-photon production holds approximately within the multiplicity interval covered by the measurement.

The inverse slope parameter of the direct-photon spectrum is affected by blue shift due to radial flow. Instead, the temperature extracted from a fit to the invariant-mass excess spectrum of dileptons in the mass region $m_\phi < m_{ll} < m_{J/\psi}$, dominated by thermal radiation from QGP, is not influenced by radial flow. The excess dilepton spectrum is obtained by subtracting from the inclusive spectrum the contribution from hadron decays, the so-called "hadronic cocktail". The latter is obtained from simulations based on parametrized p_T distributions of hadrons from data and scaling laws obtain the input distributions of hadrons that are not measured. The dilepton excess spectrum was measured by STAR in Au–Au collisions at $\sqrt{s_{\text{NN}}} = 27$ and 54.4 GeV and compared with previous results in In–In collisions at $\sqrt{s_{\text{NN}}} =$ 17.3 GeV at NA61, as shown in Fig. 6 left. Higher temperatures and yields are obtained at RHIC indicating a hotter and longer-lived QGP at RHIC than at the SPS. The excess spectrum is compared with theoretical model calculations including thermal radiation from the QGP and contributions from in-medium ρ^0 decays, characterized by a broader spectral function, as shown in Fig 6 right. A good agreement is found with different approaches in the mass region $0.2 < m_{ll} < 1.3$ GeV, while the model calculations underestimate the slope of the invariant-mass spectrum for $m_{ll} > m_\phi$, thus implying that the QGP is hotter than predicted in the models.

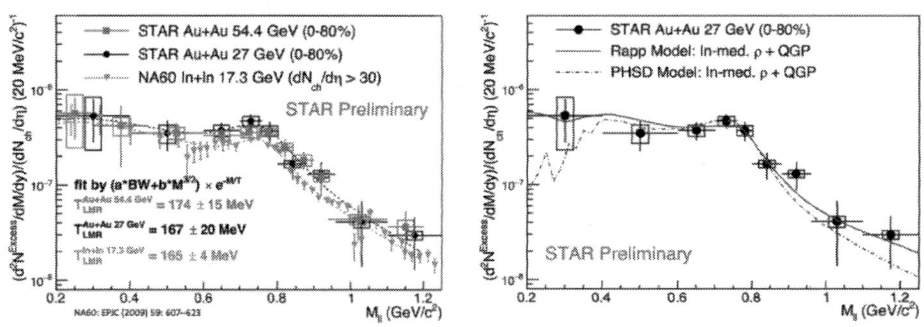

Figure 6. Dilepton excess spectra measured by STAR and NA61 at the SPS (left) and comparison with model calculations (right).

The first dilepton spectrum in Pb–Pb collisions at $\sqrt{s_{\text{NN}}} = 5.02$ TeV was measured by the ALICE Collaboration, as shown in Fig. 7 left. The measured spectrum is consistent with contributions from hadron decays, QGP radiation, and in-medium ρ^0 decays. Two different versions of the hadronic cocktail are compared to the data: one is based on $\langle N_{\text{coll}} \rangle$-scaling of correlated semi-leptonic heavy-flavor decays, while in the other, heavy-flavor suppression is implemented by scaling the heavy-flavor contributions by the measured $R_{\text{AA}}^{c/b\to e}$ [31]. The latter version of the cocktail gives a better agreement with the data. The fraction of virtual direct photons is extracted from the low-mass region in different p_{T} intervals. This fraction is then scaled by the inclusive photon spectrum measured using the photon-conversion technique to obtain the first direct-photon spectrum at LHC. This is shown in Fig. 7 right in comparison with theoretical model calculations including prompt, pre-equilibrium, and thermal photons. All theoretical approaches give a good description of the measured spectrum considering the experimental uncertainties.

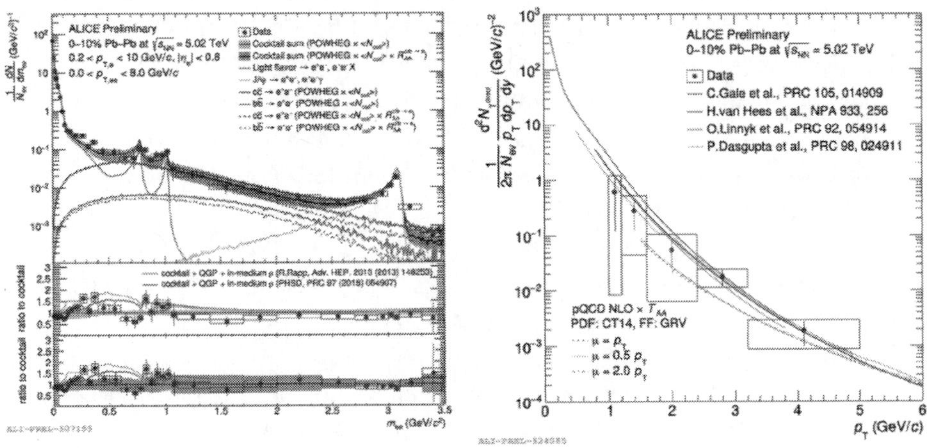

Figure 7. Dilepton spectrum measured in central Pb–Pb collisions at $\sqrt{s_{\mathrm{NN}}}$ = 5.02 TeV (left) and direct-photon spectrum (right).

2 Conclusions

Significant progress has been made over the last few years in the field of high-energy nuclear physics and hadronic physics at particle colliders. This includes a detailed characterization of the QGP properties and an in-depth understanding of a large variety of phenomena related to hadron production and hadronic interactions. The study of rare events and high-precision measurements at the TeV-energy scale will be possible with the restart of the LHC. Exciting times are ahead with the start of sPHENIX at RHIC and with future detectors and upgrades.

References

[1] L. Adamczyk et al., Phys. Lett. **B771**, 13 (2017)
[2] J. Adam et al., Phys. Lett. **B797**, 134917 (2019)
[3] B. Svetitsky, Phys. Rev. **D37**, 2484 (1988)
[4] P. Braun-Munzinger, J. Stachel, Phys. Lett. **B490**, 196 (2000)
[5] R.L. Thews, M. Schroedter, J. Rafelski, Phys. Rev. **C63**, 054905 (2001)
[6] A. Andronic et al., Phys. Lett. **B797**, 134836 (2019)
[7] X. Du, R. Rapp, Nucl. Phys. **A943**, 147 (2015)
[8] S. Digal, P. Petreczky, H. Satz, Phys. Rev. **D64**, 094015 (2001)
[9] S. Acharya et al. (ALICE Collaboration), J. High Energ. Phys. **174** (2022)
[10] J. Cleymans et al., Phys. Rev. **C84**, 054916 (2011), 1105.3719
[11] A. Andronic, others., Phys. Lett. **B697**, 203 (2011), 1010.2995
[12] F. Becattini et al., Phys. Rev. **C90**, 054907 (2014), 1405.0710
[13] V. Vovchenko, H. Stöcker, Phys. Rev. **C95**, 044904 (2017), 1606.06218
[14] A. Andronic et al., Nature **561**, 321 (2018), 1710.09425
[15] N. Sharma et al., Phys. Rev. **C99**, 044914 (2019), 1811.00399
[16] V. Vovchenko, B. Dönigus, H. Stoecker, Phys. Rev. **C100**, 054906 (2019), 1906.03145
[17] S.T. Butler, C.A. Pearson, Phys. Rev. **129**, 836 (1963)
[18] J.I. Kapusta, Phys. Rev. **C21**, 1301 (1980)
[19] R. Scheibl, U.W. Heinz, Phys. Rev. **C59**, 1585 (1999), nucl-th/9809092
[20] W. Zhao et al., Phys. Rev. **C98**, 054905 (2018), 1807.02813
[21] K. Blum, M. Takimoto, Phys. Rev. **C99**, 044913 (2019), 1901.07088
[22] K.J. Sun, C.M. Ko, B. Dönigus, Phys. Lett. **B792**, 132 (2019), 1812.05175
[23] M. Kachelrieß, S. Ostapchenko, J. Tjemsland, Eur. Phys. J. **A56**, 4 (2020), 1905.01192
[24] V. Vovchenko, B. Dönigus, H. Stoecker, Phys. Lett. **B785**, 171 (2018), 1808.05245
[25] ALICE Collaboration (2022), https://arxiv.org/abs/2204.10166
[26] ALICE Collaboration (2022), https://arxiv.org/abs/2206.03343
[27] A. Pérez-Obiol, D. Gazda, E. Friedman, A. Gal, Phys. Lett. **B811**, 135916 (2020)
[28] J. Adam et al., Nature Physics **13**, 535 (2017)
[29] A. Motornenko et al., Phys. Rev. **C102**, 024909 (2020)
[30] U.A. Acharya, others. (2022), https://arxiv.org/abs/2203.17187
[31] S. Acharya et al., J. High Energ. Phys. **2018** (2018)

Charm hadronisation measurements with the ALICE experiment

Marco Giacalone[1,2,*], on behalf of the ALICE Collaboration

[1]Dipartimento di Fisica e Astronomia, University of Bologna
[2]Istituto Nazionale di Fisica Nucleare (INFN), sezione di Bologna

Abstract. Charm quarks are produced in hard scatterings at the early stages of the hadronic collisions due to the heavy mass of the quark. The production cross section of charm can be described by pQCD calculations based on the factorisation approach. This description traditionally assumes universal fragmentation functions of charm into hadrons among different collision systems.
The ALICE experiment measured charmed-baryon production in pp collisions for different baryon species, observing a significant enhancement of the charm baryon-to-meson yield ratios, for $p_{\rm T} < 10$–12 GeV/c, with respect to measurements performed at e^+e^- and ep colliders. Measurements of charmed baryon production in pp collisions also provide a fundamental reference for heavy-ion collisions, in which an enhancement of the baryon-to-meson ratio could be due to an additional hadronisation mechanism via coalescence of charm quarks with lighter quarks in a deconfined medium.
This contribution discusses the latest measurements of charm production performed by the ALICE Collaboration in pp collisions at the LHC via D mesons, Λ_c^+, $\Sigma_c^{0,+,++}$, $\Xi_c^{0,+}$, and Ω_c^0 baryons at midrapidity at $\sqrt{s} = 5.02$ and 13 TeV. The most recent measurements of the Λ_c^+/D^0 ratio in Pb–Pb collisions at $\sqrt{s_{\rm NN}} = 5.02$ TeV are also presented, together with the first measurement of the charm fragmentation fraction into hadrons in pp and p–Pb collisions at the same energy.

1 Introduction

Heavy flavour production in proton–proton (pp) collisions can be described via pQCD calculations which employ the factorisation approach [1]. In particular, the cross section of charm (and beauty) hadrons is calculated as a convolution of three main terms:

1. Parton distribution functions, which account for the probability to find a specific parton inside the colliding protons carrying a given fraction of the proton momentum (Bjorken-x).
2. cross section of a quark–antiquark pair creation via parton interactions (hard scattering)
3. fragmentation functions, which describe the probability of the hadronisation of a heavy quark into a given hadron, which carries a specific fraction z of the original quark momentum.

*e-mail: marco.giacalone@bo.infn.it

In particular, the fragmentation functions are traditionally assumed to be universal [2] and are constrained from e^+e^- and ep collision measurements. This assumption was observed to be not valid, given the many results reported at the LHC [3, 4] showing an increased heavy-flavour baryon production relative to mesons, with respect to measurements in e^+e^- and ep collisions.
Moreover, heavy flavours are an important tool to characterise the quark–gluon plasma, the state of matter created in relativistic heavy-ion collisions. Charm and beauty quarks are produced at the early stages of the collisions and they can interact with the medium during its entire lifetime.
This contribution presents an overview of the recent charm measurements performed by the ALICE Collaboration in pp collisions at $\sqrt{s} = 5.02$ and 13 TeV and in p–Pb and Pb–Pb collisions at $\sqrt{s_{\rm NN}} = 5.02$ TeV.
The reconstruction of different charmed particles is performed via an invariant mass analysis of fully reconstructed decay topologies, selected with criteria based on geometrical and kinematic variables related to the displaced charmed-hadron decay vertex and on the particle identification of the decay products. The heavy-flavour hadron reconstruction strategies exploit the track impact parameter resolution provided by the ITS detector: better than 50 μm for $p_{\rm T} > 1$ GeV/c.
The production of D mesons, Λ_c^+, $\Sigma_c^{0,+,++}$, $\Xi_c^{0,+}$, and Ω_c^0 baryons at midrapidity was measured down to very low $p_{\rm T}$ in pp collisions at $\sqrt{s} = 5.02$ and 13 TeV.
This contribution focuses on the discussion of the baryon-to-meson ratio measurements, since they show an intriguing dependence on the collision system, which is not observed for the yield ratios of different meson species.
In Fig. 1 (left) the D^+/D^0 yield ratios for prompt and non-prompt (i.e. originating from beauty-hadron decays) mesons are shown as a function of $p_{\rm T}$, presenting a flat trend. They are in agreement within uncertainties with the measurements performed in e^+e^- and ep collisions [5], and they are well described by pQCD calculations (FONLL [6] and GM-VFNS [7]). On the other hand, the Λ_c^+/D^0 ratio measured in pp collisions at the LHC show a strong decreasing trend as a function of $p_{\rm T}$, as shown in Fig. 1 (right).

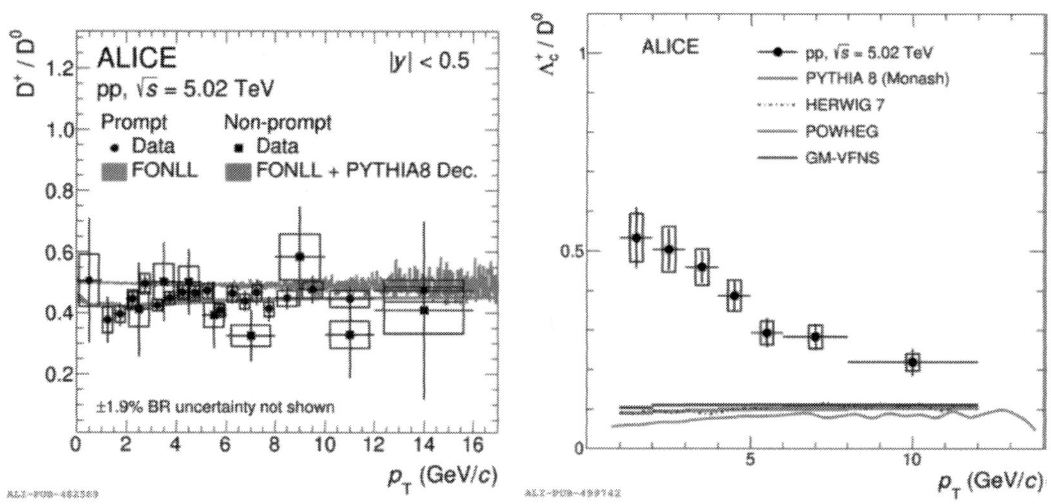

Figure 1. Comparison of measured production cross section ratios with Monte Carlo generators using D^+/D^0 meson ratios (left) [5] and Λ_c^+/D^0 baryon-to-meson ratio in pp collisions at 5.02 TeV (right) [3].

The baryon-to-meson measurements performed in pp collisions at midrapidity by the ALICE Collaboration show a different behaviour both with respect to results from e^+e^- collisions at LEP ($\Lambda_c^+/D^0 \approx 0.11$ [2]) and to pQCD calculations and event generators, in which the charm-quark fragmentation is constrained using e^+e^- and ep results [3]. The underestimation of the baryon-to-meson ratio might be due to modified hadronisation mechanisms in the parton-rich environment created in pp collisions at the LHC, which enhance the production of baryons relative to mesons, thus breaking the universality of fragmentation.

2 Recent results and discussion

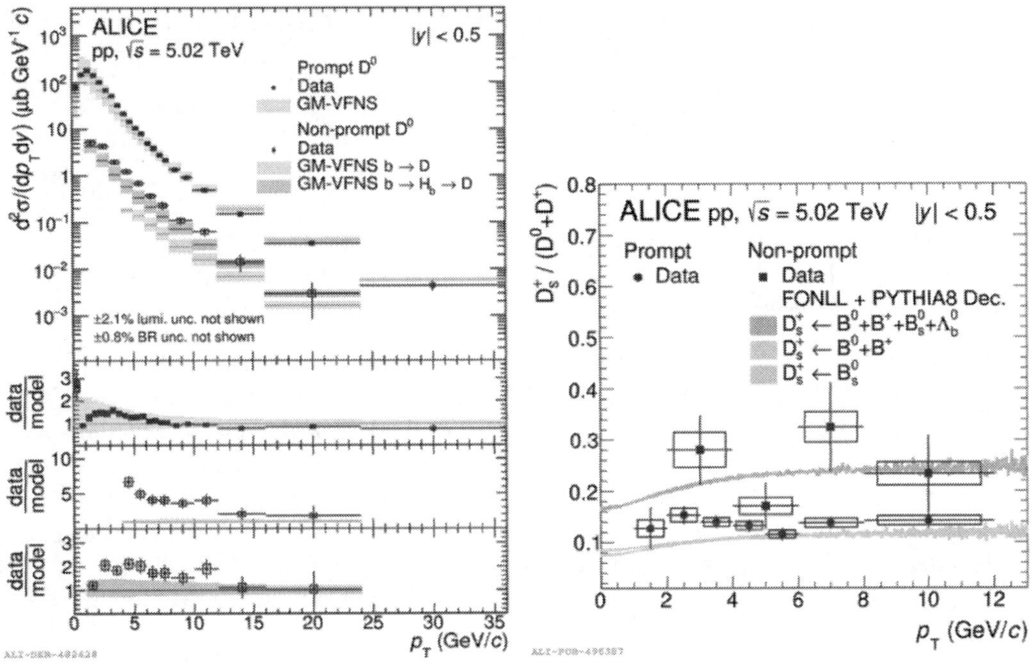

Figure 2. Prompt and non-prompt production of D^0 mesons in pp collisions at $\sqrt{s} = 5.02$ TeV compared to pQCD calculations [5] down to $p_T = 0$ (left) and ratio between D_s^+ and the sum of D^+ and D^0 meson yields (right).

The production of D mesons in pp collisions at $\sqrt{s} = 5.02$ TeV has been measured down to $p_T = 0$ for D^+ and D^0 mesons. In Fig. 2 (left) the measured production cross sections of prompt and non-prompt D^0 mesons is compared with pQCD calculations based on the factorisation approach, namely GM-VFNS. Results for prompt D mesons are described well by GM-VFNS calculations, while the measurements of non-prompt D^0 production are underestimated: this is especially true for the case in which the fragmentation and decay are done in one step (b $\rightarrow$ D curve), while the two-steps approach (b $\rightarrow$ H_b $\rightarrow$D) describes better the data. On the other hand, the ratio between the yields of non-prompt charmed strange D_s^+ mesons and the sum of non-prompt D^0 and D^+, reported in Fig. 2 (right), shows a good agreement within the uncertainties with predictions based on FONLL calculations for beauty-hadron production and PYTHIA 8 for the B$\rightarrow$D decays. As for the D^+/D^0 ratio shown in Fig. 1 (right), no trend with p_T is observed within current uncertainties.
The Λ_c^+/D^0 ratio, now measured down to $p_T = 0$ in pp collisions at $\sqrt{s} = 13$ and 5.02 TeV, is found to be compatible at the two considered centre-of-mass energies (Fig. 3 left).

As seen in the previously published results, we notice that PYTHIA8 with Monash tune [8] and HERWIG7 [10] results are underestimating the measured ratio (because the parton fragmentation in these event generators is tuned on e^+e^- measurements at LEP), while the

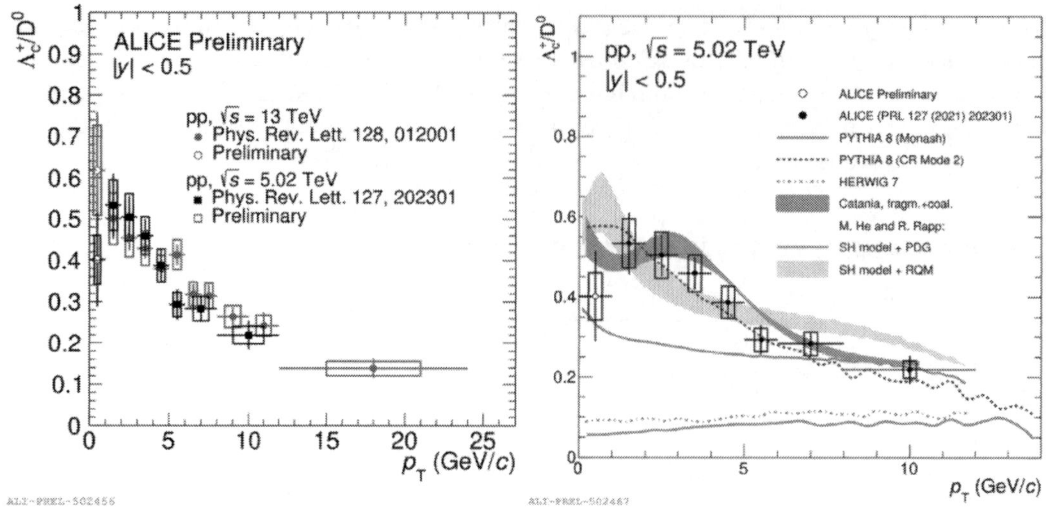

Figure 3. Prompt Λ_c^+/D^0 production ratio in pp collisions at $\sqrt{s} = 13$ and 5.02 TeV (left) and comparison of the ratio with various event generators and model calculations including the ones with enhanced baryon production mechanisms (right).

PYTHIA8 event generator with enhanced colour reconnection mechanisms [9], the Catania model [12], and the Statistical Hadronisation Model (SHM) [11] provide a better description of the data. The presence of junction topologies in PYTHIA8 with enhanced colour reconnection mechanisms increases the production of baryons; the Catania model assumes the presence of a medium of thermalised quarks and gluons, in which hadron formation occurs both with fragmentation and quark coalescence (recombination); in the SHM, instead, it is assumed that hadron abundances follow the equilibrium populations of a hadron–resonance gas including also not yet measured baryon resonances foreseen by the RQM models [11].

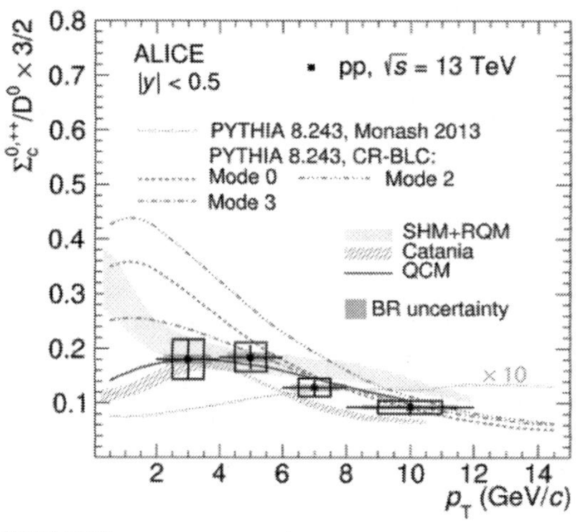

Figure 4. Prompt $\Sigma_c^{0,+,++}/D^0$ production ratio in pp collisions at $\sqrt{s} = 13$ compared to different models [14].

Another result is shown in Fig. 4, in which the $\Sigma_c^{0,+,++}/D^0$ baryon-to-meson ratios are compared with the already discussed models, as well as with the Quark Recombination Model (QCM) [13], in which charm quarks are recombined with lighter quarks sharing their same

velocity. Even for this heavier charmed baryon species, the measured baryon-to-meson ratio is larger than the expected values from models using fragmentation fractions as measured at LEP or a fragmentation process tuned on e^+e^- collisions, while the models with enhanced baryon production are in better agreement with the measurements. Compared to e^+e^- measurements an enhancement up to a factor 10 is found.
Additional insight can be obtained from the results of the ratio between the charmed strange baryon $\Xi_c^{0,+}$ and the D^0 meson yields shown in Fig. 5, for which the measured values are underestimated also by the models with enhanced baryon production. When considering the $\Xi_c^{0,+}/\Sigma_c^{0,+,++}$ ratio, the PYTHIA8 generator with Monash 2013 tune is able to describe the data, thanks to the fact that a similar enhancement in pp with respect to e^+e^- collisions might be happening for the two baryon species. However, all the ratios involving the $\Xi_c^{0,+}$ baryon are in better agreement with the Catania model compared to the other calculations.

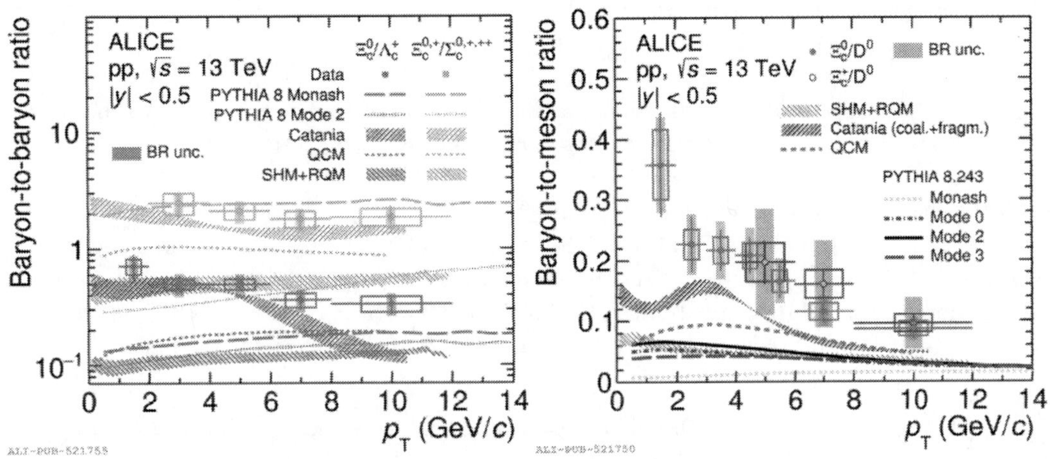

Figure 5. Prompt $\Xi_c^{0,+}$ baryon-to-baryon (left) and baryon-to-meson (right) ratio in pp collisions at $\sqrt{s}$ = 13 TeV compared to different models [15].

The last measurement performed in pp collisions and discussed in this contribution is the production cross section of the doubly strange charmed baryon Ω_c^0, whose first measurement at the LHC was performed by the ALICE Collaboration [16]. Figure 6 (left) shows the Ω_c^0/D^0 ratio as a function of p_T. The ratio is scaled by the branching ratio of the decay $\Omega_c^0 \to \Omega^-\pi^+$, which is not yet measured, but it can be estimated with theoretical calculations [17]. The measured Ω_c^0/D^0 ratio is larger than the theoretical predictions, with the Catania model being the one closer to the experimental results when adding higher mass resonance decays.

The cross section of the Ω_c^0 multiplied by the branching ratio is about 5 times larger than results obtained at BELLE in e^+e^- collisions, which might hint to the possibility of a larger than expected fraction of charm quarks fragmenting into Ω_c^0 baryons at LHC energies.
The measurements of all the ground states of charm hadrons down to very low p_T allowed ALICE to compute the fragmentation fractions in pp, and in p–Pb collisions, as shown in Fig. 6 (right), where they are compared with LEP, HERA and B factory measurements. The measured larger fraction of Λ_c^+ is compensated by a relative decrease of D^0 (and less evidently of $D^\pm$) fraction. The charm fragmentation fractions in p–Pb are compatible with those computed in pp collisions for all the particle species.
The Λ_c^+/D^0 yield ratios in different centrality intervals for Pb–Pb collisions and in pp collisions at $\sqrt{s}$ = 5.02 TeV are shown in Fig. 7. The measurements are compared to the same models used for pp collisions (Catania and SHM), and to the TAMU transport model predictions [20], in which heavy quarks are transported through the QGP phase, where they interact

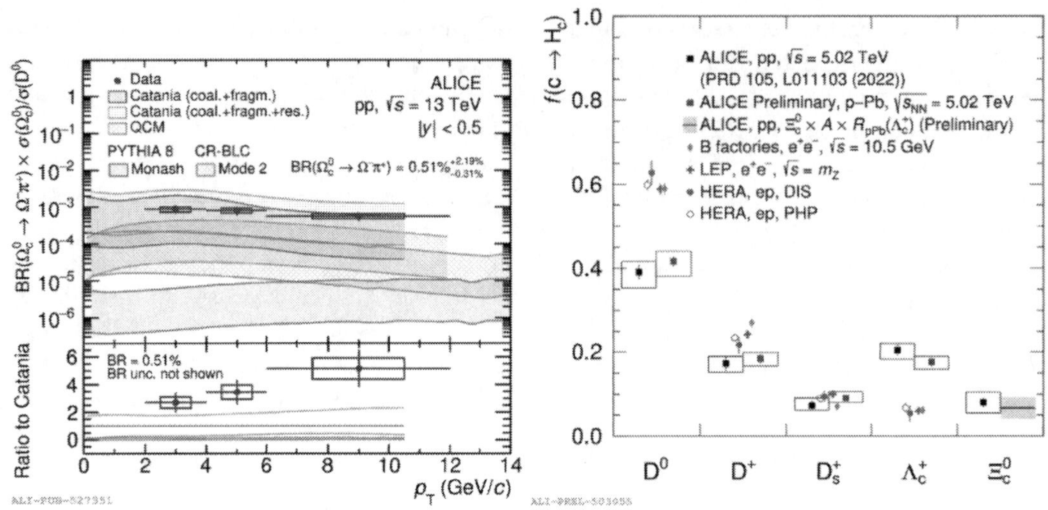

Figure 6. Ω_c^0/D^0 ratio in pp collisions at $\sqrt{s}$ = 13 TeV (left) [16] and charm fragmentation fraction results by ALICE compared to other experiments in different collision systems (right).

with the medium constituents and they can hadronise via quark recombination. The ratio presents different trends as a function of p_T for pp and Pb–Pb collisions, with the centrality interval 30-50% being closer to pp results. This might depend on the presence of radial flow, which modifies in a different way the Λ_c^+ and D^0 p_T shapes in Pb–Pb collisions at different centralities [19]. In this case, TAMU is the model that best describes the results.

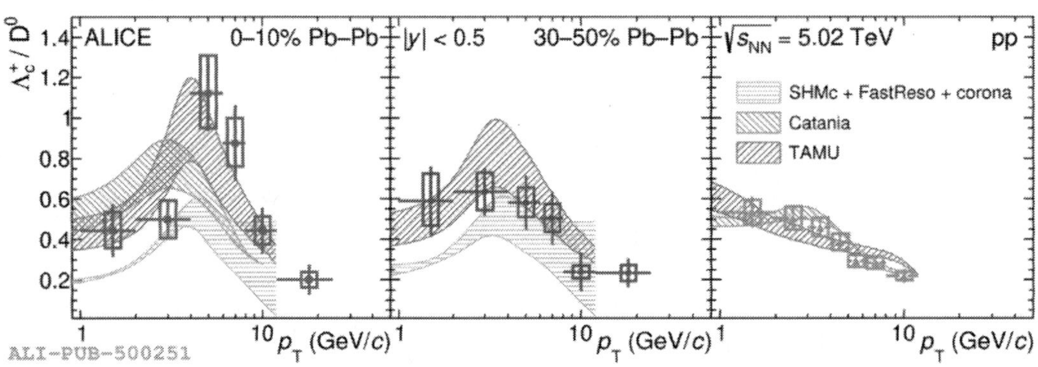

Figure 7. Λ_c^+/D^0 ratio in pp collisions and in Pb–Pb collisions in different centrality intervals compared to models [19].

3 Conclusions

The results presented in this contribution provide novel insights into the mechanisms of charm hadronisation in hadronic collisions. The comparison of the measurements of charm baryon-to-meson ratios with different models suggests that additional mechanisms for the hadronisation of charm into baryons could take place in pp collisions with respect to e^+e^- and ep collisions. More studies are needed to discriminate among different theoretical descriptions. Higher-accuracy measurements will be performed by the ALICE Collaboration exploiting the larger data samples that will be collected in Run 3 and the improved tracking resolution provided by the upgraded experiment after the Long Shutdown 2.

References

[1] J. C. Collins et al, *Factorisation of Hard Processes in QCD*. Adv. Ser. Direct. High Energy Phys., 5:1–91 (1989)

[2] L. Gladilin, *Fragmentation fractions of c and b quarks into charmed hadrons at LEP*. Eur. Phys. J. C, 75(1) 19 (2015)

[3] S. Acharya et al (ALICE Collaboration), Λ_c^+ *production in pp and p–Pb collisions at* $\sqrt{s_{NN}}$ *= 5.02 TeV*, Phys. Rev. C 104, 054905 (2021)

[4] A.M. Sirunyan et al. (CMS Collaboration), *Production of* Λ_c^+ *baryons in proton–proton and lead-lead collisions at* $\sqrt{s}$ *= 5.02 TeV*, Phys. Lett. B 803, 135328 (2020)

[5] S. Acharya et al (ALICE Collaboration), *Measurement of beauty and charm production in pp collisions at* $\sqrt{s}$ *= 5.02 TeV via non-prompt and prompt D mesons*, JHEP 05, 220 (2021)

[6] M. Cacciari et all, *Theoretical predictions for charm and bottom production at the LHC*, JHEP 10, 137 (2012)

[7] I. Helenius and H. Paukkunen, *Revisiting the D-meson hadroproduction in general-mass variable flavour number scheme*, JHEP 2018, 196 (2018)

[8] P. Skands et al, *Tuning PYTHIA8.1: the Monash 2013 Tune*, Eur. Phys. J. C 74, 8, 3024 (2014)

[9] J.R. Christiansen and P. Skands, *String formation beyond leading colour*, JHEP 08, 003 (2015)

[10] J. Bellm et al, *Spacetime colour reconnection in Herwig 7*, Eur. Phys. J. C 79, 12, 1003 (2019)

[11] M. He and R. Rapp, *Charm-Baryon Production in Proton-Proton Collisions*, Phys. Lett. B 795, 117–121 (2019)

[12] V. Minissale et al, *Charm Hadrons in pp collisions at LHC energy within a Coalescence plus Fragmentation approach*, Phys. Lett. B 821, 136622 (2021)

[13] J. Song, *New feature of low* p_T *charm quark hadronisation in pp collisions at* $\sqrt{s}$ *= 7 TeV*, Eur. Phys. J. C 78, 4, 344 (2018)

[14] S. Acharya et al (ALICE Collaboration), *Measurement of Prompt* D^0, Λ_c^+ *and* $\Sigma_c^{0,+,++}$ *(2455) production in Proton-Proton Collisions at* $\sqrt{s}$ *= 13 TeV*, Phys. Rev. Lett. 128, 012001 (2022)

[15] S. Acharya et al (ALICE Collaboration), *Measurement of the cross sections of* Ξ_c^0 *and* Ξ_c^+ *baryons and of the branching-fraction ratio* $BR(\Xi_c^0 \to \Xi^- e^+ \nu_e)/BR(\Xi_c^0 \to \Xi^- \pi^+)$ *in pp collisions at* $\sqrt{s}$ *= 13 TeV*, Phys. Rev. Lett. 127, 272001 (2021)

[16] S. Acharya et al (ALICE Collaboration), *First measurement of* Ω_c^0 *production in pp collisions at* $\sqrt{s}$ *= 13 TeV*, arXiv:2205.13993 (2022)

[17] Y.K. Hsiao, L. Yang, C.C. Lih et al., *Charmed* Ω_c^0 *weak decays into* Ω *in the light-front quark model*. Eur. Phys. J. C 80, 1066 (2020)

[18] M. Niiyama et al (BELLE Collaboration), *Production cross sections of hyperons and charmed baryond from* e^+e^- *annihilation near* $\sqrt{s}$ *= 10.52 GeV*, Phys. Rev. D 97, 072005 (2018)

[19] S. Acharya et al (ALICE Collaboration), *Constraining hadronisation mechanisms with* Λ_c^+/D^0 *production ratios in Pb–Pb collisions at* $\sqrt{s_{NN}}$ *= 5.02 TeV*, arXiv:2112.08156 (2021)

[20] M. He and R. Rapp, *Hadronisation and Charm-Hadron Ratios in Heavy-Ion Collisions*, Phys. Rev. Lett. 124, 042301 (2020)

First bent wafer-scale sensor in truly-cylindrical geometry for the ALICE ITS3 detector

Arianna G. Torres Ramos[1,*], on behalf of the ALICE Collaboration

[1]Università e INFN, Bari, Italy

Abstract. An upgrade of the three innermost layers of the ALICE ITS, intended to be installed during the LHC Long Shutdown 3, is under development. The detector concept foresees the usage of curved, wafer-scale Monolithic Active Pixel Sensors. Extensive characterization studies of bent single ALPIDE chips (used for the current ITS), have been carried out to evaluate their performance under the mechanical stress involved in the bending process. These tests on small sensors have opened the way to the investigation of a large scale sensor: a full size demonstrator of a half-layer in a truly cylindrical shape based on so called super-ALPIDE chips. Such activity has required the development of special tools and procedures dedicated to bend and read out the new pixel matrix.

1 Introduction

A pillar of the ALICE upgrade program is the improvement of the Inner Tracking System (ITS2) performance by the replacement of its three innermost layers during the next/third long shutdown of the LHC (LS3) . The proposal is based on a vertex detector consisting of three cylindrical layers composed by curved wafer-scale silicon sensors, with the first layer positioned at 18 mm radial distance from the interaction point [1]. The new detector will present a significant reduction of the material budget, down to only 0.03% of an interaction length, thus improving the spatial resolution of the reconstructed charged tracks.

Currently ALPIDE MAPS sensors, the same as used in the ITS2, are being used for preliminary bending studies. The primary goal is testing the chips performance under the mechanical strain involved in the bending process. Since silicon is a piezoresistive material, when a mechanical stress is applied on the crystal, the internal energy structure changes, causing a change in electron mobility and, in turn, electrical resistance [2]. Moreover it has to be verified that the electronic circuitry on the chips preserve its functionalities after the mechanical stress of the bending.

Under these premises, extensive test campaigns, devoted to the electrical characterization of bent single ALPIDE chips and the evaluation of their performance in a test-beam environment, have been carried out.

*e-mail: arianna.grisel.torres.ramos@cern.ch

2 Electrical characterization in laboratory tests

The setup for the electrical tests was composed by a single ALPIDE chip wire bonded to a flexible printed circuit (FPC), this component is extended by coupling it to an I-board, used both as noise filter and to release mechanical stress from the setup. The last is connected through a cable to an adapter board (Flex2daq) for the communication between the chip and the DAQ board (Fig.1). The whole setup was placed inside a black box in order to isolate it from any light source.

Fig. 1. Setup assembled for the electrical tests with flat ALPIDE.

The tests were performed with the ALPIDE both in flat and bent positions. For the bending procedure a cylindrical jig made of Aluminum was designed and realized to assure a good control of the shape and it is attached to the platform holding the setup (Fig. 2 left). The radius of the jig is r=18 mm, this corresponds to the intended radius for the innermost layer of the ITS3. The chip is positioned on top of this tool, such that the bent axis is parallel to its long side and is bent carefully forcing it to follow the curved shape by using a piece of kapton foil. Starting at one of the short sides and then moving on towards the other one in a rolling fashion, the chip bends gradually and is fixed in the final position by screwing the kapton to the jig at both extremes (Fig. 2 right).

Fig. 2. (left) Setup assembled for the electrical tests with the ALPIDE bent over a cylindrical jig. (right) Closed view if the ALPIDE bent along its long side over the cylindrical jig and held in position with a kapton foil screwed at the edges.

The ALPIDE chip was characterized mainly in terms of threshold and fake-hit rate.

The charge threshold is the key parameter leading the chip performance. It is defined for every pixel by three front-end parameters: ITHR, VCASN and IDB. The effective value is increased by increasing ITHR or IDB and decreased by augmenting VCASN. For a given set of these parameters, pixel charge threshold can be determined using the analogue pulsing. The measurement consists in injecting N_{inj} times a range of test charges Q_{inj} via pulsing capacitance C_{inj} such that

$$Q_{inj} = C_{inj}\,(VPLSE\_HIGH - VPLSE\_LOW) \qquad (1)$$

VPLSE_HIGH and VPLSE_LOW maximum value is 1.8 V and are set via on-board 8-bit DACs. For each test charge Q_{iinj} , out of N_{inj} injections, N^{i}_{hit} pass the threshold i.e. result in pixel hits. The plot of N^{i}_{hit} / N_{inj} as a function of Q^{i}_{inj} is often referred to as s-curve measurement as its shape resembles the letter S. The threshold is defined as charge at which pixel fires in 50% of cases (injections).

The fake-hit rate of a sensor is defined as the number of hits per pixel per event in absence of any external stimulus. It is measured by sending a number of triggers (N_{trg}) to the chip without providing any charge injection, and recording the number of hits N_{hits}

$$FHR = N_{hits} \,/\, (N_{pix} \cdot N_{trg}) \qquad (2)$$

where N_{pix} is the number of pixels in the sensor.

A similar test was done by bending the chip along its short side [3]. In this procedure, the long edge, hosting the bonding pads and the periphery logic, was glued onto a carrier board. The bonding area remained flat, and the rest of the sensor was left unattached and lightly compressed between two layers of polyimide film, which were tied to two lateral wheels that could be moved along the bending direction (Fig. 3). Following this method the chip was bent up to curvature radius of less than 18mm.

Here, the sensor was also characterized in terms of threshold and fake hit rate before and after bending. The distribution of thresholds over the pixel matrix, showed only negligible variations between both positions.

Fig. 3. ALPIDE sensor bent along the short edge to a radius of 18 mm.

3 Performance of bent ALPIDEs in a test-beam

In a previous study [3], curved ALPIDE MAPS have been characterized in a test beam environment, using 5.4 GeV electrons. In this case, a sensor was bent along the short side as described above and taken as device under test. It was placed between reference flat detector planes in a telescope configuration.

An important component of the analysis was the measurement of the detection efficiency as a function of the beam position and local inclination angle. The inefficiency as a function of threshold for different rows and incident angles can be seen in figure 4.

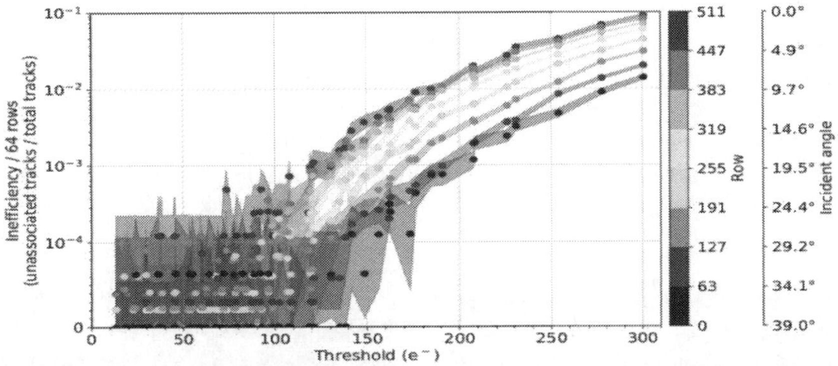

Fig. 4. Inefficiency as a function of threshold for different rows and incident angles. The dark circles represent the calculated efficiency and the shaded areas the statistical uncertainty. Each data point corresponds to at least 8k tracks.

For thresholds above 100 electrons, which is the nominal operating point of ALPIDE, the efficiency increases with increasing beam incident angle, this is, decreasing row number. Below 100 electrons, the inefficiency is generally lower than 10^{-4}, showing that the sensors maintain their excellent performance after bending to radii of about 2cm, with detection efficiencies above 99.9% at typical operating conditions.

More recently, during a beam test campaign realized at CERN SPS facility in July 2021 a beam telescope featuring six flat ALPIDE chips as reference detectors was used to measure the performance of other six ALPIDE chips, bent at the foreseen ITS3 layer radii, composing the so called micro-ITS3. A picture of the resulting DUTs, forming the μ-ITS3 and a top view of the DUTs mounted within the ALPIDE beam telescope can be seen in figure 5.

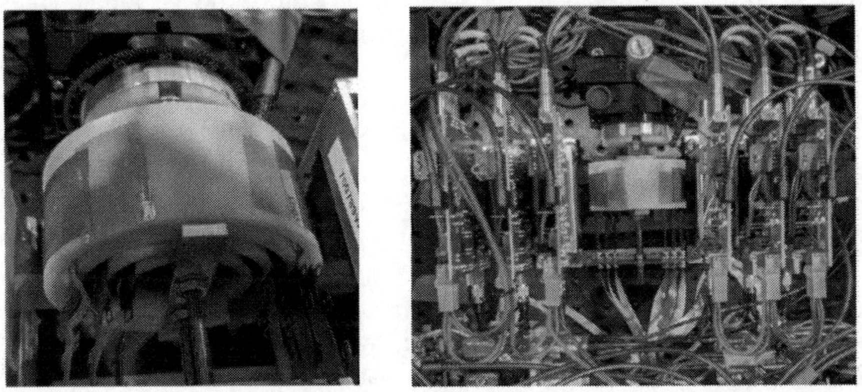

Fig. 5. (left) Image of the six DUTs used in the test-beam campaign, bent over three cylindrical jigs at different radii, forming the μ-ITS3. (right) Top view of the bent DUTs mounted within the ALPIDE beam telescope.

4 Beyond the single-ALPIDE

An important part of the R&D activities dedicated to the detector performance upgrade of the Inner Tracking System is focused on the prototyping and testing of bending techniques, the mechanics, and the integration into the detector.

A full size (1:1 scale) demonstrator of a half-layer in a truly cylindrical shape has been produced, based on so called super-ALPIDE chips. Wafer-scale bending tests have also been carried out on dummy silicon wafers. A bending technique was developed and bending down to a radius of 18 mm can be carried out in a reliable and repeatable way. For this purpose, a dedicated tool was designed (Fig. 6). In this case a tensioned mylar foil is attached by one extreme to a cylindrical mandrel of the desired radius. The silicon wafer is placed between both objects, then the mandrel is rotated, allowing the mylar to wrap it down and to bend the wafer around it. The mechanical components for the bending and electrical assembly of this demonstrator, as shown in figure 6, are being completed in 2022. Once assembled, the detectors will be characterized in the laboratory and eventually on a test-beam.

Fig. 6. Mechanical tool developed for bending the super-ALPIDE (left) and main components designed for the electrical assembly (right).

Forward activities will be dedicated to the testing and characterization of the newest chip designs, specifically the test chips contained in the MLR1, which stands for Multi Layer Reticle 1. This is the first submission of a chip fabricated with the TowerJazz 65nm technology, featuring many devices for technology exploration, including transistor test structures for radiation hardness, various diode matrices for charge collection studies, and both digital and analogue test structures.

5 Conclusions

Looking towards the realization of the new ITS3, which envisages a detector consisting of three truly-cylindrical layers, composed only by wafer-scale silicon sensors, several test with bent ALPIDE chips have been carried out, to evaluate the consequences of bending. These studies have included the electrical characterization of a bent single ALPIDE chip, through the measurement of the main parameters steering the chip functioning. On the other hand, the evaluation of in-beam performance has been realized in different test-beams, including the one carried out at CERN SPS with a telescope of six flat ALPIDE chips as reference detectors used to measure the efficiency of other six ALPIDE chips, bent at the foreseen ITS3 layer radii, composing the so called micro- ITS3.

The results obtained up to now, exhibit the expected behavior of the bent chips with respect to that when they are flat, demonstrating that the functionalities of the ALPIDE are not affected by the curvature effect.

References

1. ALICE Collaboration, *Letter of Intent for an ALICE ITS Upgrade in LS3*, Tech. Rep. CERN- LHCC-2019-018. LHCC-I-034, CERN, Geneva, (Dec. 2019)

2. *Y.* Kanda, *Piezoresistance effect on silicon*, Sensors and Actuators A: Physical 28 (July 1991) 83-91
3. ALICE ITS project, *First demonstration of in-beam performance of bent Monolithic Active Pixel Sensors*, Nuclear Inst. and Methods in Phys. Research, A **1028** (2022)

Electromagnetism in quark matter at intermediate densities

Efrain J. Ferrer[1]

[1]Department of Physics and Astronomy, University of Texas Rio Grande Valley, Texas, USA

Abstract. Several anomalous electromagnetic effects that can take place in quark matter at low temperatures and intermediate densities will be discussed. The anomalous transport properties of the spatially inhomogeneous phase of quark matter known as the Magnetic Dual Chiral Density Wave (MDCDW) phase will be reviewed. Going beyond mean-field approximation, it will be shown how linearly polarized electromagnetic waves that penetrate the MD-CDW medium mix with the phonon fluctuations to give rise to two hybridized modes of propagation called axion polaritons. Finally, some possible implications of these results for the astrophysics of neutron stars will be indicated.

1 Introduction

The QCD phase diagram has two regions that are very well understood: The region of extremely high temperature and the one with a very high density. Its feasibility is due to the weakening of the strong coupling by the phenomenon of asymptotic freedom. They are respectively described by the quark-gluon plasma (QGP) phase at high temperature and low density and by the color superconducting color-flavor locked (CFL) phase at asymptotically large density and low-temperature [1]. More challenging, nonetheless, is to determine the phases in the intermediate density-temperature regions, where lattice QCD is not applicable due to the sign problem, so one has to rely on nonperturbative methods and effective theories. Nevertheless, it is believed that precisely one of the denser objects that we have in nature, neutron stars (NS), should have densities which are not extremely high but in that intermediate region.

It has long been considered, on the other hand, that the region of intermediate density and relatively low temperature may feature inhomogeneous phases, many of which have spatially inhomogeneous chiral condensates favored over the homogeneous ones. Such phases have been found in the large-N limit of QCD [2, 3], in NJL models [4]-[7], and in quarkyonic matter [8, 9]. In all the cases, chiral condensates with single-modulation are energetically favored over higher-dimensional modulations. However, single-modulated phases in three spatial dimensions are known to be unstable against thermal fluctuations at any finite temperature, a phenomenon known in the literature as Landau-Peierls instability [10]. In dense QCD models, the Landau-Peierls instability occurs in the periodic real kink crystal phase [11]; in the Dual Chiral Density Wave (DCDW) phase [12], and in the quarkyonic phase [13].

Another factor that plays an important role in the physics of these inhomogeneous phases and which also has relevance in the astrophysics of compact objects, as well as in heavy-ion collisions (HIC), is the presence of a strong magnetic field. In off-central HIC, quark-matter degrees of freedom become relevant and are known to produce large magnetic fields ($eB \simeq$

10^{18} G at RHIC, $eB \simeq 10^{19}$ G, at the LHC [14, 15].). Likewise, NS can have strong inner magnetic fields. Estimates based on the scalar virial theorem give inner fields for magnetars of order 10^{18} G for nuclear matter [16] and up to 10^{20} G for quark matter [17]. Even inner fields, one to three orders of magnitude smaller, would be significant and should not be ignored in NS studies [18]. The magnetic field can noticeably enhance the window for inhomogeneous phases [9, 19, 20]; and activate attractive channels producing new condensates, as it occurs with chiral condensate [19], color superconductivity [20] and quarkyonic matter [9].

In the context of the DCDW phase, the presence of an external magnetic field favors the formation of a spatially inhomogeneous condensate with a modulation along the field direction [21]. This phase has a new symmetry pattern determined by the breaking of the rotational and isospin symmetries. This new magnetized phase is called the magnetic dual chiral density wave (MDCDW) phase [22, 23]. Another important result is that the presence of a magnetic field removes the Landau-Peierls instability in this phase [24]. Moreover, the constant external magnetic field produces Landau momentum quantization, giving rise to an asymmetric quark energy spectrum in the lowest Landau level (LLL). This asymmetry, together with the activation of the chiral anomaly in the presence of an electric field with a component along the magnetic field direction give the system a non-trivial topology. The topology of the MDCDW phase manifests in the effective electromagnetic action by the presence of a dynamical axion field coupled to the electromagnetic field. This coupling in turn leads to several topological effects [22, 23] that we will discuss later.

It is timely to comment that in addition to the properties of the MDCDW phase that were already pointed out, this phase exhibits other important characteristics that make it more attractive as a possible candidate for the inner matter phase of NS. First, it has been shown in [25] that the temperature needed to evaporate the inhomogeneous condensate for fields $\sim 10^{18}$ G is higher than the characteristic stellar temperatures for the whole range of densities characteristic of NS; second, it was proved in [26] that the maximum stellar mass of a hybrid star with a quark-matter core in the MDCDW phase satisfies the maximum mass observation constraints [27, 28]; and finally, in [29] it was shown that if the NS core is formed by quarks in the MDCDW phase, the heat capacity will be well above the lower limit expected for NS ($C_V \gtrsim 10^{36}(T/10^8)$ erg/K) [30]. This limit was established by long-term observations of NS temperatures in the range from months to years after accretion outburst together with continued observations on timescales of years. On the other hand, this lower-limit value put out of the game the matter components that exhibit superfluidity and/or superconductivity of any kind, since as showed in [29] all these cases are exponentially damped. This will lead to the striking conclusion that, if the only quark matter state to be realized in NS interior is the color superconducting phases, quark will be ruled out from forming part of the NS core [30].

The rest of the paper is organized in two sections. In Section 2, it is expose how the presence of the chiral anomaly in the MDCDW phase produces a peculiar electromagnetism with field equations know as axion electrodynamics, which exhibit magnetoelectricity. In Section 3, by going beyond mean-field approximation, we show that the photon-phonon interaction in this medium produces a couple of hybridized modes, one of those being gapped, and which are known as axion polaritons (AP). Possible astrophysical applications are indicated.

2 Magnetoelectricity in the MDCDW phase

The action of the MDCDW phase is [22, 23]

$$
\begin{aligned}
S_{eff} &= \int d^4x\{\bar{\psi}[i\gamma^{\mu}(\partial_{\mu} + iQA_{\mu} + i\tau_3\gamma_5\partial_{\mu}\theta) + \gamma_0\mu - m]\psi - \frac{m^2}{4G} \\
&+ \frac{\kappa}{4}\theta(x)F_{\mu\nu}\tilde{F}^{\mu\nu} - \frac{1}{4}F_{\mu\nu}F^{\mu\nu}\}, \qquad (1)
\end{aligned}
$$

Here m is the condensate magnitude, b its modulation and $\theta(x)$ is the axion field whose coupling with the electromagnetic tensor is given by $\frac{\kappa}{4} = \frac{3(e_u^2-e_d^2)}{8\pi^2} = \frac{e^2}{8\pi^2} = \frac{\alpha}{2\pi}$. In (1), $Q = \text{diag}(e_u, e_d) = \text{diag}(\frac{2}{3}e, -\frac{1}{3}e)$ denotes the electric-charge matrix in flavor.

The electromagnetic effective action $\Gamma(A)$ in the MDCDW phase is obtained as usual from

$$\Gamma = -i \log Z, \tag{2}$$

where the partition function Z is

$$Z = e^{i\Gamma} = \int \mathcal{D}\bar{\psi}(x)\mathcal{D}\psi(x)e^{iS_{eff}} \tag{3}$$

Integrating in the fermion fields, performing the Matsubara sum, taking the zero-temperature limit, and expanding Γ in powers of the fluctuation field A_μ we obtain

$$\begin{aligned}\Gamma(A) &= -V\Omega + \int d^4x\left[-\frac{1}{4}F_{\mu\nu}F^{\mu\nu} - \kappa\int d^4x\epsilon^{\mu\alpha\nu\beta}A_\alpha\partial_\nu A_\beta\partial_\mu\theta\right] \\ &- \int d^4xA_\mu(x)J^\mu(x),\end{aligned} \tag{4}$$

where we integrated by parts the axion term and cut the series in power of A_μ in the linear term. The four-current $J^\mu(x) = (J^0, \mathbf{J})$ represents the contribution of the ordinary (non-anomalous) electric four-current, obtained from the one-loop tadpole diagrams [22, 23]. In (4), V is the system volume and Ω the thermodynamic potential which does not depend on A_μ.

From (4) we can obtain the source depending equations of axion electrodynamics, which in terms of the D and H fields, are given by

$$\nabla\cdot\mathbf{D} = J^0, \quad \nabla\times\mathbf{H} - \frac{\partial\mathbf{D}}{\partial t} = \mathbf{J} \tag{5}$$

where

$$\mathbf{D} = \mathbf{E} - \kappa\theta\mathbf{B}, \quad \mathbf{H} = \mathbf{B} + \kappa\theta\mathbf{E} \tag{6}$$

Equations (6) show that a magnetic field induces an electric polarization $\mathbf{P} = -\kappa\theta\mathbf{B}$ and an electric field induces a magnetization $\mathbf{M} = -\kappa\theta\mathbf{E}$, a phenomenon known as magnetoelectricity. The linear magnetoelectricity of the MDCDW medium is a direct consequence of the chiral anomaly. It reflects the fact that the ground state of the MDCDW medium breaks parity and time inversion symmetries. The magnetoelectricity in the MDCDW phase is different from the one found in the magnetic-CFL phase of color superconductivity, where parity was not broken and the effect was a consequence of an anisotropic electric susceptibility [31], thus it was not linear. It also follows from Eq. (5) that there is an anomalous Hall current $\mathbf{J}_{anom} = -\frac{e^2}{4\pi^2}\mathbf{q}\times\mathbf{E}$, which is a medium-induced magnetic current density $\nabla\times\mathbf{M}$. This medium-induced magnetic current density is produced by the space-dependent anomalous magnetization coming from the axion term and it indicates that if there exists an electric field perpendicular to the magnetic field it will produce a dissipationless current in the medium. Such a current could act as a propellant of the large magnetic fields existing in magnetars in case that their core are formed by quark matter in the MDCDW phase.

3 Axion-Polariton modes in the MDCDW phase

Considering a small phonon fluctuation $\theta(x)$ on the order parameter, which is associated with the breaking of the translational symmetry by the single-modulated inhomogeneous condensate, and expanding it about the condensate solution up to quadratic order in the fluctuation, it can be obtained the fluctuation low-energy Lagrangian density as [24, 32]

$$\mathcal{L}_\theta = \frac{1}{2}[(\partial_0\theta)^2 - v_z^2(\partial_z\theta)^2 - v_\perp^2(\partial_\perp\theta)^2], \tag{7}$$

with coefficients given by

$$v_z^2 = a_{4.2} + \bar{m}^2 a_{6.2} + 6\bar{q}^2 a_{6.4} + 3\bar{q} b_{5.3} \tag{8}$$

$$v_\perp^2 = a_{4.2} + \bar{m}^2 a_{6.2} + 2\bar{q}^2 a_{6.4} + \bar{q} b_{5.3} - a_{4.2}^{(1)} - \bar{m}^2 a_{6.2}^{(1)} \tag{9}$$

These coefficients represent the squares of the parallel and transverse group velocities respectively.

Let's consider now the propagation of electromagnetic waves in the MDCDW phase by going beyond the mean-field approximation to study the effects of the phonon-photon interactions in the MDCDW medium.

When an electromagnetic wave penetrates the MDCDW medium, the low-energy theory of the fluctuations acquires the following additional contributions

$$\mathcal{L}_{A-\theta} = -\frac{1}{4}F_{\mu\nu}F^{\mu\nu} + J^\mu A_\mu + \frac{\kappa}{8}\theta_0(x)F_{\mu\nu}\tilde{F}^{\mu\nu} + \frac{\kappa}{8}\theta(x)F_{\mu\nu}\tilde{F}^{\mu\nu}, \tag{10}$$

The first two terms are the conventional Maxwell and ordinary 4-current contributions respectively, the latter obtained after integrating out the fermions in the original MDCDW effective action [22, 23], as it was explained above. The last two terms are the axial anomaly with background axion field $\theta_0(x) = mqz$ and its (phonon-induced) fluctuation $\theta(x)$. Here $\kappa = 2\alpha/\pi m$.

The combined Lagrangian $\mathcal{L} = \mathcal{L}_\theta + \mathcal{L}_{A-\theta}$ effectively describes the low-energy theory of an axion field $\theta(x)$ interacting nonlinearly with the photon via the chiral anomaly. Let us now assume that a linearly polarized electromagnetic wave, with its electric field **E** parallel to the background magnetic field $\mathbf{B}_0$, propagates in the MDCDW medium [33]. From (7) and (10) we obtain the field equations

$$\nabla \cdot \mathbf{E} = J^0 + \frac{\kappa}{2}\nabla\theta_0 \cdot \mathbf{B} + \frac{\kappa}{2}\nabla\theta \cdot \mathbf{B}, \tag{11}$$

$$\nabla \times \mathbf{B} - \partial\mathbf{E}/\partial t = \mathbf{J} - \frac{\kappa}{2}(\frac{\partial\theta}{\partial t}\mathbf{B} + \nabla\theta \times \mathbf{E}), \tag{12}$$

$$\nabla \cdot \mathbf{B} = 0, \quad \nabla \times \mathbf{E} + \partial\mathbf{B}/\partial t = 0 \tag{13}$$

$$\partial_0^2\theta - v_z^2\partial_z^2\theta - v_\perp^2\partial_\perp^2\theta + \frac{\kappa}{2}\mathbf{B}\cdot\mathbf{E} = 0, \tag{14}$$

which contains terms where the axion field and the photon are coupled. In (11)-(14), **B** is the total magnetic field, meaning the background field plus the wave magnetic field.

Since we are interested in applications to NS, we should consider a neutral medium, hence we assume that J^0 contains an electron background charge that ensures overall neutrality

$$J^0 + \frac{\kappa}{2}\nabla\theta_0 \cdot \mathbf{B} + \frac{\kappa}{2}\nabla\theta \cdot \mathbf{B} = 0. \tag{15}$$

Hence, from (11)-(14) and (15), the linearized field equations can be given as

$$\partial^2\mathbf{E}/\partial t^2 = \nabla^2\mathbf{E} + \frac{\kappa}{2}(\partial^2\theta/\partial t^2)\mathbf{B}_0 \tag{16}$$

$$\partial^2\theta/\partial t^2 - v_z^2(\partial^2\theta/\partial z^2) - v_\perp^2(\partial^2\theta/\partial x^2 + \partial^2\theta/\partial y^2) + \frac{\kappa}{2}\mathbf{B}_0\cdot\mathbf{E} = 0. \tag{17}$$

The solutions of (16)-(17) describe two hybridized propagating modes of coupled axion and photon fields that we call AP [33], borrowing the term from condensed matter. In general, polaritons are hybridized propagating modes that emerge when a collective mode like phonons, magnons, etc., couples linearly with light.

The energy spectrum of the hybrid modes are

$$\omega_0^2 = A - B, \tag{18}$$

$$\omega_m^2 = A + B \tag{19}$$

with

$$A = \frac{1}{2}[p^2 + q^2 + (\frac{\kappa}{2}B_0)^2], \tag{20}$$

$$B = \frac{1}{2}\sqrt{[p^2 + q^2 + (\frac{\kappa}{2}B_0)^2]^2 - 4p^2q^2}, \tag{21}$$

and

$$q^2 = v_z^2 p_z^2 + v_\perp^2 p_\perp^2 \tag{22}$$

where we used the notation $p_\perp^2 = p_1^2 + p_2^2$.

From (18)-(21) we identify ω_0 as the gapless mode and ω_m as the gapped mode with field-dependent gap

$$\omega_m(\vec{p} \to 0) = m_{AP} = \alpha B_0/\pi m, \tag{23}$$

which is proportional to the external applied magnetic field and inversely proportional to the inhomogeneous condensate amplitude.

Similarly coupled modes of axion and photon have been found in topological magnetic insulators [34], underlining once again the striking similarities between MDCDW quark matter and topological materials in condensed matter.

This result implies that if a linearized electromagnetic wave can reach the MDCDW medium, it will split into the two AP modes, which are the real eigenmodes of this medium. Hence, since one of the modes is massive, this can serve as a mechanism to transfer photon energy into rest mass.

This mechanism can be of interest for astrophysics, since if the interior of magnetars host quarks in the MDCDW phase, γ-photons that penetrate and reach the quark medium with sufficient energy can be converted into gapped AP's that will increase the star mass. We should call attention to the fact that although the AP mass for fields of order $\sim 10^{17}$ G is not too large (i. e. ~ 0.5 MeV), extragalactic sources of gamma ray bursts (GRB) show an isotropic distribution over the whole sky flashing with a rate of 1000/year. The energy output of these events is $\sim 10^{56} - 10^{59}$ MeV, with photon energies of order $0.1 - 1$ MeV [35]. Hence, each one of these events can produce at least $10^{56} - 10^{59}$ γ-photons that could be converted into AP's, with half of them being gapped. If we assume that about only 10% of these photons reach the star, which is a conservative estimate if the star is in the narrow cone of a GRB beam, then at least about $10^{55} - 10^{58}$ of those photons can reach the star per each GRB event. This scenario can serve to give an alternative explanation [32] to the astronomical puzzle called the missing pulsar problem, which refers to the failed expectation to observe a large number of pulsars within the distance of 10 pc of the galactic center. Theoretical predictions have indicated that there should be more than 10^3 active radio pulsars in that region [36], but these numbers have not been observed.

Acknowledgments: This work was supported in part by NSF grant PHY-2013222.

References

[1] M. Buballa, Phys. Rep. **407**, 205 (2005). K. Fukushima and T. Hatsuda, Rept. Prog. Phys. **74**, 014001 (2011); M. G. Alford, A. Schmitt, K. Rajagopal and T. Schafer, Rev. Mod. Phys. **80**, 1455 (2008)

[2] D. V. Deryagin, D. Y. Grigoriev and V. A. Rubakov, Int. J. Mod. Phys. A **7**, 659 (1992); E. Shuster and D. T. Son, Nucl. Phys. B **573**, 434 (2000); B.-Y. Park, M. Rho, A. Wirzba and I. Zahed, Phys. Rev. D **62**, 034015 (2000)

[3] R. Rapp, E. Shuryak, and I. Zahed, Phys. Rev. D **63**, 034008 (2001); N. V. Gubina, K. G. Klimenko, S.G. Kurbanov, and V.Ch. Zhukovsky, Phys. Rev. D **86**, 085011 (2012)

[4] E. Nakano and T. Tatsumi, Phys. Rev. D **71**, 114006 (2005)

[5] D. Nickel, Phys. Rev. D **80**, 074025 (2009); Phys. Rev. Lett. **103**, 072301 (2009)

[6] S. Carignano, D. Nickel, and M. Buballa, Phys. Rev. D **82**, 054009 (2010)

[7] H. Abuki, D. Ishibashi, and K. Suzuki, Phys. Rev. D **85**, 074002 (2012)

[8] T. Kojo, Y. Hidaka, L. McLerran and R. D. Pisarski, Nucl. Phys. A **843**, 37 (2010); T. Kojo, Y. Hidaka, K. Fukushima, L. McLerran and R. D. Pisarski, Nucl. Phys. A **875**, 94 (2012); T. Kojo, R. D. Pisarski and A. M. Tsvelik, Phys. Rev. D **82**, 074015 (2010); T. Kojo, Nucl. Phys. A **877**, 70 (2012)

[9] E. J. Ferrer, V. de la Incera and A. Sanchez, Acta Phys. Polon. Supp. **5**, 679 (2012)

[10] R. Peierls, Helv. Phys. Acta **7**, 81 (1934); Annales de l'Institut Henri Poincaré **5**, 177 (1935); L. D. Landau, Phys. Z. Sowjet Union **11**, 26 (1937); Zh. Eksp. Teor Fiz. **7**, 627 (1937)

[11] Y. Hidaka, K. Kamikado, T. Kanazawa, and T. Noumi, Phys. Rev. D **92**, 034003 (2015)

[12] T-G. Lee, E. Nakano, Y. Tsue, T. Tatsumi and B. Friman, Phys. Rev. D **92**, 034024 (2015)

[13] R. D. Pisarski, V. V. Skokov, and A. M. Tsvelik, Phys. Rev. D **99**, 074025 (2019)

[14] D. E. Kharzeev, L. D. McLerran and H. J. Warringa, Nucl. Phys. A **803**, 227 (2008)

[15] V. Skokov, A.Yu. Illarionov and V. D. Toneev, Int. J. Mod. Phys. A **24**, 5925 (2009)

[16] D. Lai and S. L. Shapiro, ApJ. **383**, 745 (1991)

[17] E. J. Ferrer, V. de la Incera, J. P. Keith, I. Portillo and P. L. Springsteen, Phys. Rev. C **82**, 065802 (2010)

[18] E. J. Ferrer and A. Hackebill, Phys. Rev. C **99**, 065803 (2019); Universe **5**, 104 (2019)

[19] E. J. Ferrer, V. de la Incera, I. Portillo and M. Quiroz, Phys. Rev. D **89**, 085034 (2014); E. J. Ferrer and V. de la Incera, Phys. Rev. Lett. **102**, 050402 (2009); Nucl. Phys. B **824**, 217 (2010)

[20] B. Feng, E. J. Ferrer and V. de la Incera, Nucl. Phys. B **853**, 213 (2011); E. J. Ferrer and V. de la Incera, *Magnetism in Dense Quark Matter* (Lect. Notes in Phys. vol. 871) ed. by D. Kharzeev, K. Landsteiner, A. Schmitt and H.-U.Yee (Springer Berlin Heidelberg) 2013, pp. 399-432

[21] I. E. Frolov, V. Ch. Zhukovsky and K. G. Klimenko, Phys. Rev. D **82**, 076002 (2010)

[22] E. J. Ferrer and V. de la Incera, Phys. Lett. B **769**, 208 (2017)

[23] E. J. Ferrer and V. de la Incera, Nucl. Phys. B **931**, 192 (2018)

[24] E. J. Ferrer and V. de la Incera, Phys. Rev. D **102** , 014010 (2020)

[25] W. Gyory and V. de la Incera, Phys. Rev. D **106** , 016011 (2022)

[26] S. Carignano, E. J. Ferrer, V. de la Incera, and L. Paulucci, Phys. Rev. D **92**, 105018 (2015)

[27] P. Demorest et al., Nature **467**, (2010) 1081; Z. Arzoumanian, et al. Astrophys. J. Suppl. **235**, 37 (2018)

[28] J. Antoniadis et al., Science **340**, 6131 (2013)
[29] E. J. Ferrer, V. de la Incera and P. Sanson. Phys. Rev. D **103**, 123013. (2021)
[30] A. Cumming, E. F. Brown, F. J. Fattoyev, C. J. Horowitz, D. Page and S. Reddy, Phys. Rev. C **95**, 025806 (2017)
[31] B. Feng, E. J. Ferrer and V. de la Incera, Phys. Lett. B **706**, 232 (2011)
[32] E. J. Ferrer and V. de la Incera, Universe **7**, 458 (2021)
[33] E. J. Ferrer and V. de la Incera, V. Axion-Polariton in dense quark matter: a solution to the missing pulsar problem. arXiv: 010.02314 [hep-ph]
[34] R. Li, J. Wang, X-L Qi and S-C Zhang, Nature **6**, 284 (2010)
[35] T. Piran, Phys. Rep. **314**, 575 (1999); I. Bombaci and B. Datta, Astropys. J. **530**, L69 (2000); P. Meszaros, Rept. Prog. Phys. **69**, 2259 (2006); S. Razzaque, Journal of Physics: Conference Series **60**, 111 (2007)
[36] E. Pfahl and A. Loeb, Astropys. J. **615**, 253 (2004); J. -P. Macquart, N. Kanekar, D. Frail and S.Ransom, Astropys. J. **715**, 939 (2010); J. Dexter and R. M. O'Leary, Astropys. J. **783**, L7 (2014)

Chiral restoration of strange baryons

Eduardo S. Fraga[1], *Rodrigo* da Mata[1], *Savvas* Pitsinigkos[2,*], and *Andreas* Schmitt[2]

[1]Instituto de Física, Universidade Federal do Rio de Janeiro, Caixa Postal 68528, 21941-972, Rio de Janeiro, RJ, Brazil
[2]Mathematical Sciences and STAG Research Centre, University of Southampton, Southampton SO17 1BJ, UK

Abstract. We review the results of a phenomenological model for cold and dense nuclear matter exhibiting a chiral phase transition. The idea is to model the quark-hadron phase transition under neutron star conditions within a single model, but without adding quark degrees of freedom by hand. To this end, strangeness is included in the form of hyperonic degrees of freedom, whose light counterparts provide the strangeness in the chirally restored phase. In the future, the model can be used for instance to compute the surface tension at the (first-order) chiral phase transition and to study the possible existence of inhomogeneous phases.

1 Introduction and main results

At high temperatures or high baryon densities, Quantum Chromodynamics (QCD) is expected to form a quark-gluon plasma, while the low-temperature and low-density regime is inhabited by hadrons due to confinement. The intermediate transition regions pose a great theoretical challenge due to the difficulty of first-principle calculations, which are currently limited to the regime of very small baryon densities. However, it is this intermediate region that holds the details of quark confinement and the quark-hadron phase transition, along with a microscopic description for the physics of compact stars.

We explore this interesting region within a phenomenological approach, focusing on zero temperature and large baryon densities. Hadron and quark phases can be described by individual models, but in order to predict the phase transition and its properties we need more than the sum of the two parts. This is why in this work, rather than stitching together two individual models (which can have other benefits [1–6]), we choose a single model approach (as done, in other variants, before [7–11]). In particular, we include strangeness in the form of hyperonic degrees of freedom, to account for a more realistic "quark matter" phase, extending previous studies, where the only baryonic degrees of freedom in the Lagrangian were neutrons and protons [12–15]. The idea of a unified model for the chiral phase transition is illustrated in Fig. 1.

Our model exhibits chiral symmetry restoration at large baryon densities. This is possible because, by construction, the baryonic mass is entirely generated by the chiral condensate. The chirally restored phase is made of baryonic – confined – degrees of freedom, but, due to their lightness, shares some properties with dense quark matter in QCD. For instance, we

*Speaker

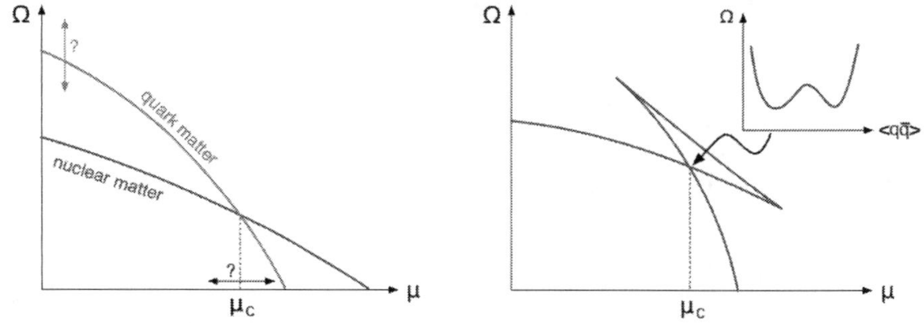

Figure 1. Schematic representation of a two-model (left) and unified (right) approach, showing the quark-hadron transition in the plot of the free energy Ω vs. the chemical potential μ. Using two models, the critical chemical potential varies as the vacuum pressure of either of the models is changed and thus it is a parameter of the approach, not a prediction. Moreover, a unified approach naturally provides an effective potential connecting the stationary points (inset), which is relevant for instance to compute the surface tension. And, in principle, the single-model allows for a smooth crossover if the two phases are not distinguished by an exact symmetry.

find that we can easily force the system to reproduce the asymptotic limit of the speed of sound, although in our model this limit is approached from above. This is different from perturbative QCD, which is the correct description at asymptotically large baryon densities. We also find that our model parameters can be chosen such that asymptotically dense matter has nonzero strangeness. However, we have not found parameters which reproduce flavor symmetry asymptotically, i.e., a strangeness fraction of 1/3, and at the same time respect astrophysical constraints. Since we include a small explicit chiral symmetry breaking term, our model allows for a chiral crossover, but, again, the constraint of producing realistic masses of compact stars excludes the corresponding parameter regime. We also find that realistic parameter sets do not allow for hyperons to appear in the chirally broken phase, i.e., strangeness is only found beyond the chiral phase transition. Finally, we show that putting together all constraints, the model can be used to predict poorly known properties of low-density nuclear matter, such as the slope of the symmetry energy at saturation. These results, and this entire contribution to the proceedings, is based on our previous work [16].

2 Lagrangian and approximations

We begin by writing down the Lagrangian of the model. The baryonic part (containing the full baryon octet, $i = n, p, \Sigma^0, \Sigma^-, \Sigma^+, \Lambda, \Xi^0, \Xi^-$) is

$$\mathcal{L}_B = \sum_i \bar{\psi}_i(i\gamma^\mu\partial_\mu + \gamma^0\mu_i)\psi_i \, , \tag{1}$$

where the chemical potentials of the baryons are denoted by μ_i. Three-flavor QCD allows for three independent chemical potentials, and in our main applications this is reduced to one by the constraints of electric neutrality and electroweak equilibrium. The mesonic part is

$$\begin{aligned} \mathcal{L}_M \;=\;& \frac{1}{2}\partial_\mu\sigma\partial^\mu\sigma - U(\sigma) - \frac{1}{4}\omega_{\mu\nu}\omega^{\mu\nu} - \frac{1}{4}\phi_{\mu\nu}\phi^{\mu\nu} - \frac{1}{4}\rho^0_{\mu\nu}\rho_0^{\mu\nu} + \frac{m_\omega^2}{2}\omega_\mu\omega^\mu + \frac{m_\phi^2}{2}\phi_\mu\phi^\mu \\ & + \frac{m_\rho^2}{2}\rho^0_\mu\rho_0^\mu + \frac{d}{4}(\omega_\mu\omega^\mu + \rho^0_\mu\rho_0^\mu + \phi_\mu\phi^\mu)^2 \, , \end{aligned} \tag{2}$$

where we have only written the mesons that will form a condensate, where $d > 0$ is the quartic self-coupling of the vector mesons, and where the vacuum potential for the chiral condensate is

$$U(\sigma) = \sum_{n=1}^{4} \frac{a_n}{n!} \frac{(\sigma^2 - f_\pi^2)^n}{2^n} - \epsilon(\sigma - f_\pi), \tag{3}$$

with parameters $a_1, a_2, a_3, a_4, \epsilon$, and the pion decay constant f_π. The baryon-meson interactions are given by

$$\mathcal{L}_I = -\sum_i \bar{\psi}_i (g_{i\sigma}\sigma + g_{i\omega}\gamma^\mu\omega_\mu + g_{i\rho}\gamma^\mu\rho_\mu^0 + g_{i\phi}\gamma^\mu\phi_\mu)\psi_i, \tag{4}$$

with Yukawa coupling constants $g_{i\sigma}, g_{i\omega}, g_{i\rho}, g_{i\phi}$. We add non-interacting electrons and muons to the system, characterized by the electron chemical potential μ_e, such that the total Lagrangian is $\mathcal{L} = \mathcal{L}_B + \mathcal{L}_M + \mathcal{L}_I + \mathcal{L}_{\rm leptons}$. Importantly, there are no explicit mass terms for the baryons. Their medium-dependent masses M_i are dynamically generated via interactions with the condensate of the scalar sigma meson $\bar{\sigma}$, which is interpreted as the (non-strange) chiral condensate, $M_i = g_{i\sigma}\bar{\sigma}$, with $\bar{\sigma} \sim \langle u\bar{u} + d\bar{d}\rangle$.

Due to the phenomenological nature of the model it is sensible to explore the parameter space, searching for qualitatively different scenarios within the experimentally given constraints. We fix our parameters in terms of:

- Vacuum masses of baryons and mesons and the vacuum value of the chiral condensate.
- Saturation density $n_0 = 0.153\ \text{fm}^{-3}$ and corresponding binding energy $E_B = -16.3$ MeV of isospin-symmetric nuclear matter.
- Incompressibility $K \simeq (200 - 300)$ MeV, symmetry energy $S \simeq (30.2 - 33.7)$ MeV, and slope of the symmetry energy with respect to the baryon density $L \simeq (40 - 140)$ MeV at saturation. We fix $S = 32$ MeV and $K = 250$ MeV, having checked that variations of neither of these quantities within the allowed range change our results dramatically. The slope parameter is not fixed a priori but turns out to be constrained indirectly by all our other conditions, as we shall discuss later, see Fig. 4.
- Effective nucleon mass at saturation $M_0 \simeq (0.7 - 0.8)\, m_N$. Our results are very sensitive with respect to variations of this quantity, and we go beyond the empirical range to explore qualitatively different behaviors, see also Fig. 4.
- Hyperon potential depths in symmetric nuclear matter at saturation, $U_\Lambda^{(N)} = -30$ MeV, and $U_{\Sigma,\Xi}^{(N)} \simeq -70$ MeV to $+30$ MeV. We use these (poorly known) potentials as a guidance to fix our Yukawa couplings, together with theoretical input from a chiral SU(3) approach, see Ref. [16] for more details.

We restrict ourselves to zero temperature and employ the mean-field and no-sea approximations. We are only interested in thermodynamic equilibrium, and thus our main calculation concerns the free energy density Ω, from which all other relevant quantities can be computed. This is done by solving the stationarity equations for the meson condensates together with the equation of local charge neutrality, all under the condition of beta equilibrium. This results in the condensates and μ_e as (multi-valued) functions of the neutron chemical potential μ_n. Inserting these functions back into the action yields the free energy density $\Omega(\mu_n)$. With standard thermodynamic relations this yields the equation of state, which can be inserted into the Tolman-Oppenheimer-Volkoff equations to compute mass-radius curves of compact stars.

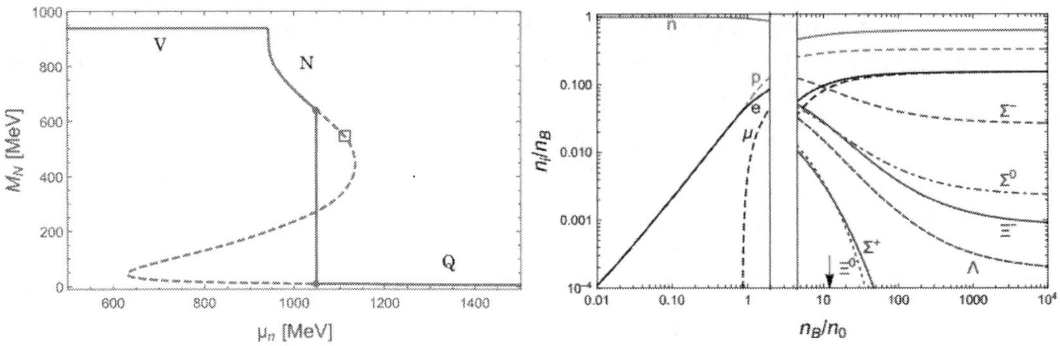

Figure 2. *Left panel:* Effective nucleon mass as a function of the neutron chemical potential. Solid lines represent the stable phases, dashed lines the metastable/unstable phases. The open square marks the appearance of the first hyperonic degree of freedom. Dots mark the chiral phase transition, and the three stable branches are labelled by vacuum (V), nuclear matter (N) and quark matter (Q). *Right panel:* Density fractions for each degree of freedom as functions of the total baryon density (normalized by nuclear saturation density). The gap in densities reflects the first-order transition. The arrow marks the density in the core of the heaviest star for this particular parameter choice, which is for both panels $M_0 = 0.8\, m_N$, $d = 21$, $U^{(N)}_{\Sigma,\Xi} = -50\,\text{MeV}$ and all other parameters as discussed in Sec. 2.

3 Results

The typical behavior of the system in the presence of a first-order phase transition is best explained with the help of Fig. 2. On the left, the effective nucleon mass is plotted as a function of the chemical potential. The shape of the curve is valid for all baryons, as their mass is proportional to the nucleon mass. The plot demonstrates that with increasing chemical potential we go from the vacuum (V) via nuclear matter (N) to a phase with very low nucleon mass, i.e., a phase with approximately restored chiral symmetry, which we interpret as an approximation to quark matter (Q). The first-order nature of the phase transition is manifest in the discontinuity of the effective mass, whose location must be determined from the free energy density. The plot also shows the critical chemical potential for the appearance of the first hyperon (open square). For the particular parameters used here, hyperons only appear in the metastable nuclear branch, not in a stable segment of the chirally broken phase. It is only after the chiral symmetry restoration that they can be found in the system, having a very low mass, and thus giving rise to nonzero strangeness of the quark matter branch. This can also be seen in the right panel, which shows that fermions with strangeness (blue) emerge right after the chiral phase transition. The particle fractions are shown up to densities much larger than present in neutron stars. This allows us to check whether our model exhibits strangeness asymptotically, as expected from QCD. The curves show that the asymptotic strangeness fraction is nonzero, but it turns out to be smaller than 1/3, in contradiction to beta-equilibrated, charge neutral QCD, which asymptotes to weakly interacting quark matter with the same number of up, down and strange quarks. We were able to find parameter sets where this flavor symmetry is recovered, but these do not produce sufficiently massive neutron stars.

In Fig. 3 we plot the speed of sound squared c_s^2 for 4 different parameter sets. In each case we see that $c_s^2 \to 1/3$ asymptotically, which can be traced back to the presence of the quartic meson coupling d [16]. Our model is of course not asymptotically free and thus interactions still play a role at asymptotic densities. Therefore, the speed of sound does not approach its asymptotic value from below, as predicted by perturbative QCD. The most prominent feature of the speed of sound is the large discontinuity from the chiral phase transition in 3 of the

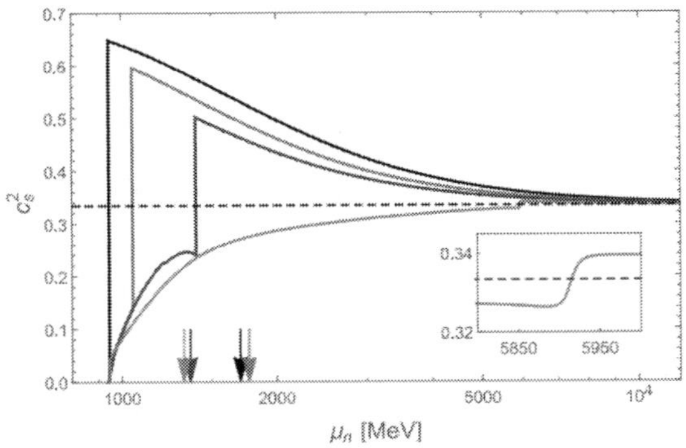

Figure 3. Speed of sound squared as a function of the neutron chemical potential for 4 parameter sets. The arrows indicate the chemical potential in the center of the heaviest star for each case. The dashed line marks the asymptotic limit $c_s^2 = 1/3$. The red curve corresponds to the parameters of Fig. 2, while the others are obtained by varying $M_0/m_N = 0.72$ (black), 0.85 (blue), 0.92 (green).

4 cases, showing a larger speed of sound in the chirally restored phase. Putting together the predictions from perturbative QCD (c_s^2 approaching 1/3 from below) and the constraints of heavy neutron stars (requiring $c_s^2 > 1/3$), it might be more natural to expect a jump *down* as we go up in density. However, there is no first-principle argument why this should be the case. It is conceivable that in a more realistic approach than ours, and in QCD itself, c_s^2 is non-monotonic in the quark matter phase, such that it can approach 1/3 from below even after having jumped *up* at the chiral transition. The plot includes a parameter choice that leads to a chiral crossover. As a remnant of the phase transition, the speed of sound still changes rapidly, yet continuously, as shown in the inset. However, as the smallness of the speed of sound already suggests, this case does not allow for the observed large masses of neutron stars.

To study the parameter space systematically we enforce the following requirements: (*i*) The model should predict compact stars compatible with the current maximum mass limit set by observations, $M_{\max} = 2.1\,M_\odot$ [17]. (*ii*) The asymptotic flavor content of chirally restored matter must include strangeness, as expected in QCD. (*iii*) Nuclear matter must be absolutely stable at zero pressure. It is conceivable that nuclear matter is metastable ("strange quark matter hypothesis" [18, 19]), but in this case quark matter will be thermodynamically preferred at all pressures and there is no quark-hadron transition to study, which is the main motivation of our work.

These 3 criteria can be translated into 3 curves in the M_0 - d plane and, equivalently and more physically, the M_0 - L plane, see Fig. 4. Together, the curves define an "allowed" region, which is shaded in both panels. Interestingly, the allowed region spans a very narrow range in the slope parameter L range, much narrower than constrained by current experiments, given their uncertainties. We see from the right panel that this region is not significantly enhanced by varying the hyperon potentials $U^{(N)}_{\Sigma,\Xi}$, and as a result we conclude that our model predicts $L \simeq (88-92)$ MeV. The left panel confirms that the need for heavy stars requires that the chiral phase transition happens before the hyperon onset, and as a result we can only find hyperons in their chirally restored form. This illustrates – within a single model – the "hyperon puzzle" [20] *and* its solution: Hyperons should appear at some point as the density is increased but at the same time they seem to prevent stars to become massive. Our model shows that indeed they do appear, but if strongly coupled quark matter takes over at the chiral phase transition, the hyperon onset can be relegated to a metastable or even unstable branch, while the stable quark matter core renders the star sufficiently massive. While the quantitative predictions of our phenomenological model should be taken with care, it is remarkable to see the hyperon puzzle and its solution being laid out dynamically within a single model.

Acknowledgments. E.S.F. is partially supported by CAPES (Finance Code 001), CNPq, FAPERJ, and INCT-FNA (Process No. 464898/2014-5). R.M. thanks CNPq for financial support.

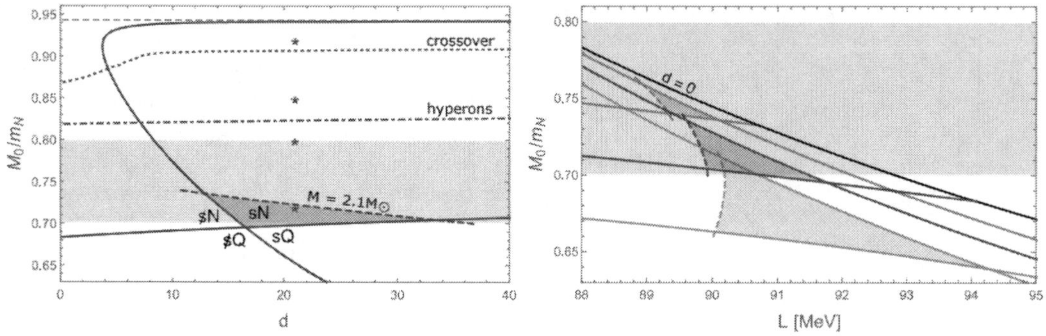

Figure 4. *Left panel*: Allowed region in the M_0-d parameter space for $U_{\Sigma,\Xi}^{(N)} = -50$ MeV. In the 4 areas defined by the solid blue lines, the label ("$\not{s}$") "s" indicates (no) strangeness asymptotically, while N (Q) indicates that nuclear (quark) matter is stable at zero pressure. The blue dashed curve is the 2.1 solar mass limit. Above the dash-dotted line hyperons appear before the chiral phase transition, and above the dotted line the quark-hadron transition is a crossover. The gray shaded area is the empirical range for M_0. The asterisks mark the 4 choices of Fig. 3, while the blue shaded area obeys the three criteria listed in the text. *Right panel*: The blue curves and area correspond to the left panel, now translated into the M_0-L plane. Two additional cases are shown: $U_{\Sigma,\Xi}^{(N)} = -30$ MeV (green) and $U_{\Sigma,\Xi}^{(N)} = -70$ MeV (red). The $d = 0$ line marks the boundary of the parameter space.

References

[1] M.G. Alford, S. Han, M. Prakash, Phys. Rev. D **88**, 083013 (2013)
[2] J.P. Pereira, M. Bejger, N. Andersson, F. Gittins, Astrophys. J. **895**, 28 (2020)
[3] D. Blaschke, A. Ayriyan, D.E. Alvarez-Castillo, H. Grigorian, Universe **6**, 81 (2020)
[4] N. Jokela, M. Järvinen, G. Nijs, J. Remes, Phys. Rev. D **103**, 086004 (2021)
[5] M. Ferreira, R. Câmara Pereira, C. Providência, Phys. Rev. D **103**, 123020 (2021)
[6] L.L. Lopes, C. Biesdorf, D.P. Menezes, Mon. Not. Roy. Astron. Soc. **512**, 5110 (2022)
[7] M. Marczenko, D. Blaschke, K. Redlich, C. Sasaki, Astron. Astrophys. **643**, A82 (2020)
[8] V. Dexheimer, R.O. Gomes, T. Klähn, S. Han, M. Salinas, Phys. Rev. C **103**, 025808 (2021)
[9] K. Bitaghsir Fadafan, F. Kazemian, A. Schmitt, JHEP **03**, 183 (2019)
[10] T. Ishii, M. Järvinen, G. Nijs, JHEP **07**, 003 (2019)
[11] N. Kovensky, A. Schmitt, JHEP **09**, 112 (2020)
[12] M. Drews, T. Hell, B. Klein, W. Weise, Phys.Rev. **D88**, 096011 (2013)
[13] M. Drews, W. Weise, Phys. Rev. **C91**, 035802 (2015)
[14] E.S. Fraga, M. Hippert, A. Schmitt, Phys. Rev. D **99**, 014046 (2019)
[15] A. Schmitt, Phys. Rev. D **101**, 074007 (2020)
[16] E.S. Fraga, R. da Mata, S. Pitsinigkos, A. Schmitt (2022), `2206.09219`
[17] E. Fonseca et al., Astrophys. J. Lett. **915**, L12 (2021)
[18] E. Witten, Phys. Rev. **D30**, 272 (1984)
[19] A.R. Bodmer, Phys. Rev. **D4**, 1601 (1971)
[20] L. Tolos, L. Fabbietti, Prog. Part. Nucl. Phys. **112**, 103770 (2020)

The axion potential in quark matter

M. Ruggieri[1], *D.E.A.* Castillo[2], *A.G.* Grunfeld[3], and *Bonan* Zhang[1]

[1]School of Nuclear Science and Technology, Lanzhou University, 222 South Tianshui Road, Lanzhou 730000, China
[2]Institute of Nuclear Physics, Polish Academy of Sciences, Radzikowskiego 152, 31-342 Cracow, Poland
[3]CONICET, Godoy Cruz 2290, C1425FQB Ciudad Autónoma de Buenos Aires, Argentina *and* Departamento de Física, Comisión Nacional de Energía Atómica, Av. Libertador 8250, C1429 BNP, Ciudad Autónoma de Buenos Aires, Argentina

Abstract. We study the QCD axion potential in hot and dense quark matter, within an NJL-like model that includes the coupling of the axion to quarks. Firstly we compute the effect of the chiral QCD crossover on the axion mass and self-coupling. Then, we compute the axion potential and study the domain walls. We find that the energy barrier between two adjacent vacuum states decreases in the chirally restored phase: this results in a lower surface tension of the walls. Finally we comment on the possibility of abundant production of walls in hot and dense quark matter.

1 Introduction

Axions are hypothetical particles whose existence was suggested by Peccei and Quinn to solve the strong CP problem [1, 2]. The existence of this particle has not been either proven or disproven, therefore it is still interesting to study them in extreme environments of large density and high temperature matter. This is the purpose of the study presented here. Within an NJL-like model [3–6] we study the coupling of axions to quarks in quark matter, like the one that could be present in the dense interior of compact stars or that produced in the very young universe before the QCD phase transition happened. In particular, we focus on the determination of the axion potential and its response to the QCD phase transition at large density and/or high temperature. Then, we comment on the possibility of production of axion walls in these extreme conditions.

2 The model

We work in the grand canonical ensemble formalism, using T and μ as state variables, where μ denotes the quark number chemical potential. We consider two flavor quark matter with Lagrangian density given by

$$\mathcal{L} = \bar{q}\,(i\partial\!\!\!/ + \hat{\mu}\gamma_0 - m_0)\,q + \bar{e}\,(i\partial\!\!\!/ + \mu_e\gamma_0)\,e + \mathcal{L}_{\mathrm{int}}, \qquad (1)$$

Here q denotes the quark field carrying Dirac, color and flavor indices,while e is the electron field. m_0 is the current quark mass, that we take to be equal for u and d quarks for simplicity.

The quark chemical potential matrix is diagonal in flavor space, $\hat{\mu} = \text{diag}(\mu_u, \mu_d)$, with

$$\mu_u = \mu - \frac{2}{3}\mu_e, \quad \mu_d = \mu + \frac{1}{3}\mu_e; \tag{2}$$

$\mu_d = \mu_u + \mu_e$ in agreement with the requirement of β−equilibrium. The interaction term is

$$\mathcal{L}_{\text{int}} = G_1\left[(\bar{q}\tau_a q)(\bar{q}\tau_a q) + (\bar{q}\tau_a i\gamma_5 q)(\bar{q}\tau_a i\gamma_5 q)\right] + 8G_2\left[e^{i\frac{a}{f_a}}\det(\bar{q}_R q_L) + e^{-i\frac{a}{f_a}}\det(\bar{q}_L q_R)\right]. \tag{3}$$

In the above equation τ_a are matrices in the flavor space with $a = 0, \ldots, 3$; τ_0 is the identity and τ_i with $i = 1, 2, 3$ are the Pauli matrices, normalized as $\text{tr}(\tau_i \tau_i) = \delta_{ij}/2$. The coupling constant G_1 governs the $U(1)_A$−invariant interaction. Similarly, G_2 regulates the strength of the $U(1)_A$−breaking term; the determinant in the latter is understood in the flavor space.

The thermodynamic potential at one loop has been discussed in the literature, see [3] and references therein; it reads

$$\Omega = \Omega_{\text{mf}} + \Omega_{1-\text{loop}} + \Omega_e. \tag{4}$$

Here we have put

$$\Omega_{\text{mf}} = -G_2(\eta^2 - \sigma^2)\cos(a/f_a) + G_1(\eta^2 + \sigma^2) - 2G_2\sigma\eta\sin(a/f_a), \tag{5}$$

that represents the mean field contribution to Ω, with $\sigma = \langle\bar{q}q\rangle$, $\eta = \langle\bar{\eta}i\gamma_5\eta\rangle$. Moreover,

$$\Omega_e = -2T\frac{4\pi}{8\pi^3}\left(\frac{7\pi^4}{180}T^3 + \frac{\pi^2\mu_e^2 T}{6} + \frac{\mu_e^4}{12T}\right) \tag{6}$$

is the contribution of the free, massless electrons. Finally, $\Omega_{1-\text{loop}}$ corresponds to the quark loop contribution, given by

$$\Omega_q = -4N_c\sum_{f=u,d}\int\frac{d^3p}{(2\pi)^3}\left[\frac{E_p}{2} + \frac{1}{2\beta}\log(1 + e^{-\beta(E_p-\mu_f)})(1 + e^{-\beta(E_p+\mu_f)})\right], \tag{7}$$

with $\beta = 1/T$. The dispersion laws of quarks are given by

$$E_p = \sqrt{p^2 + \Delta^2}, \quad \Delta^2 = (m_0 + \alpha_0)^2 + \beta_0^2, \tag{8}$$

with

$$\alpha_0 = -2\left[G_1 + G_2\cos(a/f_a)\right]\sigma + 2G_2\eta\sin(a/f_a), \tag{9}$$

$$\beta_0 = -2\left[G_1 - G_2\sin(a/f_a)\right]\eta + 2G_2\sigma\sin(a/f_a). \tag{10}$$

The electron chemical potential is fixed for each value of the pair (μ, T) by imposing the electrical neutrality condition

$$\frac{\partial\Omega}{\partial\mu_e} = 0. \tag{11}$$

Moreover, the condensates are computed self-consistently by solving the gap equations

$$\frac{\partial\Omega}{\partial\sigma} = 0, \quad \frac{\partial\Omega}{\partial\eta} = 0, \tag{12}$$

being sure that the solution $\sigma = \bar{\sigma}$, $\eta = \bar{\eta}$ corresponds to the global minimum of Ω. We notice that the first integral in the right hand side of Eq. (7) is ultraviolet divergent: we regulate this divergence by cutting the integration at $p = \Lambda$ and treat Λ as a parameter that we fix by the value of the chiral condensate in the vacuum. The set of parameters we use is $\Lambda = 590$ MeV, $G_0\Lambda^2 = 2.435$, $G_1 = (1 - c)G_0$, $G_2 = cG_0$ with $c = 0.2$, $m_0 = 6$ MeV [3].

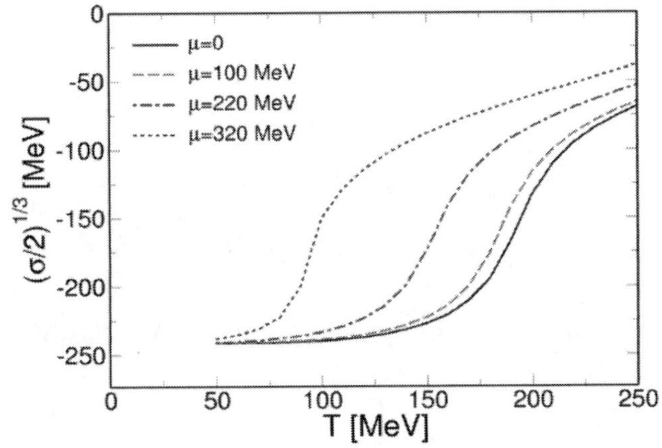

Figure 1. $(\sigma/2)^{1/3}$ versus T for several values of μ in the neutral ground state.

3 Results

In Fig. 1 we plot $(\sigma/2)^{1/3} = (\langle\bar{u}u + \bar{d}d\rangle/2)^{1/3}$ versus T for several values of μ. The chiral condensate drops down in a narrow range of temperature, signaling the approximate restoration of chiral symmetry. This allows us to define a pseudo-critical temperature, T_c, as the temperature where σ has its largest variation. T_c drops as the chemical potential increases. In addition to this, we notice that the variation of σ becomes sharper with μ: the smooth crossover at $\mu = 0$ becomes a sharp transition at at large μ.

The axion potential near the origin is characterized by the mass and the self-coupling,

$$m_a^2 = \left.\frac{d^2\Omega}{da^2}\right|_{a=0}, \quad \lambda_a = \left.\frac{d^4\Omega}{da^4}\right|_{a=0}. \tag{13}$$

At $T = \mu = 0$ we find

$$m_a f_a = 6.38 \times 10^3\ \mathrm{MeV}^2, \quad \lambda_a f_a^4 = -(55.63\ \mathrm{MeV})^4, \tag{14}$$

in agreement with previous estimates [3, 7]. In Fig. 2 we plot the axion mass (left panel) and the self-coupling (right panel) versus T for several values of μ. In correspondence of the QCD crossover the axion mass drops significantly. Moreover, increasing μ results in a sharper drop of the axion mass, similarly to what happens to the chiral condensate.

The fact that $\lambda_a < 0$ means that the quartic interaction is attractive. We notice that in correspondence of the chiral crossover, the quartic coupling experiences a kink that becomes more pronounced when the crossover becomes sharper, namely when the critical endpoint is approached. Thus, despite the fact that λ_a tends to become smaller with T, the chiral crossover enhances the axion self-coupling and this enhancement is very pronounced in proximity of the critical endpoint.

Next we turn to the full axion potential (4). In the left panel of Fig. 3 we plot the axion potential versus a/f_a for several values of μ and for $T = 10$ MeV; this has been computed along the neutrality line (11). Increasing μ results in the lowering of the barrier between the two degenerate vacua $a = 0$ and $a/f_a = 2\pi$.

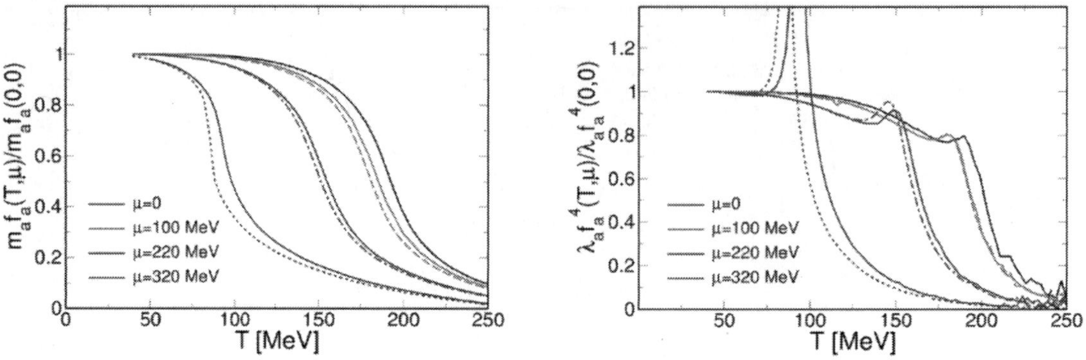

Figure 2. $m_a f_a$ versus T (left panel) and $\lambda_a f_a^4$ (right panel) for several values of μ. Solid lines correspond to the calculations with electrical neutrality while dashed lines denote the results for $\mu_e = 0$.

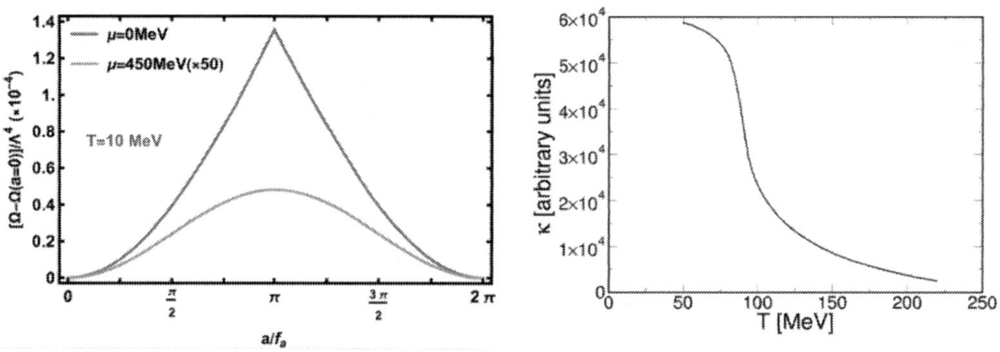

Figure 3. Axion potential $T = 10$ MeV (left panel), and surface tension of the axion walls versus T at $\mu = 320$ MeV (right panel).

The potential shown in Fig. 3 gives rise to domain walls that interpolate between two successive vacua, because the potential is invariant under the discrete symmetry transformation $\theta \to \theta + 2\pi n$ with $\theta \equiv a/f_a$ and $n \in Z$, while this symmetry is broken spontaneously by choosing one value of θ, for example $\theta = 0$. Since the energy barrier between two adjacent vacua decreases in the chirally restored phase, the energy stored in the domain walls also decreases in this phase. The domain wall solution can be built up following well known procedure of classical field theory, starting with the lagrangian density

$$\mathcal{L} = \frac{1}{2}\partial^\mu a \partial_\mu a - V(a/f_a), \quad V(\theta) = \Omega(\theta) - \Omega(0). \tag{15}$$

We checked that in the chirally restored phase a good approximation to the potential is given by

$$V(\theta) = V_0(1 - \cos\theta) = m_a^2 f_a^2 (1 - \cos\theta); \tag{16}$$

for this potential, the solution representing a domain wall at rest is

$$\theta_{\pm}(\xi) = 4 \arctan \exp(\pm m_a x) . \tag{17}$$

m_a corresponds to the in-medium axion mass computed self-consistently within the NJL model. In the chirally broken phase, the potential and the domain wall solutions have to be computed numerically solving the equation of motion arising from the lagrangian density (15). The energy per unit of transverse area of the time-independent domain wall is

$$\kappa \equiv \frac{E}{L^2} = \int_{-\infty}^{+\infty} dx \left[\frac{1}{2} \left(\frac{da}{dx} \right)^2 + V(a/f_a) \right] . \tag{18}$$

The surface tension versus temperature is shown for $\mu = 320$ MeV in the right panel of Fig 3. Around the phase transition, κ decreases as anticipated. It is useful to notice that the energy cost of adding one axion wall to bulk quark matter is κL^2. This has to be compared with the free energy of the bulk quark matter. In the thermodynamic limit, $L \to \infty$, the energy of the background of quark matter is $\propto L^3$. Accordingly, the free energy cost of adding one of these solitons to the bulk of quark matter is zero in this limit. Consequently, we expect an abundant production of axion walls in dense/hot quark matter. The study of the properties of these walls, as well as of their potential observable effects, will be the subject of future studies.

4 Conclusions

We studied the potential of axions on hot and/or dense quark matter, within an NJL-like model; this model has a phase transition to a chirally restored phase at high temperature and/or large density, hence it allows us to study the coupling of axions to quarks both in the confinement and in the deconfinement phases, and the feedback of the QCD crossover on the axion potential. We found that the potential is very sensitive to the crossover, in particular for large values of μ where the crossover is sharp and eventually becomes a real second order phase transition at the critical endpoint. We also analyzed the surface tension, κ, of the axion walls in the hot/dense quark matter. We found that κ decreases in the chirally restored phase. Moreover, we noticed that in the thermodynamic limit, adding a wall to the bulk of quark matter has a zero energy cost, therefore we expect abundant walls production in this environment.

M. R. acknowledges the organizers of QCD@Work for their kind hospitality; moreover, he acknowledges discussions with Mark Alford, Efrain Ferrer and Vivian Incera during the conference, John Petrucci for inspiration, Z. Y. Lu for numerous discussions and S. S. Wan for his support on the initial part of this project. M. R. is supported by the National Science Foundation of China (Grants No.11805087 and No. 11875153). A. G. Grunfeld would like to acknowledge CONICET for financial support under Grant No.PIP17-700.

References

[1] R. D. Peccei and H. R. Quinn, Phys. Rev. Lett. **38**, 1440-1443 (1977) doi:10.1103/PhysRevLett.38.1440

[2] R. D. Peccei and H. R. Quinn, Phys. Rev. D **16**, 1791-1797 (1977) doi:10.1103/PhysRevD.16.1791

[3] Z. Y. Lu and M. Ruggieri, Phys. Rev. D **100**, no.1, 014013 (2019) doi:10.1103/PhysRevD.100.014013 [arXiv:1811.05102 [hep-ph]]

[4] S. B. Ruester, V. Werth, M. Buballa, I. A. Shovkovy and D. H. Rischke, Phys. Rev. D **72**, 034004 (2005) doi:10.1103/PhysRevD.72.034004 [arXiv:hep-ph/0503184 [hep-ph]]

[5] D. Blaschke, S. Fredriksson, H. Grigorian, A. M. Oztas and F. Sandin, Phys. Rev. D **72**, 065020 (2005) doi:10.1103/PhysRevD.72.065020 [arXiv:hep-ph/0503194 [hep-ph]]

[6] D. N. Aguilera, D. Blaschke, M. Buballa and V. L. Yudichev, Phys. Rev. D **72**, 034008 (2005) doi:10.1103/PhysRevD.72.034008 [arXiv:hep-ph/0503288 [hep-ph]]

[7] G. Grilli di Cortona, E. Hardy, J. Pardo Vega and G. Villadoro, JHEP **01**, 034 (2016) doi:10.1007/JHEP01(2016)034 [arXiv:1511.02867 [hep-ph]]

Holographic Study of the $Q\bar{Q}$ Chaotic Dynamics in General Thermal Background

Nicola Losacco[1,2,*]

[1]Istituto Nazionale di Fisica Nucleare, Sezione di Bari, Via Orabona 4, I-70126 Bari, Italy
[2]Dipartimento Interateneo di Fisica "M. Merlin", Università e Politecnico di Bari, via Orabona 4, 70126 Bari, Italy

Abstract. The holographic approach is applied to study the chaotic behaviour of a strongly coupled $Q\bar{Q}$ pair in general thermal background. We consider two different backgrounds, one with finite temperature and baryon density, and one with finite temperature and constant magnetic field along a fixed direction. The results allow us to understand the dependence of the chaotic dynamics on the background, to test the universal bound on chaos conjectured by Maldacena, Shenker and Standford (MSS).

1 Introduction

We study the effects of chemical potential, related to a baryon reservoir, and of a magnetic field on the chaotic behaviour of a strongly coupled $Q\bar{Q}$ pair in a finite temperature background [1, 2]. Such systems can be investigated through holographic methods analyzing the dynamics of an open string hanging in the bulk in presence of a black hole (BH). The BH properties, as Hawking temperature and charge, are related to properties of the boundary gauge theory. In [3, 4] it is shown that BH belong to a set of systems called fast "scramblers", in particular BH are the fastest scramblers in nature: the time needed for a system near a BH horizon to loose information depends logarithmically on the number of the system degrees of freedom. As shown in [5] these systems present a bounded chaotic dynamics, with upper bound (MSS bound) on the largest Lyapunov exponent λ characterizing the chaotic behaviour of the thermodynamic quantum system with temperature T:

$$\lambda \leqslant 2\pi T. \tag{1}$$

The connections between chaotic quantum systems and gravity have been investigated in [6–10]. A generalization of the bound Eq. (1), proposed in the literature for systems presenting a global symmetry [11]:

$$\lambda \leqslant \frac{2\pi T}{1-\left|\frac{\mu}{\mu_c}\right|} \qquad \mu \ll \mu_c \tag{2}$$

where μ is the chemical potential and μ_c a critical value, can be checked. To test the bounds we consider the $Q\bar{Q}$ pair in a finite temperature and baryon density background [1], and with a constant and uniform magnetic field at finite temperature [2]. This allows us to test the generalized bound Eq. (2), and the features of phenomenological contexts such as in heavy-ion collisions [12–14].

*e-mail: nicola.losacco@ba.infn.it

2 Geometry

The gravity dual of a strongly coupled $Q\bar{Q}$ pair is an open string hanging in a 5-dimensional metric obtained solving Einstein equations with suitable boundary conditions. The endpoints of the string are on the boundary, they stand for the heavy quarks in the gauge boundary theory. For finite baryon density the geometry is described by a Reissner–Nordstrom metric:

$$\begin{aligned} ds^2 &= -f(r)\, r^2 dt^2 + r^2 d\mathbf{x}^2 + \frac{1}{r^2 f(r)} dr^2, \\ f(r) &= 1 - \frac{r_h^4}{r^4} - \frac{\mu^2 r_h^2}{r^4} + \frac{r_h^4 \mu^2}{r^6}, \end{aligned} \tag{3}$$

where r_h is the external BH horizon and the chemical potential μ is related to the BH charge. For the case with magnetic field, an approximate solution of the Einstein equations is used [15]:

$$ds^2 = -f(r)\, r^2 dt^2 + r^2 h(r)(dx^1)^2 + r^2 h(r)(dx^2)^2 + r^2 q(r)(dx^3)^2 + \frac{1}{r^2 f(r)} dr^2, \tag{4}$$

$$\begin{aligned} f(r) &= 1 - \frac{r_h^4}{r^4} - \frac{2B^2}{3r^4} \log \frac{r}{r_h} \\ q(r) &= 1 - \frac{2B^2}{3r^4} \log r \\ h(r) &= 1 + \frac{B^2}{3r^4} \log r\,. \end{aligned} \tag{5}$$

B is related to the magnetic field in the x^3 direction. The magnetic field breaks rotational invariance, hence $h(r) \neq q(r)$.

3 Exploring Chaos

Let us consider the general metric

$$ds^2 = g_{tt} dt^2 + g_{11}(dx^1)^2 + g_{22}(dx^2)^2 + g_{33}(dx^3)^2 + g_{rr} dr^2\,. \tag{6}$$

The string profile is obtained from the Nambu-Goto (NG) action:

$$S = -\frac{1}{2\pi\alpha'} \int dt\, d\ell\, \sqrt{-h} \tag{7}$$

where α' is the string tension and h the determinant of the induced metric $h_{ij} = g_{MN} \frac{\partial X^M}{\partial \xi_i} \frac{\partial X^N}{\partial \xi_j}$, with $\xi_{i,j}$ the worldsheet coordinates and g the metric tensor. The static solution depends on the parameters of the metric and on the proximity of the string to the BH horizon, quantified by the string tip position r_0. Analyzing the energy of the static configuration dependency on r_0, we found that it has a maximum near the black hole horizon r_h. This unstable configuration enhances the chaotic behaviour of the system, therefore we set our static solution near the BH horizon. We perturb the static configuration with a time dependent fluctuation orthogonal in each point of the string [16]:

$$\begin{aligned} r(t,\ell) &= r_{BG}(\ell) + \xi(t,\ell)\, n^r(\ell)\,, \\ x(t,\ell) &= x_{BG}(\ell) + \xi(t,\ell)\, n^x(\ell)\,, \end{aligned} \tag{8}$$

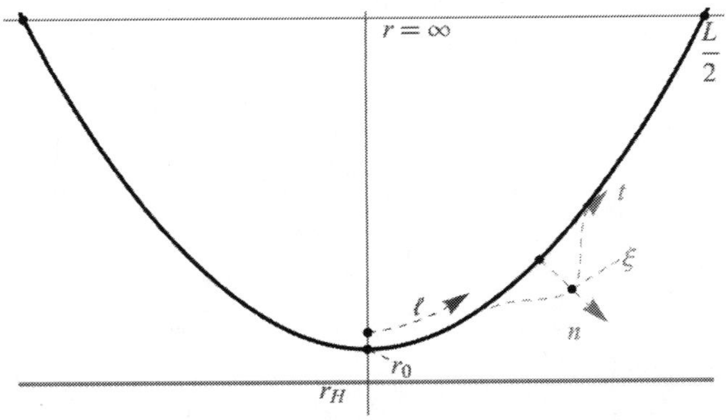

Figure 1. Static string profile and perturbation $\xi(t,\ell)$ along the direction orthogonal to the string in each point with coordinate ℓ.

leaving unperturbed the endpoints, Fig. 1. *BG* stands for background solution obtained from the action in Eq. (7). We expand the action Eq. (7) to the second order in the perturbation ξ:

$$S^{(2)} = \frac{1}{2\pi\alpha'}\int dt \int_{-\infty}^{\infty} d\ell(C_{tt}\dot{\xi}^2 + C_{\ell\ell}\acute{\xi}^2 + C_{00}\xi^2). \tag{9}$$

C_{tt}, $C_{\ell\ell}$ and C_{00} depend on ℓ and on the parameters of the metric. From the action Eq. (9) we obtain the equation of motion for the perturbation. Factorizing the variables $\xi(t,l) = \xi(l)e^{i\omega t}$, this becames a Sturm-Liouville equation $\partial_\ell\left(C_{\ell\ell}\acute{\xi}\right) - C_{00}\,\xi = \omega^2 C_{tt}\,\xi$. We solve it for the first two eigenvalues. We write the perturbation as a linear combination of the first two eigenfunctions with time dependent coefficients describing the dynamics of the system:

$$\xi(t,\ell) = c_0(t)\,e_0(\ell) + c_1(t)\,e_1(\ell)\,. \tag{10}$$

This allow us to evaluate the third order action using Eq. (10):

$$S^{(3)} = \frac{1}{2\pi\alpha'}\int dt\Big[\sum_{n=0,1}\left(\dot{c}_n^2 - \omega_n^2 c_n^2\right) + K_1 c_0^3 + K_2 c_0 c_1^2 + K_3 c_0 \dot{c}_0^2 + K_4 c_0 \dot{c}_1^2 + K_5 \dot{c}_0 c_1 \dot{c}_1\Big]. \tag{11}$$

The coefficients $K_{1,\dots,5}$ are obtained from an integration over ℓ and depend on r_0, the tip position of the hanging string, and on the parameters of the metric. The potential described by Eq. (11) is a trap potential that confines the dynamics of the unstable string configurations. In the trap region the kinetic term is negative. As shown in [1, 16], we can substitute $c_{0,1} \to \tilde{c}_{0,1}$ in the action, with $c_0 = \tilde{c}_0 + \alpha_1\tilde{c}_0^2 + \alpha_2\tilde{c}_1^2$ and $c_1 = \tilde{c}_1 + \alpha_3\tilde{c}_0\tilde{c}_1$, neglecting $O(\tilde{c}_i^4)$ terms, setting the constants α_i ensuring the positivity of the kinetic term. This replacement does not affect the dynamics of the system, and a chaotic behaviour shows up in the transformed system. The chaotic dynamics can be studied by analyzing the dynamics of $\tilde{c}_0(t)$ and $\tilde{c}_1(t)$ governed by the action Eq. (11) after the substitution.

4 Results

The chaotic dynamics is analyzed through Poincaré plots and Lyapunov exponents, evaluated at $r_0 = 1.1$ and $r_h = 1$ for different μ and B. In Fig. 2 we see that at $B = 0$ increasing the chemical potential the trajectories in the Poincaré section become more stable, hence the chemical potential stabilizes the system. In the case of the magnetic field, either parallel or

EPJ Web of Conferences **270**, 00025 (2022) https://doi.org/10.1051/epjconf/202227000025
QCD@Work 2022

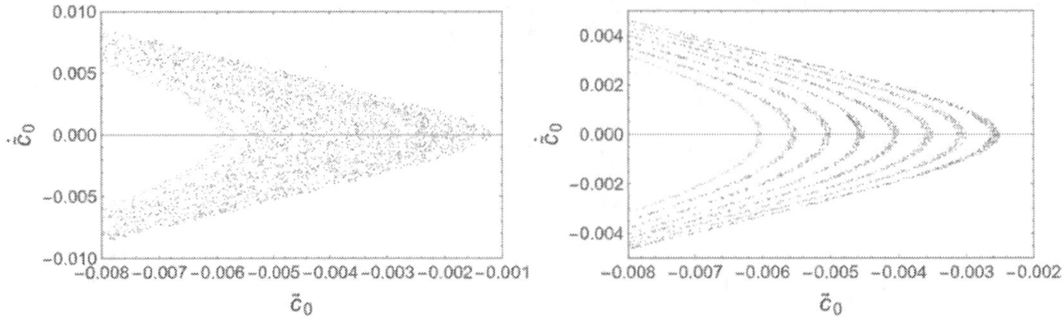

Figure 2. Poincaré sections for a time-dependent perturbed string, obtained changing the initial conditions, and increasing the chemical potential $\mu = 0.3$ (left) and $\mu = 0.9$ (right), for $\tilde{c}_1 = 0$ and $\dot{\tilde{c}}_1 \geq 0$.

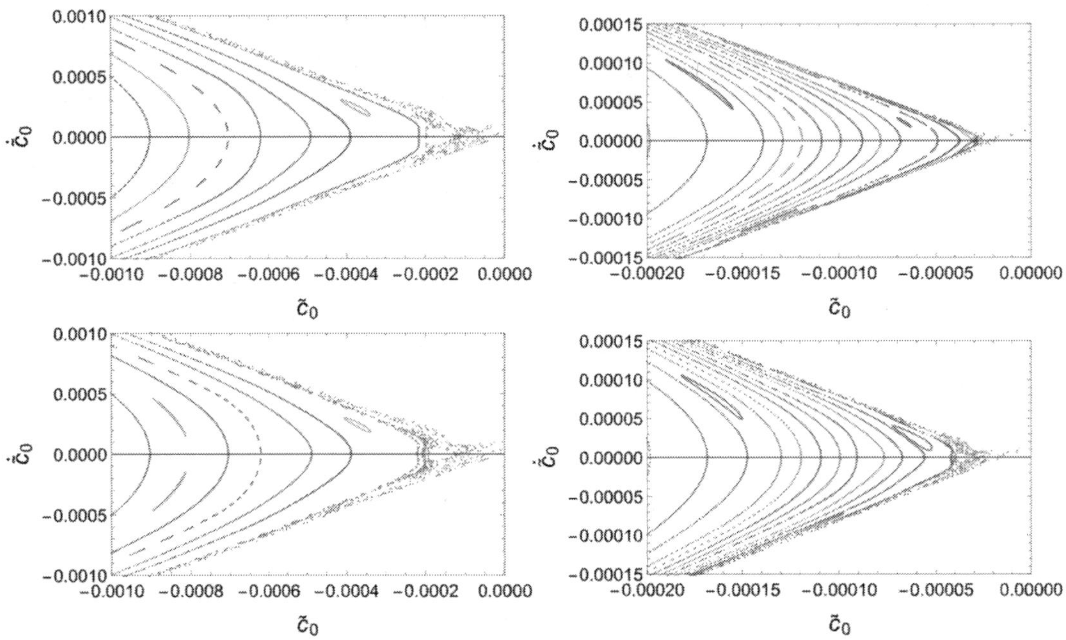

Figure 3. Poincaré sections for a time-dependent perturbed string along the magnetic field (top panels) and orthogonal to the magnetic field (bottom panels), obtained changing the initial conditions, and increasing the magnetic field $B = 0.3$ (left) and $B = 1$ (right), for $\tilde{c}_1 = 0$ and $\dot{\tilde{c}}_1 \geq 0$.

orthogonal to the $Q\bar{Q}$ system, we observe a stabilization of the orbits increasing the magnetic field from $B = 0.3$ to $B = 1$, Fig. 3. The anisotropy of the system is manifested in the different chaotic behaviour in the two directions. When the $Q\bar{Q}$ system is orthogonal to the magnetic field, the region in which the trajectories become scattered is wider than the case in which the string is along the magnetic field, as we can see in Fig. 3. It is necessary to approach the origin of the phase space to find chaotic trajectories when the string is along the magnetic field. All these properties are shown by the largest Lyapunov exponents for different values of chemical potential and magnetic field, Fig. 4. Increasing both chemical potential and magnetic field, the systems become less chaotic since the largest Lyapunov exponent decreases. Moreover, the configuration along the magnetic field stabilizes faster. In all cases

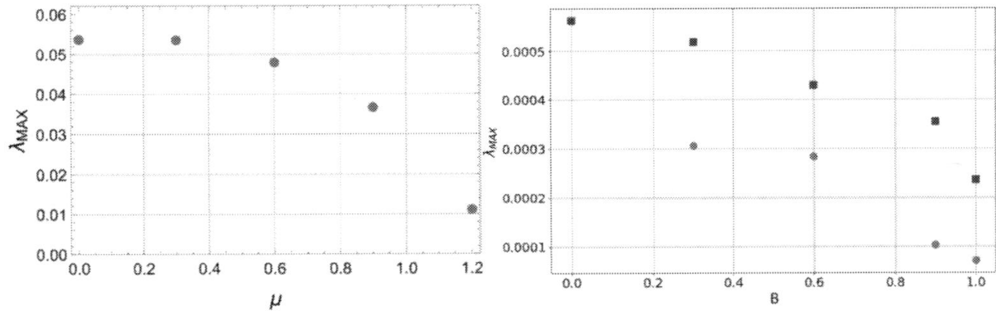

Figure 4. Largest Lyapunov exponent λ_{MAX} versus μ (left) and B (right) for $r_0 = 1.1$. In the plot at right the results for the string configurations orthogonal (blue squares) and along (red points) the magnetic field are shown.

the largest Lyapunov exponents satisfy the MSS bound Eq. (1), even in the case of baryon density where the generalized bound Eq. (2) is less stringent.

5 Conclusions

Chaos for the string dual to the $Q\bar{Q}$ system has been observed in the Poincaré plots, characterized by scattered points in the region where the tip of the string is close to the black hole horizon. The system becomes less chaotic increasing μ and B. For the magnetic field case, an anisotropy effect in two different orientations of the string is found. The MSS bound is satisfied for the largest Lyapunov exponent and remains universal.

Acknowledgements. I thank P. Colangelo, F. De Fazio and F. Giannuzzi, co-authors of the works on which this proceeding is based. This study has been carried out within the INFN project (Iniziativa Specifica) QFT-HEP.

References

[1] P. Colangelo, F. De Fazio, N. Losacco, Phys. Rev. D **102**, 074016 (2020), 2007.06980
[2] P. Colangelo, F. Giannuzzi, N. Losacco, Phys. Lett. B **827**, 136949 (2022), 2111.09441
[3] Y. Sekino, L. Susskind, JHEP **10**, 065 (2008), 0808.2096
[4] L. Susskind (2011), 1101.6048
[5] J. Maldacena, S.H. Shenker, D. Stanford, JHEP **08**, 106 (2016), 1503.01409
[6] S.H. Shenker, D. Stanford, JHEP **03**, 067 (2014), 1306.0622
[7] S.H. Shenker, D. Stanford, JHEP **05**, 132 (2015), 1412.6087
[8] A. Kitaev, Breakthrough Prize Fundamental Physics Symposium, 11/10/2014, KITP seminar (2014)
[9] J. Polchinski (2015), 1505.08108
[10] D. Giataganas (2021), 2112.02081
[11] I. Halder (2019), 1908.05281
[12] I.Y. Aref'eva, K. Rannu, P. Slepov, JHEP **07**, 161 (2021), 2011.07023
[13] I.Y. Aref'eva, A. Ermakov, K. Rannu, P. Slepov (2022), 2203.12539
[14] K. Rannu, I.Y. Aref'eva, P.S. Slepov, Rev. Mex. Fis. Suppl. **3**, 0308126 (2022)
[15] D. Li, M. Huang, Y. Yang, P.H. Yuan, JHEP **02**, 030 (2017), 1610.04618
[16] K. Hashimoto, K. Murata, N. Tanahashi, Phys. Rev. D **98**, 086007 (2018), 1803.06756

Quantum Field Theory and its Anomalies for Topological Matter

Claudio Corianò[1,*], *Mario* Cretì[1,2,**], *Stefania* D'Agostino[1,2,3,***]

[1]Dipartimento di Matematica e Fisica, Università del Salento, INFN-Lecce, Via Arnesano, 73100 Lecce, Italy
[2]Center for Biomolecular Nanotechnologies, Istituto Italiano di Tecnologia, Arnesano (LE), Italy
[3]CNR Institute of Nanotechnology (NANOTEC), SP Lecce-Monteroni, Lecce, Italy

Abstract. Topology enters in quantum field theory (qft) in multiple forms: one of the most important, in non-abelian gauge theories, being in the identification of the θ vacuum in QCD. A very relevant aspect of this connection is through the phenomenon of chiral and conformal qft anomalies. It has been realized that a class of materials, comprising topological insulators and Weyl semimetals, also exhibit the phenomenon of anomalies, which are responsible for several exotic phenomena, such as the presence of edge currents, resilient under perturbations and scattering by impurities. Another example comes from the response functions of these materials under thermal and mechanical stresses, that may be performed using correlation functions of stress energy tensors in General Relativity. In this case the conformal anomaly plays an important role. We briefly illustrate some salient features of this correspondence, and the effective action describing the long-range interactions that may account for such topological effects.

1 Introduction

Quantum anomalies have been discovered in particle physics long ago but, more recently, they have found applications in solid state physics as well, in the case of topological materials, such as Dirac and Weyl semimetals [1–5], with important implications in anomalous transport [6–8] in condensed matter and heavy ion collisions [9]. Gravity enters the field via the conformal anomaly [10] (see [11] for a review) and the Luttinger relation [12]. Dirac semimetals, for example, are 3D crystals whose low-lying excitations are described by the massless Dirac equation. The relativistic description is emergent, in the sense that a linear dispersion relation in the band structure of such materials is associated with the Fermi velocity v_F rather than with the speed of light c. In a Dirac semimetals the valence and conductance bands touch each other in a single point, called a Dirac point. At such specific point the energy/momentum dispersion relation is linear, similarly to the relativistic case. In Weyl semimetals a Dirac point could also split into a pair of so-called Weyl points, at which the dispersion relation is exactly that of a Weyl Hamiltonian $H^{\pm}_{Weyl} = \pm\vec{p}\cdot\vec{\sigma}$.

*e-mail: claudio.coriano@le.infn.it
**e-mail: mario.creti@unisalento.it
***e-mail: stefania.dagostino@iit.it

The response function of such materials may exhibit an effective anomaly if we engineer an axial-vector-vector (AVV) interaction, in which left- and right-handed Weyl modes of the electrons in the material couple with different strengths to an external chiral current, generated via a chiral chemical potential (μ_5).
The **AVV** anomaly vertex, for example, which is responsible for the chiral magnetic effect, is generated when the material is subjected to a background of a magnetic field as well as to a non-zero chiral chemical potential (the time-like component of the remaining axial current), with the generation of an electric current. Therefore, a combined study of this interaction from the qft and condensed matter theory point of views is crucial in order to understand the phenomenology of such materials. It is expected that the interaction involves an intermediate pseudoscalar state whose properties resemble those of an axion. The analysis of such state, in the long wavelength limit, in which the description of the material can be performed using ordinary chiral/conformal qft actions, raises some interesting issues concerning the type of effective actions which are most suitable for this purpose. Understanding the main features of this interaction and its difference/similarity with respect to the ordinary axion action of the QCD theta-vacuum is the point which we will briefly concentrate upon.
All anomalies are associated with effective actions characterised by a specific nonlocal structure, in the form of anomaly poles [13][11, 14, 15]. A beautiful unification of such phenomenon is in the case of supersymmetric theories [16], where the conformal and chiral anomalies, together with a supersymmetric anomaly, are part of a multiplet. Nonlocal exchanges in the anomaly effective action are identified in all the components of the multiplet.
In qft, chiral anomalies are a relativistic phenomenon connected with the presence of Weyl fermions, whose couplings to conserved external currents is different for left (L) and right handed (R) components. A chiral anomaly can be generated at trilinear level - the counting refers to the number of external currents - and anomaly cancelation is necessary if the corresponding interaction violates the Ward identities that result from the original quantum symmetry of the action.
If we expand the partition function of a certain theory

$$\mathcal{Z}(\chi_i) \equiv \int D\psi D\bar{\psi} e^{i\mathcal{S}_0(\psi,\bar{\psi},B)} \tag{1}$$

in the case of the chiral anomaly, the underlying interaction is that of an **AVV** (axial-vector/vector/vector) anomaly diagram, appearing at $O(e^2)$ in the expansion of (1), when $\mathcal{S}_0$ is the fundamental action of axial QED

$$\mathcal{S}_0 = \bar{\psi}\gamma^\mu(\partial_\mu - ieA_\mu - i\gamma_5 B_\mu)\psi - m\bar{\psi}\psi \tag{2}$$

and B_λ is the axial-vector source. In the presence of propagating axial-vector and vector interactions, in the massless limit, such anomaly interactions have to vanish, by summing over all the fermions species that are coupled to the corresponding gauge fields. In the case of the Standard Model (SM) of the elementary particles, this is realized by summing over each fermion generation of quarks and leptons, though such cancelation may occur, in other models, by a suitable choice of the fermion charges, in the form of algebraic costraints involving several generations. The constraints take the form of linear, quadratic and cubic Diophantine equations. In general, in the case of the AVV diagram, a generic parameterization of this contribution leads to a correlator that violates the ordinary vector and axial vector Ward identities in the form

$$k_{1\mu}\Delta^{\lambda\mu\nu}(k_1,k_2) = a_1\epsilon^{\lambda\nu\alpha\beta}k_1^\alpha k_2^\beta,\ \ k_{2\nu}\Delta^{\lambda\mu\nu}(k_1,k_2) = a_2\epsilon^{\lambda\mu\alpha\beta}k_2^\alpha k_1^\beta,\ \ k_\lambda\Delta^{\lambda\mu\nu}(k_1,k_2) = a_3\epsilon^{\mu\nu\alpha\beta}k_1^\alpha k_2^\beta,$$

where $a_1 = -\frac{i}{8\pi^2}$ $a_2 = -\frac{i}{8\pi^2}$ $a_3 = -\frac{i}{4\pi^2}$. In the correlator $\Delta^{\lambda\mu\nu}$, μ and ν are vector indices and λ an axial-vector one. Notice that $a_1 = a_2$, as expected from the Bose symmetry of the two **V** lines. We have denoted with k_1 and k_2 the momenta of the two photons. k is the momentum of the axial-vectors source. It is also well known that the total anomaly $a_1 + a_2 + a_3 \equiv a_n$ is regularization scheme independent ($a_n = -\frac{i}{2\pi^2}$). A general parameterization involves 6 form factors (A_i) in the expression of the diagram

$$\begin{aligned}\Delta_0^{\lambda\mu\nu} &= A_1(k_1,k_2)\epsilon[k_1,\mu,\nu,\lambda] + A_2(k_1,k_2)\epsilon[k_2,\mu,\nu,\lambda] + A_3(k_1,k_2)\epsilon[k_1,k_2,\mu,\lambda]k_1{}^{\nu} \\ &+A_4(k_1,k_2)\epsilon[k_1,k_2,\mu,\lambda]k_2^{\nu} + A_5(k_1,k_2)\epsilon[k_1,k_2,\nu,\lambda]k_1^{\mu} + A_6(k_1,k_2)\epsilon[k_1,k_2,\nu,\lambda]k_2^{\mu},\end{aligned} \tag{3}$$

where A_1 and A_2 are divergent by power counting. The parameterization will give conserved vector Ward identites if we allow a Chern-Simons (CS) contribution with a parameter β, introduced by the momentum reparametrization in the loop

$$\Delta^{\lambda\mu\nu}(\beta,k_1,k_2) = \Delta^{\lambda\mu\nu}(k_1,k_2) - \frac{i}{4\pi^2}\beta\epsilon^{\lambda\mu\nu\sigma}(k_{1\sigma} - k_{2\sigma}). \tag{4}$$

The CS term (the Bardeen counterterm) will affect only A_1 and A_2, and corresponds to a finite renormalization of the diagram. A_1 and A_2 will shift by the same counterterm under such reparametrization, while their difference, obviously, will be invariant. As shown in [17], the pole is associated with the difference of these two form factors and is, therefore, unambiguous. The pole shows up once we reformulate (3) in the W_L/W_T (longitudinal/transverse or L/T) decomposition of the same vertex, performed respect to the momentum of the axial-vector current, k. The presence of such unique interactions and its contribution to the longitudinal Ward identity of the axial-vector current, is the most direct proof that the 1PI (1-particle irreducible) effective action is affected by these types of massless exchanges. The transition from the representation in (3) to the W_L/W_T one, requires repeated use of the Schouten relations, with the generation of massless poles also in the transverse part. The effective action, obtained by integrating out the fermions, in the W_L/W_T for is given by [17, 18]

$$\begin{aligned}\mathcal{S}_{eff} &= -\int d^4x\frac{1}{4}F_{\mu\nu}F^{\mu\nu} + a_n\int d^4x\partial\cdot B\,\Box^{-1}(x,y)F_{\mu\nu}\tilde{F}^{\mu\nu}(y) \\ &+\int d^4x\,d^4y\,d^4z\,W_T^{\lambda\mu\nu}(z,x,y)\,B_\lambda(z)\,A_\mu(x)\,A_\nu(y).\end{aligned} \tag{5}$$

where A_μ is the photon. In the equation above, beside the kinetic term of the photon field ($F_{\mu\nu}$), the second contribution has a specific nonlocal structure with a $1/\Box$ interaction. In momentum space it is identified by the 2-particle cut in the variable $s \equiv k^2$, the virtuality of the axial-vector line. It corresponds to an interaction term of the form

$$\partial\cdot B(x)\,\Box^{-1}(x,y)F_{\mu\nu}\tilde{F}^{\mu\nu}(y), \tag{6}$$

between the longitudinal component of B_λ and the anomaly ($F\tilde{F}$). The third contribution is divergenceless and corresponds to the transverse component W_T. Its expression can be found in [17]. Notice that for on-shell photons, the amplitude reduces just to the anomaly pole-like behaviour given in (6) with no extra term. A similar separation is introduced in the reconstruction program of the graviton vertices in conformal field theory [19], since this is also based on a similar decomposition, with the inclusion also of traceless and longitudinal sectors. The generation of such a pole is reminiscent of a collinear interaction with an intermediate fermion-antifermion pair in the loop [13, 20], generated by the decay of an off-shell axial-vector current. It can be understood via the Coleman-Norton interpretation of a singularity in a certain diagram, as due to a physical process. Note that in the massive case, the pole of the spectral density turns into the 2-particle cut of a (single) anomaly form factor, whose spectral density can be shown to satisfy a sum rule related to the anomaly. The same behaviour is present for the conformal and superconformal anomaly. The pole reappears from the cut once we perform the conformal (zero fermion mass) limit, and the area under the cut is preserved, being mass-independent.

2 The nonlocal conformal anomaly effective action

Non conserved currents carry significant implications in relativistic hydrodynamics. Indeed, the anomalous charge and energy flows can be described in terms of the chiral/conformal/gravitational anomaly actions, depending on the specific case. Certain quantum anomalies, for instance conformal and mixed axial-gravitational anomalies, appear in curved backgrounds and involve the energy-momentum tensor [11]. In condensed matter systems, these anomalies are realized in an off-equilibrium regime using the Luttinger theory of thermal transport coefficients [12](see also [21]). In this theory, a temperature gradient $\mathbf{\nabla} T$ – drives a system out of equilibrium, and it is compensated by a non-uniform gravitational potential Φ, at linear order in its fluctuation

$$\frac{1}{T}\mathbf{\nabla} T = -\frac{1}{c^2}\mathbf{\nabla}\Phi\,, \tag{7}$$

where c is the speed of light. In the case of a weak gravitational field the gravitational potential Φ,

$$g_{00} = 1 + \frac{2\Phi}{c^2}\,, \tag{8}$$

turns Newtonian, with g_{00} the time/time component of the metric. The other components of the metric tensor are unmodified. Luttinger's observation finds its origin in the Tolman–Ehrenfest effect [22, 23]. Tolman concluded that systems in thermal equilibrium under the action of a gravitational field do not have a constant temperature. The local temperature distribution is position dependent, a result that is well-known to relativists and cosmologists and used in several applications in these areas. In other words, they stated that heat has weight, while Luttinger identified a connection between gravitational fields and thermal transport. The temperature depends on the spatial coordinates as

$$T(x) = T_0/\sqrt{g_{00}(x)}, \tag{9}$$

where T_0 denotes a reference temperature, selected at a point with $g_{00} = 1$, i.e. at asymptotic distance from a gravitational source. The Luttinger formula (7) may be derived from simple thermodynamic considerations.

Once the connection between thermal and mechanical stresses of such materials and gravity is made explicit, the analysis inherits a great deal of the formalism used in the context of general relativity, opening the way to the use of correlation functions in external metric backgrounds in the study of these materials. Renormalization of the matter sector generates a back-reaction of the quantum corrections on the external metric [24] that is described by an anomaly action, which is, in part, topological. Its topological nature is evident from the operatorial structure of the equation connecting the trace of the stress energy tensor of a conformal theory with topological invariants such as the Euler-Poincarè density, according to the quantum relation

$$g_{\mu\nu}\langle T^{\mu\nu}\rangle = bE + b'C^2 + c_1\Box R \tag{10}$$

where E and C^2 denote the Euler Poincarè density (Gauss Bonnet term) and the Weyl tensor squared

$$E_4 \equiv R_{\mu\nu\alpha\beta}R^{\mu\nu\alpha\beta} - 4R_{\mu\nu}R^{\mu\nu} + R^2 \tag{11}$$

$$(C^{(4)})^2 \equiv R_{\mu\nu\alpha\beta}R^{\mu\nu\alpha\beta} - 2R_{\mu\nu}R^{\mu\nu} + \frac{1}{3}R^2, \tag{12}$$

where the first two terms on the rhs of (10) originate from a nonlocal effective action. The E invariant can be generalized to any even dimension, becoming cubic, quartic, and so on, in the Riemann tensor and its contractions

$$E_d = \frac{1}{2^{d/2}}\delta^{\nu_1\cdots\nu_d}_{\mu_1\cdots\mu_d}R^{\mu_1\mu_2}{}_{\nu_1\nu_2}\cdots R^{\mu_{d-1}\mu_d}{}_{\nu_{d-1}\nu_d}\,, \tag{13}$$

at d=4,6, etc. At $d = 2$ this role is taken by the Einstein-Hilbert action. These contributions are investigated in dimensional regularization, generating a dilaton in the effective theory that can be removed by a finite renormalization of the Gauss-Bonnet term [24, 25]. The anomalous trace of the energy-momentum tensor is known to be generated by the nonlocal action [20, 26–28]

$$\begin{aligned} S_{\text{anom}}[g,A] &= \frac{1}{8}\int d^4x\sqrt{-g(x)}\int d^4y\sqrt{-g(y)}\cdot E(x)G^{(4)}(x,y)\times \\ &\quad\times\Big[2b'C^2(y) + bE(y) + 2cF_{\mu\nu}(y)F^{\mu\nu}(y)\Big], \end{aligned}$$

where $G^{(4)}(x,y)$ is the Green function the fourth-order conformally covariant differential operator

$$\Delta_4 = \nabla_\mu\left(\nabla^\mu\nabla^\nu + 2R^{\mu\nu} - \frac{2}{3}Rg^{\mu\nu}\right)\nabla_\nu. \tag{14}$$

An application of the formalism is provided by a system in a slightly off-equilibrium regime [29] with a small temperature variation along a certain (x_3) direction. We use the Luttinger relation (7) to link the time-independent temperature gradient to the gradient of the gravitational potential Φ. Then, the perturbation of the metric is nonvanishing only for the $h_{00} \equiv 2\Phi$ component, that in our case is a function of only on one spatial variable, $h_{00}(x) \equiv h_{00}(x_3)$. One finds that the anomalous part of the TTT vertex contributes to the expectation value for the energy-momentum tensor in the form

$$\langle T^{00}\rangle_{TTT} = \frac{4b'}{9}\Big[3\big(\partial_3^2\Phi\big)^2 + 4\big(\partial_3\Phi\big)\big(\partial_3^3\Phi\big) + 2\Phi\partial_3^4\Phi\Big], \tag{15}$$

$$\langle T^{11}\rangle_{TTT} = \langle T^{22}\rangle_{TTT} = \frac{4b'}{9}\Big[2\big(\partial_3\Phi\big)\big(\partial_3^3\Phi\big) + \Phi\partial_3^4\Phi\Big]. \tag{16}$$

Eqs. (15)–(16) show some striking features, one of them being their local expression in terms of the gravitational potential $\Phi \equiv h_{00}/2$, despite the fact that they have been derived from the non-local anomaly action. A second important point is that the expectation value of the energy-momentum tensor involves only the anomalous coefficient b'. This coefficient, given explicitly in (10) for the case of one-species QED, is related to the truly anomalous part of the energy-momentum tensor (the Weyl term C^2). Indeed, E is an evanescent counterterm at $d = 4$, whose appearance in the Weyl variation of the anomaly action originates by requiring that the same action satisfies the Wess-Zumino consistency condition. Consequently, there is no topological contribution coming from the Euler density to the trace of the energy-momentum tensor at cubic order. These analysis can be extended to higher point functions.

3 Conclusions

The study of anomaly effective actions, and in particular their local and nonlocal formulations, finds important applications in the context of topological materials. Here we have briefly described some of these connections. This shows the existence of an important interplay between some fundamental areas of high energy physics - anomalies are such - and condensed matter theory.

Acknowledgements

We thank M. Chernodub, E. Hankiewicz, M.M. Maglio, E. Mottola and M. Vozmediano for discussions. We thank S. Lionetti and R. Tommasi for collaborating to related analysis. The work of C.C. is partly supported by INFN Iniziativa Specifica QFT-HEP.

References

[1] M.N. Chernodub, Y. Ferreiros, A.G. Grushin, K. Landsteiner, M.A.H. Vozmediano, Phys. Rept. **977**, 1 (2022), 2110.05471
[2] M. Fruchart, D. Carpentier, Comptes Rendus Physique **14**, 779 (2013), 1310.0255
[3] R. Arouca, A. Cappelli, T.H. Hansson (2022), 2204.02158
[4] C. Tutschku, F.S. Nogueira, C. Northe, J. van den Brink, E.M. Hankiewicz, Phys. Rev. B **102**, 205407 (2020), 2007.11852
[5] G. Tkachov, E.M. Hankiewicz, Physica E **44**, 900 (2012), 1106.1059
[6] K. Landsteiner, E. Megias, F. Pena-Benitez, Lect. Notes Phys. **871**, 433 (2013), 1207.5808
[7] M.N. Chernodub, E. Kilincarslan, Phys. Rev. D **103**, 036019 (2021), 2010.02104
[8] M.N. Chernodub, A. Cortijo, M.A.H. Vozmediano (2017), 1712.05386
[9] D.E. Kharzeev, *Chiral magnetic effect in heavy ion collisions and beyond* (2022), 2204.10903
[10] M.J. Duff, Class. Quant. Grav. **11**, 1387 (1994), hep-th/9308075
[11] C. Corianò, M.M. Maglio, Phys. Rept. **952**, 2198 (2022), 2005.06873
[12] J.M. Luttinger, Phys. Rev. **135**, A1505 (1964)
[13] M. Giannotti, E. Mottola, Phys. Rev. **D79**, 045014 (2009), 0812.0351
[14] R. Armillis, C. Corianò, L. Delle Rose, Phys. Lett. **B682**, 322 (2009), 0909.4522
[15] C. Corianò, M.M. Maglio, Phys. Lett. **B781**, 283 (2018), 1802.01501
[16] C. Corianò, A. Costantini, L. Delle Rose, M. Serino, JHEP **06**, 136 (2014), 1402.6369
[17] R. Armillis, C. Corianò, L. Delle Rose, M. Guzzi, JHEP **12**, 029 (2009), 0905.0865
[18] R. Armillis, C. Corianò, L. Delle Rose, Physics Letters B **682**, 322 (2009), 0909.4522
[19] A. Bzowski, P. McFadden, K. Skenderis, JHEP **03**, 111 (2014), 1304.7760
[20] R. Armillis, C. Corianò, L. Delle Rose, Phys. Rev. **D81**, 085001 (2010), 0910.3381
[21] B. Bermond, M. Chernodub, A.G. Grushin, D. Carpentier (2022), 2206.08784
[22] R.C. Tolman, Phys. Rev. **35**, 904 (1930)
[23] R. Tolman, P. Ehrenfest, Phys. Rev. **36**, 1791 (1930)
[24] C. Corianò, M.M. Maglio, D. Theofilopoulos (2022), 2203.04213
[25] C. Corianò, M.M. Maglio, Phys. Lett. B **828**, 137020 (2022), 2201.07515
[26] R.J. Riegert, Phys. Lett. **134B**, 56 (1984)
[27] P.O. Mazur, E. Mottola, Phys.Rev. **D64**, 104022 (2001), hep-th/0106151
[28] E. Mottola, R. Vaulin, Phys. Rev. **D74**, 064004 (2006), gr-qc/0604051
[29] M.N. Chernodub, C. Corianò, M.M. Maglio, Phys. Lett. **B802**, 135236 (2020), 1910.13727

Tracking performance studies for the Experimental Setup at the Electron-Ion Collider

Shyam Kumar[1,3,*], *Annalisa* Mastroserio[2,3], and *Domenico* Elia[3]

[1]Università degli Studi di Bari Aldo Moro, Italy
[2]Università degli Studi di Foggia, Italy
[3]INFN, sezione di Bari, Italy

Abstract. The US Electron-Ion Collider (EIC) is a future facility to be built at the Brookhaven National Laboratory (BNL) to study the collisions of polarized electrons with polarized protons and ions. It will provide the answers to fundamental questions on quark and gluon interactions such as: how colored partons and colorless jets travel in the nuclear medium, the properties of the state of matter in high gluon density (low-x) regime, distribution of quarks, gluons, and their spin inside a nucleon. EIC is expected to run at the luminosity of 10^{32}-10^{34} cm^{-2} sec^{-1} and center-of-mass energy 20-140 GeV [1]. ATHENA (A Totally Hermetic Electron Nucleus Apparatus) is one of the detector designs provided as the response to the call for the proposal issued by the EIC Project in 2021. It consists of a tracking system with wide pseudorapidity coverage ($|\eta|<3.5$), high granularity (pixel size of inner layers ~10 μm), and low material budget (0.05% of X_0 per layer) for the innermost silicon layers to achieve good tracking and vertexing performances.

1 Introduction

ATHENA is one of the experimental designs proposed for experiments at the US EIC, providing the basis for the further optimization of tracking and vertexing performance currently ongoing. Within this context, we performed the simulation in Fun4All [2] and Detector Description Toolkit for High Energy Physics (DD4hep) [3] frameworks. We first simulate a single particle using an event generator uniform in pseudorapidity (η) and momentum then transport it to the detector material using a particle gun in GEANT4 [4]. The dedicated full digitization and hit reconstruction is performed considering the response of each detector. After the hit reconstruction, tracking is performed using a Kalman filter to estimate the tracking performance of the detector system. The tracking performances are given by the momentum resolution and distance of closest approach (DCA) resolution [5,6]. The momentum resolution affects the RMS of invariant mass spectra of the identified particles and DCA resolution is very important for the reconstruction of secondary particles mostly originated by the weak decays of particles that decay close to the interaction point, e.g. D^0 meson ($c\tau$ = 123 μm). In this paper, we present the study of tracking performances

*e-mail: shyam.kumar@ba.infn.it

of the ATHENA detector [7] in each η bin and compare them with the requirements dictated by the physics goals of the experiment.

2 Design of the Tracker

The ATHENA tracker consists of detectors in the central region, forward region, and the backward region as shown in Fig. 1 and detailed in the following sections.

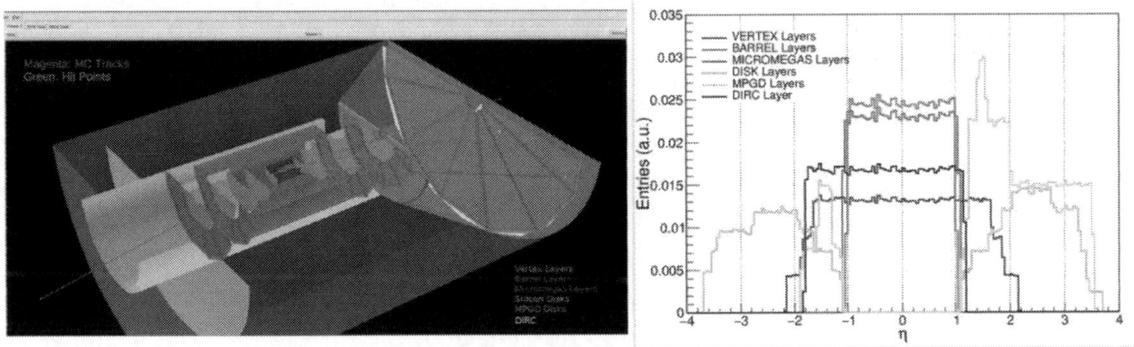

Fig. 1. (Left) Event Display of ATHENA showing the detector volumes with embedded pion MC tracks (magenta) and hits (green) using Fun4All simulation framework; (Right) Pseudorapidity coverage for each detector from the hit points on the detector surface.

1. Central Tracker

The central part of the ATHENA tracking system consists of three vertex layers and two barrel sagitta layers based on silicon Monolithic Active Pixel Sensors (MAPS) in 65 nm CMOS technology currently being developed by the ALICE Collaboration for the upgrade of the inner tracking system in view of Run 4 (ITS3): a schematic picture of the ALICE ITS3 system is shown in Fig. 2. Each of the silicon vertex layers consists of pixel cells with a size ~ 10 μm. It also consists of four micromegas layers with a resolution of $\sigma_{r\varphi} = \sigma_z =$ 150 μm. The innermost layers of silicon sensors with an extremely small material budget and excellent spatial resolution will be key for achieving the physics requirements. The configuration details of the central tracking layers are described in Tables 1, 2, and 3.

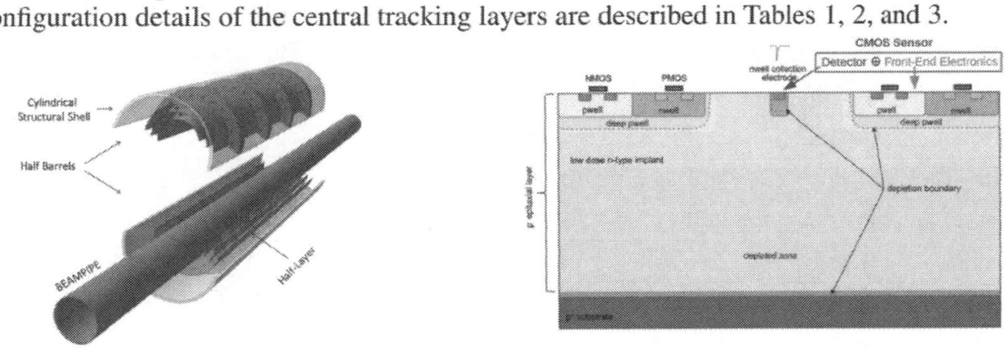

Fig. 2. (Left) The layout of the ALICE ITS3 detector [8]; (Right) ALICE ITS3 MAPS Sensor [9].

Table 1. The geometric parameters of Silicon Vertex Layers [10].

Radius (cm)	Length (cm)	Pixel Pitch	(X/X_0 %)
3.3	28.0	10 μm	0.05
4.35	28.0	10 μm	0.05
5.4	28.0	10 μm	0.05

Table 2. The geometric parameters of Silicon Barrel (Sagitta) Layers [10].

Radius (cm)	Length (cm)	Pixel Pitch	(X/X_0 %)
13.34	34.34	10 μm	0.55
17.96	46.68	10 μm	0.55

Table 3. The geometric parameters of Micromegas Barrel Layers [10].

Radius (cm)	Length (cm)	Resolution	(X/X_0 %)
47.72	127.47	150 μm ($\sigma_{r\varphi}$) x 150 μm (σ_z)	0.4
49.57	127.47	150 μm ($\sigma_{r\varphi}$) x 150 μm (σ_z)	0.4
75.61	201.98	150 um ($\sigma_{r\varphi}$) x 150 μm (σ_z)	0.4
77.46	201.98	150 μm ($\sigma_{r\varphi}$) x 150 μm (σ_z)	0.4

2. Forward and Backward Tracker

In the forward region, the ATHENA tracker consists of six silicon disks ($X/X_0 = 0.24\%$ per layer) with a pixel cell size ~ 10 μm, while in the backward region five silicon disks with the similar pixel size and material budget. The two outer silicon disks in the forward and backward region are complemented by the micro-pattern gaseous detector (MPGD) layers to avoid the acceptance gap while moving from the barrel to the forward or backward region. The setup also has an MPGD layer in the most forward region. The MPGD layers have the material budget ($X/X_0 = 0.4\%$ per layer) and spatial resolution σ_r = 250 μm, $\sigma_{r\varphi}$ = 50 μm. The configuration details of the forward and backward tracking layers are described in Tables 4 and 5.

Table 4. The geometric parameters of Forward and Backward Silicon Disks [10].

Six Forward Silicon Disks			5 Backward Silicon Disks		
R_{in} (cm)	R_{out} (cm)	Z (cm)	R_{in} (cm)	R_{out} (cm)	Z (cm)
3.18	18.62	25.0	3.18	18.62	-25.0
3.18	36.50	49.0	3.18	36.50	-49.0
3.47	43.23	73.0	3.18	43.23	-73.0
5.08	43.23	103.65	3.95	43.23	-109.0
6.58	43.23	134.33	5.26	43.23	-145.0
8.16	43.23	165.0			

Table 5. The geometric parameters of Forward and Backward MPGD Layers [10].

Forward MPGD Layers			Backward MPGD Layers		
R_{in} (cm)	R_{out} (cm)	Z (cm)	R_{in} (cm)	R_{out} (cm)	Z (cm)
44.68	76.91	105.76	44.68	76.91	-103.0
44.68	76.91	161.74	44.68	76.91	-141.74
19.34	195.5	332.0			

3 Simulation Results

To understand the performance of the tracking system negative pions were simulated and their transport to the detector material was done using GEANT4. After simulation, digitization, hit reconstruction, and tracking was performed within the framework. We evaluated the momentum resolution as a function of momentum, and DCA in the transverse plane as a function of transverse momentum in different η intervals as shown in Fig. 3 and 4. The green points are extracted from the full detector simulations, the orange curve corresponds to the results of the fit to the data points, and the blue curve represents the physics requirements as shown in Table 6.
The uncertainty in the momentum measurement comes from two sources: the finite size of pixels, namely the spatial resolution (S.R.), and the thickness of the detector material, which impacts on the multiple scattering (M.S.), experienced by the particles crossing it. These effects at η = 0 can be parameterized for the detector planes with equal spatial resolutions, spacing, and equal thickness (Gluckstern approach) with the following expressions [5,6] :

$$\left(\frac{\sigma_p}{p}\right)_{SR} \approx \frac{\sigma_{r\varphi}\, p}{0.3\, B\, L^2}\sqrt{\frac{720}{N+5}} \qquad \left(\frac{\sigma_p}{p}\right)_{MS} \approx \frac{0.0136\, GeV/c}{0.3\, \beta B\, L}\sqrt{\frac{d_{total}}{X_0}} \qquad \left(\frac{\sigma_p}{p}\right) = \sqrt{\left(\frac{\sigma_p}{p}\right)_{SR}^2 + \left(\frac{\sigma_p}{p}\right)_{MS}^2}$$

where p, $\sigma_{r\varphi}$, B, β, d_{total}, X_0, and L represent momentum, spatial resolution of the detector, magnetic field, velocity, total thickness of detector, radiation length of the traversed material, and the lever arm (distance between first and the last layer of setup) of the detector setup, respectively.
Momentum resolution due to spatial resolution varies linearly with momentum for which the slope can be reduced by increasing the number of detector planes (N), by decreasing the spatial resolution, and by increasing the magnetic field and lever arm. The momentum resolution due to multiple scattering stays constant except at low momentum where there is an increase because of the lower value of beta at low momentum. The constant term can be reduced by increasing the product between the magnetic field and lever arm (B*L) but that will also increase d_{total}. Therefore, we need to balance between two terms S.R. and M.S. to achieve an optimal detector design to minimize the uncertainty in the momentum measurement. The analytical expression used for the fit ($f(dp/p)=\sqrt{A^2 p^2+B^2}$) of the momentum resolution curves obtained by the ATHENA simulation studies consists of two terms: the first one (M.S.) is considered to be constant after neglecting the low momentum regime (β <1) while the other (S.R.) is linear in momentum.
The transverse DCA resolution (DCA_{xy} or DCA_T) also consists of two similar terms associated with S.R. and M.S. respectively, as appears from the following formulas [5,6]:

$$(\sigma DCA_{xy})_{SR} \approx \frac{3\sigma_{r\phi}}{\sqrt{N+5}}\sqrt{1+8\frac{r_0}{L}+28\frac{r_0^2}{L^2}+40\frac{r_0^3}{L^3}+20\frac{r_0^4}{L^4}}$$

$$(\sigma DCA_{xy})_{MS} \approx 0.0136\,GeV/c\,\frac{r_0}{\beta p_T}\sqrt{\frac{d}{X_0 \sin\theta}}\sqrt{1+\frac{1}{2}\left(\frac{r_0}{L}\right)+\frac{N}{4}\left(\frac{r_0}{L}\right)^2} \qquad \sigma DCA_{xy}=\sqrt{(\sigma DCA_{xy})_{SR}^2+(\sigma DCA_{xy})_{MS}^2}$$

In the case of DCA_{xy} resolution, the S.R. terms stay constant with p_T. It can be reduced by decreasing spatial resolution, increasing the number of layers, and putting the first layer very close to the interaction point (r_0 radius of the first layer). The multiple scattering term decreases with momentum because of the $1/p_T$ term: it can be further be decreased by reducing r_0 and the thickness of sensors. Therefore, the ATHENA fit ($f(DCA_T)=\sqrt{A^2/p_T^2+B^2}$) for the DCA_{xy} resolution also consists of two terms, one constant (S.R.) and the other proportional to the inverse of momentum (M.S.). The tracking performances met the physics requirements as stated in the Yellow Report (YR) [1], except for the most backward -3.5 < η < -2.5 and momenta below 2 GeV, respectively. Further technology R&D to reduce material, combined with sophisticated analysis techniques and further overall detector layout optimizations will be needed to fully achieve the EIC physics goals.

Table 6. Physics requirements on tracking in different bins [1].

Acceptance	Momentum Resolution σ(p)/p		DCA_T Resolution	
	Performances	Requirements	Performances	Requirements
-3.5 < η< -2.5	~ 0.04%×p ⊕1.5 %	~ 0.1%×p ⊕0.5 %	~ 80/p_T ⊕10 μm	~ 30/p_T ⊕50 μm
-2.5 < η< -1.0	~ 0.01%×p ⊕ 0.5 %	~ 0.05%×p ⊕ 0.5 %	~ 50/p_T ⊕5 μm	~ 30/p_T ⊕20 μm
-1.0 < η< 1.0	~ 0.05%×p ⊕ 0.4 %	~ 0.05%×p ⊕ 0.5 %	~ 30/p_T ⊕5 μm	~ 20/p_T ⊕5 μm
1.0 < η< 2.5	~ 0.01%×p ⊕ 0.5 %	~ 0.05%×p ⊕ 1.0 %	~ 50/p_T ⊕5 μm	~ 30/p_T ⊕20 μm
2.5 < η< 3.5	~ 0.02%×p ⊕ 1.5 %	~ 0.1%×p ⊕ 2.0 %	~ 80/p_T ⊕10 μm	~ 30/p_T ⊕50 μm

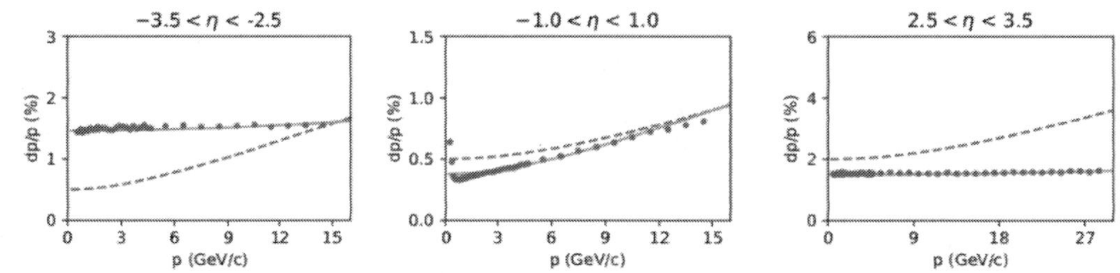

Fig. 3. Momentum resolutions versus momentum for the ATHENA setup corresponding to simulated pions- uniform in η momentum compared to the physics requirements in YR for selected η bins [7].

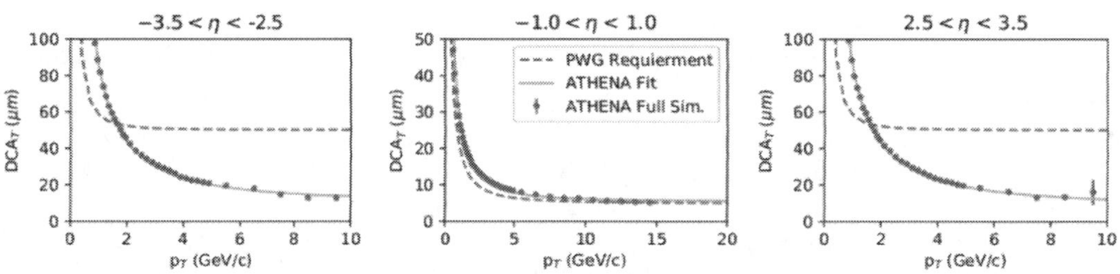

Fig. 4. Transverse DCA resolutions versus transverse momentum for the ATHENA setup corresponding to simulated pions- uniform in η momentum compared to the physics requirements in YR for selected η bins [7].

4 Conclusions

EIC is a future facility with unique physics programme due to its unique characteristics (e.g. energy, luminosity, polarized beams). It will require an excellent tracking detector system that can reconstruct particles produced in the collision with very small uncertainties. As described above, we performed the full simulation of the ATHENA tracker in Fun4All and DD4hep simulation frameworks. The tracking performances are estimated using the above geometry and the physics requirement are satisfied everywhere except that in few η bins where the tracker configuration still needs to be optimized both in geometry and material budget. The EIC community is currently working in Electron Proton Ion Collider (EPIC) collaboration with the aim of combining the existing studies performed for the detector proposals in 2021 (including ATHENA) and preparing a baseline detector configuration in view of the technical design report (TDR) activities in 2023.

References

1. R. Abdul Khalek, *et al.*, Science Requirements and Detector Concepts for the Electron-Ion Collider: *EIC Yellow Report*, Nucl. Phys. A **1026**, 122447 (2022)
2. Fun4All, https://wiki.bnl.gov/sPHENIX/index.php/EIC_SPHENIX_FUN4ALL
3. M. Frank, F. Gaede, C. Grefe and P. Mato, DD4hep: DD4hep: A Detector Description Toolkit for High Energy Physics Experiments, J. Phys. Conf. Ser. **513**, 022010 (2014)
4. S. Agostinelli *et al.* [GEANT4], GEANT4--a simulation toolkit, Nucl. Instrum. Meth. A **506**, 250-303 (2003)
5. R. L. Gluckstern, Uncertainties in track momentum and direction, due to multiple scattering and measurement errors, Nucl. Instrum. Meth. **24**, 381-389 (1963)
6. Z. Drasal and W. Riegler, An extension of the Gluckstern formulae for multiple scattering: Analytic expressions for track parameter resolution using optimum weights, Nucl. Instrum. Meth. A **910**, 127-132 (2018)
7. J. Adam *et al.*, ATHENA detector proposal — a totally hermetic electron nucleus apparatus proposed for IP6 at the Electron-Ion Collider, *JINST* **17** P10019 (2022)
8. ALICE Collaboration, *Letter of Intent for an ALICE ITS Upgrade in LS3*, CERN-LHCC-2019-018 , LHCC-I-034, https://cds.cern.ch/record/2703140/files/LHCC-I-034.pdf
9. W. Snoeys, *et al.*, A process modification for CMOS monolithic active pixel sensors for enhanced depletion, timing performance and radiation tolerance, Nucl. Instrum. Meth. A **871**, 90-96 (2017)
10. ATHENA Twiki, https://wiki.bnl.gov/athena/index.php/Tracking

Trigger rejection factor in the first kaonic helium run with the complete SIDDHARTA-2 setup

L. De Paolis[1,*], *M.* Bazzi[1], *D.* Bosnar[3], *M.* Bragadireanu[4], *M.* Cargnelli[2], *M.* Carminati[5], *A.* Clozza[1], *G.* Deda[5], *R.* Del Grande[6,1], *L.* Fabbietti[6], *C.* Fiorini[5], *I.* Friščić[3], *C.* Guaraldo[1], *M.* Iliescu[1], *M.* Iwasaki[7], *A.* Khreptak[1,8], *P.* King[5], *P.* Levi Sandri[1], *S.* Manti[1], *J.* Marton[2], *M.* Miliucci[1], *P.* Moskal[8], *F.* Napolitano[1], *S.* Niedźwiecki[8], *H.* Ohnishi[9], *K.* Piscicchia[10], *Y.* Sada[9], *A.* Scordo[1], *F.* Sgaramella[1], *H.* Shi[2], *M.* Silarski[8], *D.L.* Sirghi[1,10,4], *F.* Sirghi[1,4], *M.* Skurzok[8], *A.* Spallone[1], *K.* Toho[9], *M.* Tüchler[2,11], *O.* Vazquez Doce[1], *C.* Yoshida[9], *J.* Zmeskal[2], and *C.* Curceanu[1]

[1]INFN, Laboratori Nazionali di Frascati, Via E. Fermi 54, I-00044 Frascati(RM), Italy

[2]Stefan Meyer Institute for Subatomic Physics, Kegelgasse 27, 1030 Wien, Austria

[3]Department of Physics, Faculty of Science, University of Zagreb, Zagreb, Croatia

[4]Horia Hulubei National Institute of Physics and Nuclear Engineering, Str. Atomistilor No. 407, PO Box MG-6 Bucharest-Măgurele, Romania

[5]Politecnico di Milano, Dipartimento di Elettronica, Informazione e Bioingegneria and INFN Sezione di Milano, I-20133 Milano, Italy

[6]Excellence Cluster Universe, Technische Universiät München, Boltzmannstraße 2, 85748 Garching bei München, Germany

[7]RIKEN, Institute of Physical and Chemical Research, Wako, Japan

[8]Faculty of Physics, Astronomy, and Applied Computer Science, Jagiellonian University, Łojasiewicza 11, 30-348 Kraków, Poland

[9]Research Center for Electron Photon Science (ELPH), Tohoku University, Sendai, Japan

[10]Centro Ricerche Enrico Fermi—Museo Storico della Fisica e Centro Studi e Ricerche 'Enrico Fermi', Via Panisperna 89A, 00184, Roma, Italy

[11]University of Vienna, Vienna Doctoral School in Physics, Vienna, Austria

Abstract. The SIDDHARTA-2 experiment aims to perform the first measurement of the kaonic deuterium 2p $\rightarrow$ 1s x-ray transition energy. Such measurement, together with the measurement of kaonic hydrogen 2p $\rightarrow$ 1s x-ray energy transition performed by the SIDDHARTA experiment in 2011, allows the determination of kaon proton and kaon neutron scattering lengths and represents a fundamental input for the low energies QCD in the strangeness sector theory. The SIDDHARTA-2 experiment is presently installed at the DAΦNE electron-positron collider at the National Laboratories of Frascati, in Italy. In May 2022, the kaonic ^4He x-ray transitions measurement was performed by the complete SIDDHARTA-2 setup, by using a gaseous target. The result of this measurement is presented in this paper, with a specific focus on the background rejection performed by the kaon trigger system.

*e-mail: Luca.DePaolis@lnf.infn.it

1 Introduction

Hadronic atoms are atomic systems in which an electron is replaced by a different negatively charged hadron, for example a pion or a kaon, bound by its electromagnetic interaction with the nucleus. The experimental research on hadronic atoms allows the study of the strong interaction in an atomic system, at low energy, and provided a large amount of data and constraints for nuclear and particle physics. Quantum electrodynamics (QED) allows a precise description of the electromagnetic properties of the kaonic atom, providing energy levels and x-ray transitions of the atomic system. However, in the innermost level of the hadronic atom, energy shifts and broadening of the atomic levels, due to the strong hadron-nucleus interaction, can be measured with x-ray spectroscopy, thus getting fundamental information on quantum chromodynamics (QCD) and strong interaction at threshold [1].
Kaonic atoms are hadronic atoms formed whenever a K^- is captured by an atom, replacing an electron in the atomic shell. The K^- contains the strange quark s. Therefore, experiments with kaonic atoms allow the investigation of the strong interaction at low-energy in the strageness sector. In the kaonic atoms' formation, the K^- is captured in a highly excited energy level and it starts an electromagnetic cascade process until the nuclear absorption. The average lifetime of K^- is $\tau_{K^-} = 1.2 \times 10^{-8}$ s [2], whilst each of the electromagnetic cascade occurs in less than 10^{-9} s [1]. The kaon-nucleus strong interaction affects the energy orbitals of the kaonic atoms producing shifts and broadenings of the energy levels, but only for the innermost levels, these contributions become measurable. The kaonic hydrogen and kaonic deuterium offer a unique opportunity for the experimental investigation of the strong interaction between kaons and nucleons at low energies.
At low energies, the perturbative QCD models cannot be employed anymore for the description of the strong interaction among hadrons, and various phenomenological models are used. In this regime, the K^-N strong interaction is described for example with models based on Chiral Perturbation Theory (χPT). The strong coupling $\pi\Sigma$ and the resonance Λ (1405), just below the K^-N energy threshold, complicate the description of kaon-nucleon strong interaction with simple approach models in χPT [3, 4]. The state-of-art of theoretical approaches is presented with a comparative analysis in [5]. The measurement of shifts and widths of the 1s energy level in kaonic hydrogen and kaonic deuterium allows the extraction of the K^-p and K^-n scattering lengths, providing fundamental constraints for the theoretical model for the description of kaon-nucleon strong interaction at low energies [6–8].
In the last 25 years, several experiments performed the measurement of light kaonic atoms with x-ray spectroscopy. Kaonic hydrogen was investigated by the KpX experiment at J-PARC in 1997 [9], and by the DEAR experiment at the National Laboratories of Frascati (LNF) in 2005 [10]. In 2011, the SIDDHARTA experiment performed the most precise measurement of kaonic hydrogen 1s level shift and width [11]. Moreover, kaonic ^3He and ^4He measurements were performed by the E570 and the E62 experiment at J-PARC [12], in Japan, and by the SIDDHARTA experiment at LNF, in Italy [13]. In July 2022, the SIDDHARTA-2 experiment, actually installed on the Interaction Region (IR) of the DAΦNE electron-positron collider at LNF [14], started the data taking to perform the first measurement of the 1s energy level shift and width in kaonic deuterium.
In January 2021, the SIDDHARTINO experiment, a reduced version of the SIDDHARTA-2 setup with 1/6 of the x-ray Silicon Drift Detectors (SDDs), performed the best measurement of gaseous $K^{-4}He$ x-ray transitions [15]. The SIDDHARTINO $K^{-4}He$ measurement had a dual goal: to perform the tuning of the various components of the SIDDHARTA-2 setup and to provide a feedback to DAΦNE collider during its commissioning. In the second half of 2021, the installation of the SIDDHARTA-2 apparatus was completed and a new $K^{-4}He$ x-ray transitions measurement, in gas, was performed. The goal of the SIDDHARTA-2 $K^{-4}He$

measurement is to confirm the performance shown by SIDDHARTINO, in view of the kaonic deuterium measurement (2023). A background rejection factor provided by the Kaon Trigger of the same order of magnitude obtained with SIDDHARTINO (10^{-5}) was confirmed, a key ingredient for the challenging $K^{-}d$ measurement.
In this paper, the SIDDHARTA-2 setup is described, and spectra of the $K^{-4}He$ x-ray atomic transitions measurement is shown. Future perspectives for the SIDDHARTA-2 kaonic deuterium measurement, are concluding the article.

2 The SIDDHARTA-2 experimental setup

Following SIDDHARTINO, the SIDDHARTA-2 apparatus was installed on the Interaction Region (IR) of the DAΦNE $e^{+}e^{-}$ collider, at INFN-LNF, in 2021. The DAΦNE collider is a Φ factory. Electron and positron beams, both tuned at 510 MeV/c, are collided, producing Φ mesons which decay in low energy monochromatic back-to-back $K^{+}K^{-}$ pairs (48.9% of branching ratio). The kaons are produced with a momentum of $\sim 127 MeV/c$ and a low energy spread ($\Delta p/p \simeq 0.1\%$) [14]. The DAΦNE collider is a low energy kaon source, ideal for kaonic atom measurements.
The SIDDHARTA2 experimental target is a cylindrical cell of 144 mm in diameter and 125 mm in height. The walls are made of a 2-Kapton layer 150 μm thick, placed on aluminium support (see Fig. 1). The target is filled with the gas where kaonic atoms are being formed.

Figure 1. Picture of the SIDDHARTA2 target cell. The cylindrical target is surrounded by the SDD detector arrays for the x-ray spectroscopy.

In the SIDDHARTA-2 experiment, the cylindrical target is surrounded by 384 large area monolithic SDDs, specifically developed for the kaonic atom x-ray spectroscopy and providing a high collection efficiency (> 90%) for x-ray energies between 5 keV and 12 keV [16]. Each SDD is 450 μm thick and has an active area of 0.64 cm^2. The detectors are organized in arrays. Each array consists of 8 SDDs placed in two raws, with 4 SDD cells each (2×4 matrix), as is shown in Fig. 2.

The silicon wafer is glued on an alumina carrier providing the polarization of the SDD cells via an external voltage. Whenever an x-ray is absorbed within the silicon bulk, the generated charges are collected by a central anode. A C-MOS charge-sensitive amplifier is

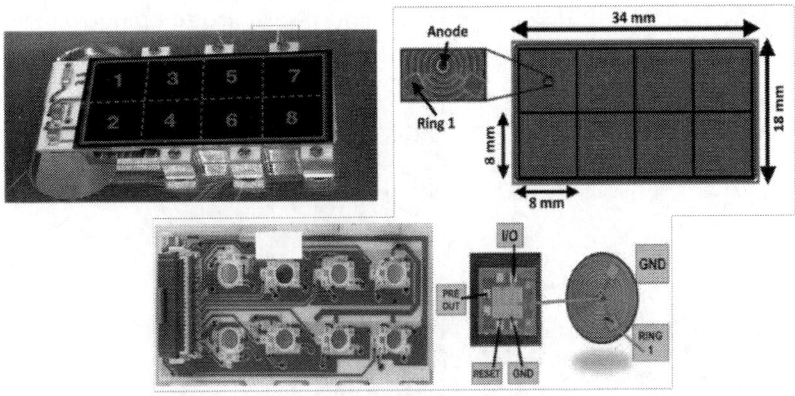

Figure 2. *Top left:* a picture of an SDD array; *Top right:* A schematic view of the SDD array; *Bottom:* schematization of the ceramic support on which the SDD cells are mounted.

closely bonded to each detector unit [17]. The signals are processed through a dedicated ASIC chip named SFERA [18, 19]. The SIDDHARTA-2 SDDs provide a resolution of $157.8 \pm 0.3(stat)^{+0.2}_{-0.2}(syst)$ eV at 6.4 keV ($K^{-4}He$ 3d → 2p transition energy) and a linearity of 2-3 eV, as tested and optimized in the commissioning phase at DAΦNE [16, 20]. In the SIDDHARTA-2 apparatus, the SDDs and the target are placed inside a vacuum chamber evacuated below 10^{-5} mbar. A closed-cycle helium refrigerator cooling system keeps the target cell cooled at about 25 K. Pressure and temperature parameters provide the best performance conditions for SDD devices and to stop kaons within the target. The SDD detectors calibration is performed *in situ* with two x-ray tubes installed outside the vacuum chamber, on the apparatus, which exciting foils of titanium and copper placed close to the detectors, inside the vacuum chamber. In the vacuum chamber, plastic scintillators read by pairs of silicon photo-multipliers (SiPMs) are placed radially outside the cylindrical target detectors for external background identification (Veto-2 system) [21, 22]. Outside the vacuum chamber, twelve plastic scintillators read by pairs of photo-multiplier tubes (PMTs), are placed radially around the target (Veto-1 system). The Veto-1 and Veto-2 systems aim to suppress the synchronous and asynchronous backgrounds coming from the pions emitted by the kaon absorption and kaon decays in kaonic atoms. A lead shielding is placed around the vacuum chamber, downstream and upstream the beam-pipe, to reduce the background from the accelerator. Moreover, two lead walls are installed and placed along the beamline in both directions, externally to the setup, to shelter the apparatus from the particles, mostly minimum ionizing particles (MIPs), mainly lost from the electron and positron rings because of the Touschek effect. A shaped Mylar degrader is placed between the target and the Interaction Point (IP) of the DAΦNE accelerator to slow down the kaons, optimizing their stopping in the target gas and the kaonic atoms formation [15]. A Kaon Trigger (KT), consisting of two pairs of plastic scintillators read by photo-multipliers placed above and below the IR, is installed in the apparatus for the identification of the K^+K^- back-to-back emitted from the decay of the ϕ meson in the IR. The KT uses the Time Of Flight (ToF) to identify the K^- directed towards the target cell, in the vertical region. The SIDDHARTA-2 luminometer (constructed based on the J-PET technology [23–25]), consisting of a pair of plastic scintillators read by photo-multipliers placed on the longitudinal plane, is installed near the IR (see Fig. 3). The SIDDHARTA-2 luminometer measures the kaons rate using the horizontal direction [26]. A GEANT4 [27] dedicated simulation was used to optimize the geometry of the entire apparatus. A schematic drawing of the SIDDHARTA-2 setup is provided in Fig. 3.

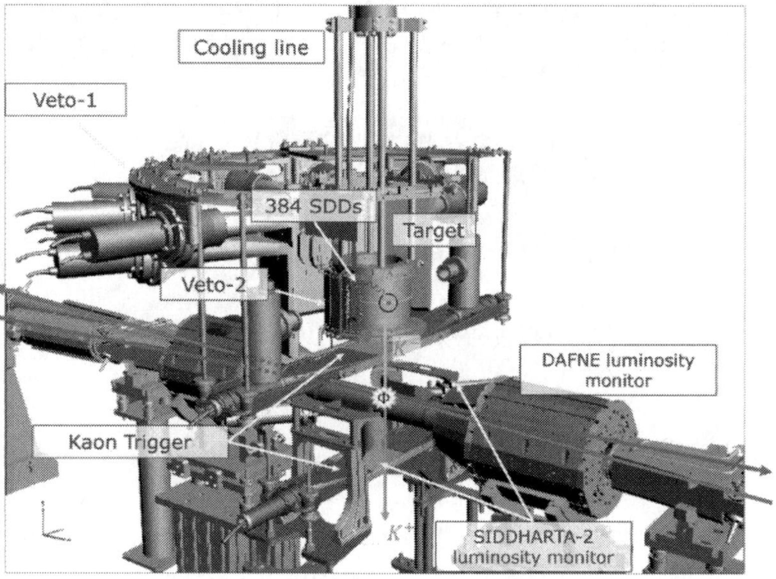

Figure 3. A schematic drawing of the SIDDHARTA-2 apparatus installed on the DAΦNE Interaction Region at the INFN National Laboratories of Frascati.

3 The Kaonic Helium 4 SIDDHARTA-2 measurement

The SIDDHARTA-2 experiment performed the $K^{-4}He$ x-ray transitions measurements in May 2022 period, collecting 28 pb^{-1} integrated luminosity aiming to characterize the full system, previous to the $K^{-}d$ data taking campaign. The ^4He gas in the target cell was kept to a density of about 1.4% the Liquid Helium one. The SDDs were cooled at $-145°C$. The Kaon Trigger was fundamental for the discrimination of signal events. For the asynchronous background cut, only the triggered events with signals detected in coincidence by the two scintillators were selected. The MIPs produced by the electromagnetic showers were discriminated from the kaons using the Time-of-Flight information (ToF). The final spectrum of the $K^{-4}He$ x-ray atomic transitions measurement performed by the SIDDHARTA-2 experiment is shown in Fig. 4. The background rejection factor by using the Kaon Trigger is 3.1×10^5, confirming the excellent performance measured by SIDDHARTINO [15]. The effect can be observed comparing the spectra shown in Fig. 4.

4 Conclusions and future perspectives

This paper presents the kaonic helium 4 measurement performed by the SIDDHARTA-2 experiment. The measurement is the first one performed with the complete SIDDHARTA-2 experimental apparatus installed on the Interaction Region of the DAΦNE collider at the National Laboratories of Frascati (INFN). The measurement was performed in May 2022. The results confirm the very good performance of the trigger system achieved during the SIDDHARTINO test run [15], performed in early 2021. In the specific, the crucial background

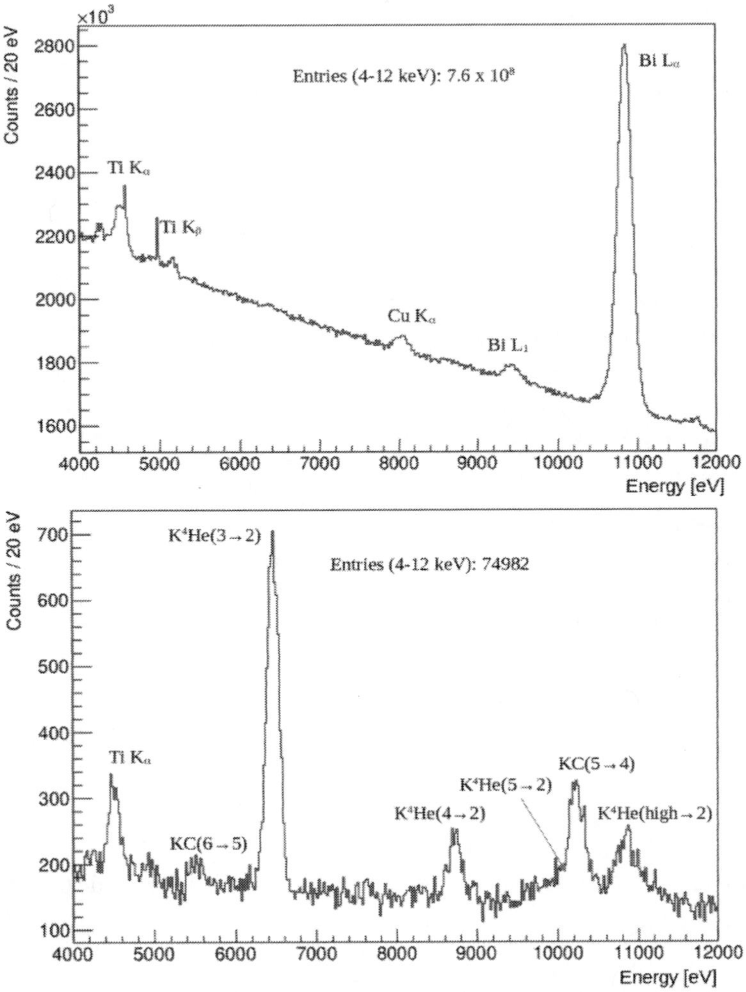

Figure 4. *Top*: The complete spectrum obtained by the SIDDHARTA-2 experiment measuring $K^{-4}He$ x-ray atomic transitions (28 pb^{-1}). Fluorescence peaks from Ti, Cu and Bi are present. *Bottom:* The final spectrum got applying the Kaon Trigger background reduction.

rejection factor is confirmed to be in the order of 10^5, thus providing optimal experimental conditions to start the first measurement of the kaonic deuterium 2p → 1s transition, which will be performed in 2023.

Acknowledgments

We thank C. Capoccia from LNF-INFN and H. Schneider, L. Stohwasser, and D. Pristauz-Telsnigg from Stefan- Meyer-Institut for their fundamental contribution in designing and building the SIDDHARTA-2 setup. We thank as well the DAΦNE staff for the excellent working conditions and permanent support. Part of this work was supported by the Austrian Science Fund (FWF): [P24756-N20 and P33037-N]; the Croatian Science Foundation under the project IP-2018-01-8570; the EU STRONG-2020 project (Grant Agreement No. 824093), the EU Horizon 2020 project under the MSCA (Grant Agreement 754496); Japan Society for the Promotion of Science JSPS KAKENHI Grant No. JP18H05402; the Polish Ministry of

Science and Higher Education grant No. 7150/E-338/M/2018; the Polish National Agency for Academic Exchange (grant no PPN/BIT/2021/1/00037); the SciMat and qLife Priority Research Areas budget under the program Excellence Initiative - Research University at the Jagiellonian University.

References

[1] C. Curceanu et al., Rev. Mod. Phys. **91**, 025006 (2019)
[2] P. A. Zyla et al. (Particle Data group), Prog. Theor. Exp. Phys. (2020)
[3] Chang-Hwan Lee, Hong Jung, Dong-Pil Min, and Mannque Rho, Phys. Lett. B **326**, 14–20 (1994)
[4] V. Bernard, Prog. Part. Nucl.Phys. **60**, 82–160 (2008)
[5] A. Cieplý, M. Mai, Ulf-G. Meißner, and J. Smejkal., Nucl. Phys. A **954**, 3 (2016)
[6] M. Döring, U. G. Meißner, Phys. Lett. B **704**, 663 (2011)
[7] U.-G. Meissner, U. Raha, A. Rusetsky, Eur. Phys. J. C **47**, 473 (2006)
[8] F. Napolitano et al., Physica Scripta **97**, 8 (2022)
[9] M. Iwasaki et al., Phys. Rev. Lett. **78**, 3067 (1997)
[10] G. Beer et al. (DEAR Collaboration), Phys. Rev. Lett. **94**, 212302 (2005)
[11] M. Bazzi et al., Phys. Lett. B **704**, 113 (2011)
[12] S. Okada et al., Phys. Lett. B **653**, 387 (2007)
[13] M. Bazzi et al., Eur. Phys. J. A **50**, 91 (2014)
[14] M. Zobov et al., Phys. Rev. Lett. **104**, 174801 (2010)
[15] D. Sirghi et al., J. Phys. G **49**, 055106 (2022)
[16] M. Miliucci et al., Meas. Sci. Technol. **33**, 095502 (2022)
[17] L. Bombelli et al., IEEE Nuclear Science Symposuim and Medical Imaging Conference, pp. 135–138 (2010)
[18] R. Quaglia et al., Nucl. Instrum. Meth. A **824**, 449 (2016)
[19] F. Schembari, R. Quaglia, G. Bellotti, C. Fiorini, IEEE Transactions on Nuclear Science **63**, 1797 (2016)
[20] Miliucci et al., Condens. Matter. **4**, 31 (2019)
[21] F. Sgaramella et al., arXiv **2201**, 12101 (2022)
[22] M. Tüchler et al., J. Phys.: Conf. Ser. **1138**, 012012 (2018)
[23] P. Moskal et al., Science Advances **7**, eabh4394 (2021)
[24] P. Moskal et al., Nature Communications **12**, 5658 (2021)
[25] S. Niedźwiecki et al., Acta Phys. Pol. B **48**, 1567 (2017)
[26] M. Skurzok et al., JINST **15**, P10010 (2020)
[27] S. Agostinelli et al., Nucl. Inst. Meth. Phys. Res. Sect. A **506**, 250–303 (2003)

Hadronic ligh-by-light contribution to the anomalous magnetic moment of the muon: The role of scalar resonances in a holographic model of QCD

Luigi Cappiello[1,2,*]

[1]Dipartimento di Fisica "Ettore Pancini", Universitá di Napoli 'Federico II', Via Cintia, 80126 Napoli, Italy
[2]INFN-Sezione di Napoli, Complesso Universitario di Monte S. Angelo, Via Cintia, Edificio 6, 80126 Napoli, Italy

Abstract. We review the evaluation of scalar mesons contribution to the hadronic light-by-light piece of the muon anomalous magnetic moment, using a holographic model of QCD. We evaluate the contributions of the lightest, sub-GeV scalars $\sigma(500)$, $a_0(980)$ and $f_0(980)$ and their associated towers of excited states. Our results point at a negative contribution, overwhelmingly dominated by the $\sigma(500)$ meson, that we estimate at $a_\mu^{\mathrm{HLbL,S}} = -9(2)\cdot 10^{-11}$, in very good agreement with recent determinations from dispersive analyses.

1 Introduction

The Fermilab measurement of the muon anomalous magnetic moment [1], based on the first data run, has confirmed the previous BNL E871 result [2], with a similar precision. Compared to the most recent theory determination [3] (based on Refs. [4–23]) the discrepancy with the theory expectation is now beyond 4σ. In few years, Fermilab precision will be improved by a factor 4, thereby reaching the projected 0.14ppm.

On the theory side, hadronic contributions are currently the focus of attention. Rather surprisingly, the most serious issue concerns now the hadronic vacuum polarization contribution (HVP), with an evident discrepancy between increasingly precise lattice simulations (see, e.g., Ref. [24]) and data-driven analyses [12, 13] and low-energy phenomenology [25–28].

The hadronic light-by-light (HLbL) shows instead a reasonable level of precision, with an overall agreement between data-driven dispersive analysis [29, 30] and lattice simulations [22, 31], still having room for more precise determinations in the near future.

2 Holographic QCD

In a parallel theoretical approach, some hadronic resonances contributions to HLbL have been evaluated using five-dimensional Lagrangian-based models using techniques inspired by the AdS/CFT correspondence [32–34]. When compactified to four dimensions, these models have a natural interpetation as hadronic models of QCD with an infinite number of states for the different meson channels, as it is expected in the large-N_c limit of QCD [35, 36]. These

*e-mail: luigi.cappiello@na.infn.it

HQCD models [37–42] are provided with the following relevant features: *(i)* at high energies, they are conformal invariant and match the QCD asymptotic behaviour of correlators, reproducing leading short distance (SD) constraints of perturbative QCD; *(ii)* spontaneous chiral symmetry breaking can be implemented, so that they contain a pion multiplet, together with all the consequences of chiral dynamics built in; *(iii)* chiral anomaly is consistently implemented at all energy scales; *(iv)* in a minimal settings, HQCD models contain very few free parameters (e.g. the size of the extradimension, the 5D gauge coupling), which can be fitted to physical quantities, making the HQCD model rather predictive (or actually over constrained); *(v)* correlators of QCD quark currents, relevant to the evaluations of hadronic contributions to a_μ can be obtained almost analytically [43–45].

Thus, HQCD models appear especially suited to address conceptual issues in simplified but consistent settings, in particular those related to the duality between hadronic contributions at high energies and QCD short-distance constraints. A clear example, shown in Refs. [46, 47], was the nontrivial combination of the Goldstone and axial-vector contributions, dictated by anomaly matching at all energy scales, that naturally resolved a long-standing puzzle [15, 20, 21] concerning the saturation of short-distance constraints with hadronic states in the HLbL. The holographic models also predict a more sizeable axial-meson contribution than previously estimated (see, e.g., Ref. [48]). This conclusion has been shown to hold even in the presence of quark-mass corrections [49]. It is worth mentioning that HQCD models allow simple expressions for the HVP contribution [50], however the ballpark of values is large, showing a strong dependence on the models and the chosen values of their parameters.

3 Scalars contributions from HQCD

The generalized consensus is that the scalar contributions are modest, largely dominated by the $\sigma(500)$ resonance and with the opposite sign with respect to the Goldstone- and axial-vector contributions. As is usual for HQCD models, apart from the contribution of the lightest, sub-GeV scalars, our model allows us to estimate the contribution of the infinite towers of excited scalar states, and thereby get an idea of the uncertainties that can be ascribed to heavier, not experimentally accessible, scalar states.

All HQCD model treat resonances as narrow-width states, and we are forced to do so even for the $\sigma(500)$ whose width is quite big. This raises the question of whether its contribution to the HLbL, as predicted by the model, is reliable. In order to estimate the $\sigma(500)$ parameters (e.g., its mass and its decay width into photons), the most rigorous approach uses dispersive methods including coupled channels applied to $\gamma\gamma \to \pi\pi$ (see, e.g., Refs. [51–53]) or, more recently, to $\gamma\gamma \to \pi\pi$ and $\gamma\gamma \to KK$ data [54, 55]. To the best of our knowledge, the set of values closest to our parametrization are the ones provided in Ref. [56] from a fit to $\gamma\gamma \to \pi\pi$ data, as we shall discuss in the numerical analysis.

3.1 The 5D action

The action of our HQCD model reads [57]:

$$S[L_M, R_M, X] = \int d^5x \sqrt{g}\, \mathcal{L}(L_M, R_M, X) + S_{\rm CS}\,, \tag{1}$$

where

$$\begin{aligned}\mathcal{L}(L_M, R_M, X) = &-\lambda\, \mathrm{tr}\Big[F^{MN}_{(L)} F_{(L)MN} + F^{MN}_{(R)} F_{(R)MN}\Big] + \rho\, \mathrm{tr}\Big[D^M X^\dagger D_M X - m_X^2 X^\dagger X\Big] \\ &+ z\,\delta(z - z_0) V(X) + \zeta_+\, \mathrm{tr}\Big[X^\dagger X F^{MN}_{(R)} F_{(R)MN} + X X^\dagger F^{MN}_{(L)} F_{(L)MN}\Big] + \zeta_-\mathrm{tr}\Big[X^\dagger F^{MN}_{(L)} X F_{(R)MN}\Big]\end{aligned} \tag{2}$$

and

$$S_{\mathrm{CS}} = c \int \mathrm{tr}[\omega_5(L) - \omega_5(R)] . \tag{3}$$

The Chern-Simons term makes sure that the model correctly reproduces the effects the axial anomaly. Its explicit expression in terms of the gauge fields reads $\omega_5(L) = LF_{(L)}^2 + \frac{i}{2}L^3 F_{(L)} - \frac{1}{10}L^5$, where the wedge product of forms is implicitly understood. The same conventions and definitions apply to the right-handed sector.

The five-dimensional background metric is anti-de Sitter,

$$g_{MN}dx^M dx^N = \frac{1}{z^2}\left(\eta_{\mu\nu}dx^\mu dx^\nu - dz^2\right) , \tag{4}$$

where $\mu, \nu = (0, 1, 2, 3)$, $M, N = (0, 1, 2, 3, z)$ and $\eta_{\mu\nu}$ has a mostly negative signature. For convenience, we have normalized the AdS curvature to unity. The fifth dimension is taken to be compact, with $\epsilon < z < z_0$.

The first term in Eq. (2) collects the Yang-Mills term for the 5D gauge fields L_M and R_M. Together with the Chern-Simons term, this action was used in Ref. [47] to describe the contribution of axials and Goldstone bosons to the HLbL, with $F_{(L)\,MN} = \partial_M L_N - \partial_N L_M - i[L_M, L_N]$, $L_M = L_M^a t^a$ and $\mathrm{tr}\,(t^a t^b) = \frac{1}{2}\delta_{ab}$, where t^a are the eight Gell-Mann matrices extended with $t^0 = \mathbf{1}_3/\sqrt{6}$. The seond term in Eq. (2) contains the complex scalar field $X = X^a t^a$ transforming as a bifundamental of $U(3)_L \times U(3)_R$, $X \to g_L X g_R^\dagger$, with $D_M X = \partial_M X - iL_M X + iXR_M$.

We have also introduced a potential term, localized on the infrared boundary [40]:

$$V(X) = \frac{1}{2}\mu^2\,\mathrm{tr}\left[X^\dagger X\right] - \eta\,\mathrm{tr}\left[(X^\dagger X)^2\right] . \tag{5}$$

The last two additional terms in Eq. (2), generate the $S\gamma\gamma$ interaction and are therefore essential to obtain a prediction for the single scalar exchange contributions to the HLbL.

4 Numerical analysis and discussion

Here we summarize some of the steps of the numerical analysis done in [57], where the explicit analytic expressions for the various physical quantities obtained in the HQCD model are presented.

The model contains nine parameters: six coefficients of the different terms in the 5D action (2) (λ, c, ρ, $\zeta_\pm$, m_X), the size of the fifth dimension z_0 and two parameters from the scalar boundary potential (μ, η). Actually, for the scalar contributions to the HLbL, only the combination $\zeta = \zeta_+ + \frac{1}{2}\zeta_-$ is relevant. The parameters of the boundary potential, can be traded for the quark condensate $\langle\bar{q}q\rangle$ and γ. The five-dimensional scalar mass m_X is fixed to the value dictated by the AdS/CFT correspondence, $m_X^2 = -3$.

We obtain the values of ζ and ρ by matching the QCD OPE short-distance constraint for the three-point function $\langle SVV\rangle$ and the decay width of the lowest-lying scalars into two photons. Experimentally, the partial decay widths of the $f_0(980)$ and $a_0(980)$ are well known and given by [58]

$$\Gamma_{\gamma\gamma}^{(f_0)} = 0.29^{+0.11}_{-0.06}\,\mathrm{keV}; \qquad \Gamma_{\gamma\gamma}^{(a_0)} = 0.30(10)\,\mathrm{keV} . \tag{6}$$

For the $\sigma(500)$, however, rescattering effects are dominant and the determination is less precise. The most recent analyses suggest $\Gamma_{\gamma\gamma}^{(\sigma)} = (1.3-2)\,\mathrm{keV}$ [52, 55, 59].

To evaluate the scalar contributions to the HLbL in a realistic way, we need to introduce flavour breaking. Following what was done in [47] for the case of Goldstone and axial-vector towers, we consider independent copies of the original Lagrangian for each of the different light scalar states, with a threefold set of parameters associated to the three neutral scalar towers, whose lightest states are the $\sigma(500)$, $a_0(980)$ and $f_0(980)$. For simplicity, we will assume a flavour-invariant quark condensate $\langle \bar{q}q \rangle = (-260\,\mathrm{MeV})^3$, which is consistent with the phenomenological determinations at 1 GeV. The size of the fifth dimension is inversely proportional to the vector meson mass scale as and the value $z_0 = (322\ \mathrm{MeV})^{-1}$, guarantees that the first vector multiplet lies at $m_\rho = 775$ MeV.

The parameters γ, ρ and ζ, will be flavour dependent, with each γ fixed by the mass of the lightest scalars. A flavour-independent short-distance contraint can be used to fix the ratio $\zeta_j/\rho_j \simeq -7.72$, while another constraint links ρ_j to γ_j and $\Gamma_{j\gamma\gamma}$

The model has limitations as a model of scalar mesons. The parameter ρ, for instance, has been determined in previous papers [38, 40, 60] through the asymptotic behaviour of the two-point correlator

$$\delta^{ab}\Pi^{SS}(q^2) = i \int d^4x e^{iq\cdot x} \langle 0|T\left\{j^a(x)j^b(0)\right\}|0\rangle\,, \tag{7}$$

where $j^a(x) = \bar{q}(x)t^a q(x)$. At large Euclidean momenta, the HQCD model successfully reproduces the QCD OPE $Q^2 \log Q^2$ behaviour. However, the coefficient would be matched with $\rho = N_c(8\pi)^{-2}$, while the constraint used predict $\rho_j \sim O(10^3)$. Such numerical mismatches at $N_c = 3$ are, however, to be expected in a minimal model. Moreover, such kind of HQCD model the predicted spectrum of excited states wich does not show a Regge behaviour, and the g_{SPP} couplings of the lightest scalars with Goldstone pairs give results of $O(10\ \mathrm{MeV})$. This is the same range that was found in Ref. [61] using a similar holographic model, and it falls short of the experimental results by 2 orders of magnitude.

Despite these limitations, the strategy to fix the parameters is meant to ensure that the model is phenomenologically sound, at least for a determination of the scalar contribution to the HLbL and it works especially well for the $a_0(980)$ and $f_0(980)$, which are rather narrow resonances. Instead, the $\sigma(500)$ cannot be considered a narrow state and one could argue that the description of this state in the model is not reliable. The most rigorous formalism to extract its parameters combines coupled Roy-Steiner equations with information from chiral perturbation theory (see, e.g., Refs. [17, 51, 52, 54, 55]). In our formalism, however, using the values for the mass and partial decay width extracted with this procedure would be inconsistent. The closest parametrization to our approach, that we have found in the literature, is the one of Ref. [56], in which a Breit-Wigner model was employed to fit $\gamma\gamma \to \pi^0\pi^0$ data. The results of two different fitting strategies gave

$$m_\sigma = 547(45)\,\mathrm{MeV}; \qquad \Gamma_{\sigma\gamma\gamma} = 0.62(19)\,\mathrm{keV}\,, \tag{8}$$

$$m_\sigma = 471(23)\,\mathrm{MeV}; \qquad \Gamma_{\sigma\gamma\gamma} = 0.33(07)\,\mathrm{keV}\,, \tag{9}$$

where in the second line constraints from chiral perturbation theory were used. The interpretation of the partial widths above should be taken with care, but for our purposes this is not relevant. The important point is that the parameters above fit the experimental data.

Using the strategy outlined above to determine the model parameters for the $\sigma(500)$ contribution, we find $a_\mu^S = -8.11 \cdot 10^{-11}$ and $a_\mu^S = -8.42 \cdot 10^{-11}$, respectively, depending on whether one takes Eq. (8) or Eq. (9) as input. These values are not only compatible with each other but are rather similar of the one obtained from a dispersion relation analysis, $a_\mu^S = -9(1) \cdot 10^{-11}$ [17, 62], a strong indication that our strategy for the $\sigma(500)$ contribution is reliable.

It is difficult, however, to come up with a reasonable estimate of the uncertainty associated to our number for the $\sigma(500)$ contribution to the HLbL. The mass and partial decay width in

Table 1. Results for the scalar contributions to $a_\mu^{\rm HLbL} \times 10^{11}$ for the set of values described in the main text.

	$n=1$	$n=2$	**Total**
$a_\mu^S(\sigma)$	-8.5(2.0)	-0.07(2)	-8.7(2.0)
$a_\mu^S(a_0)$	-0.29(13)	-0.025(10)	-0.32(14)
$a_\mu^S(f_0)$	-0.27(13)	-0.025(9)	-0.29(14)
a_μ^S	-9(2)	-0.12(4)	-9(2)

Eqs. (8) and (9) are strongly correlated. However, in Ref. [56], no information is given of the correlation matrix. Our estimate

$$a_\mu^S(\sigma) = (-8.5 \pm 2.0) \cdot 10^{-11} \tag{10}$$

is therefore orientative, but we believe that it correctly captures the right order of magnitude for the uncertainty. Instead, the contributions of $a_0(980)$ and $f_0(980)$ can be computed in a less problematic way, and we find:

$$a_\mu^S(a_0) = -0.29(13) \cdot 10^{-11}; \quad a_\mu^S(f_0) = -0.27(13) \cdot 10^{-11}\,, \tag{11}$$

in agreement with recent work [62, 63], and compatible with earlier estimates [64, 65] (see, however, the remarks in Ref. [62]).

The model also provides a prediction for the contribution of the whole tower of scalar excitations. The results for the first excitation states together with the total contributions of the three scalar towers are shown in Table 1. They amount to an overall 3% correction to the sub-GeV scalar contribution. As expected, the dynamics beyond a few GeV is of little numerical importance.

Taking all the previous points into account, we estimate the scalar contribution to the HLbL at

$$a_\mu^S = -9(2) \cdot 10^{-11}\,, \tag{12}$$

with roughly 90% of this number coming from the $\sigma(500)$. The error budget is likewise dominated by the $\sigma(500)$, whereas scalar states above the GeV scale contribute no more than 15% to the total uncertainty. This number is compatible with the one estimated with the Nambu-Jona-Lasinio model, $a_\mu^S = -7(2) \cdot 10^{-11}$ [66, 67], with dispersive analyses [62], and also in agreement with the values reported in Refs. [68–70].

I thank the organizers of the workshop and my collaborators G. D'Ambrosio and O. Catà. I was supported in part by the INFN research initiative Exploring New Physics (ENP).

References

[1] B. Abi *et al.* (Muon g-2 Collaboration), Phys. Rev. Lett. **126**, no.14, 141801 (2021) [arXiv:2104.03281 [hep-ex]]

[2] G. W. Bennett *et al.* (Muon g-2 Collaboration), Phys. Rev. D **73**, 072003 (2006) [arXiv:hep-ex/0602035 [hep-ex]]

[3] T. Aoyama, N. Asmussen, M. Benayoun, J. Bijnens, T. Blum, M. Bruno, I. Caprini, C. M. Carloni Calame, M. Cè, G. Colangelo *et al.*, Phys. Rept. **887**, 1-166 (2020) [arXiv:2006.04822 [hep-ph]]

[4] T. Aoyama, M. Hayakawa, T. Kinoshita and M. Nio, Phys. Rev. Lett. **109**, 111808 (2012) [arXiv:1205.5370 [hep-ph]]

[5] T. Aoyama, T. Kinoshita and M. Nio, Atoms **7**, no.1, 28 (2019)

[6] A. Czarnecki, W. J. Marciano and A. Vainshtein, Phys. Rev. D **67**, 073006 (2003) [erratum: Phys. Rev. D **73**, 119901 (2006)] [arXiv:hep-ph/0212229 [hep-ph]]

[7] C. Gnendiger, D. Stöckinger and H. Stöckinger-Kim, Phys. Rev. D **88**, 053005 (2013) [arXiv:1306.5546 [hep-ph]]

[8] M. Davier, A. Hoecker, B. Malaescu and Z. Zhang, Eur. Phys. J. C **77**, no.12, 827 (2017) [arXiv:1706.09436 [hep-ph]]

[9] A. Keshavarzi, D. Nomura and T. Teubner, Phys. Rev. D **97**, no.11, 114025 (2018) [arXiv:1802.02995 [hep-ph]]

[10] G. Colangelo, M. Hoferichter and P. Stoffer, JHEP **02**, 006 (2019) [arXiv:1810.00007 [hep-ph]]

[11] M. Hoferichter, B. L. Hoid and B. Kubis, JHEP **08**, 137 (2019) [arXiv:1907.01556 [hep-ph]]

[12] M. Davier, A. Hoecker, B. Malaescu and Z. Zhang, Eur. Phys. J. C **80**, no.3, 241 (2020) [erratum: Eur. Phys. J. C **80**, no.5, 410 (2020)] [arXiv:1908.00921 [hep-ph]]

[13] A. Keshavarzi, D. Nomura and T. Teubner, Phys. Rev. D **101**, no.1, 014029 (2020) [arXiv:1911.00367 [hep-ph]]

[14] A. Kurz, T. Liu, P. Marquard and M. Steinhauser, Phys. Lett. B **734**, 144-147 (2014) [arXiv:1403.6400 [hep-ph]]

[15] K. Melnikov and A. Vainshtein, Phys. Rev. D **70**, 113006 (2004) [arXiv:hep-ph/0312226 [hep-ph]]

[16] P. Masjuan and P. Sanchez-Puertas, Phys. Rev. D **95**, no.5, 054026 (2017) [arXiv:1701.05829 [hep-ph]]

[17] G. Colangelo, M. Hoferichter, M. Procura and P. Stoffer, JHEP **04**, 161 (2017) [arXiv:1702.07347 [hep-ph]]

[18] M. Hoferichter, B. L. Hoid, B. Kubis, S. Leupold and S. P. Schneider, JHEP **10**, 141 (2018) [arXiv:1808.04823 [hep-ph]]

[19] A. Gérardin, H. B. Meyer and A. Nyffeler, Phys. Rev. D **100**, no.3, 034520 (2019) [arXiv:1903.09471 [hep-lat]]

[20] J. Bijnens, N. Hermansson-Truedsson and A. Rodriguez-Sanchez, Phys. Lett. B **798**, 134994 (2019) [arXiv:1908.03331 [hep-ph]]

[21] G. Colangelo, F. Hagelstein, M. Hoferichter, L. Laub and P. Stoffer, JHEP **03**, 101 (2020) [arXiv:1910.13432 [hep-ph]]

[22] T. Blum, N. Christ, M. Hayakawa, T. Izubuchi, L. Jin, C. Jung and C. Lehner, Phys. Rev. Lett. **124**, no.13, 132002 (2020) [arXiv:1911.08123 [hep-lat]]

[23] G. Colangelo, M. Hoferichter, A. Nyffeler, M. Passera and P. Stoffer, Phys. Lett. B **735**, 90-91 (2014) [arXiv:1403.7512 [hep-ph]]

[24] S. Borsanyi, Z. Fodor, J. N. Guenther, C. Hoelbling, S. D. Katz, L. Lellouch, T. Lippert, K. Miura, L. Parato and K. K. Szabo, *et al.* Nature **593**, no.7857, 51-55 (2021) [arXiv:2002.12347 [hep-lat]]

[25] A. Crivellin, M. Hoferichter, C. A. Manzari and M. Montull, Phys. Rev. Lett. **125**, no.9, 091801 (2020) [arXiv:2003.04886 [hep-ph]]

[26] A. Keshavarzi, W. J. Marciano, M. Passera and A. Sirlin, Phys. Rev. D **102**, no.3, 033002 (2020) [arXiv:2006.12666 [hep-ph]]

[27] B. Malaescu and M. Schott, Eur. Phys. J. C **81**, no.1, 46 (2021) [arXiv:2008.08107 [hep-ph]]

[28] G. Colangelo, M. Hoferichter and P. Stoffer, Phys. Lett. B **814**, 136073 (2021) [arXiv:2010.07943 [hep-ph]]

[29] G. Colangelo, M. Hoferichter, M. Procura and P. Stoffer, JHEP **1409**, 091 (2014) [arXiv:1402.7081 [hep-ph]]

[30] G. Colangelo, M. Hoferichter, B. Kubis, M. Procura and P. Stoffer, Phys. Lett. B **738**, 6 (2014) [arXiv:1408.2517 [hep-ph]]

[31] E. H. Chao, R. J. Hudspith, A. Gérardin, J. R. Green, H. B. Meyer and K. Ottnad, Eur. Phys. J. C **81**, no.7, 651 (2021) [arXiv:2104.02632 [hep-lat]]

[32] J. M. Maldacena, Adv. Theor. Math. Phys. **2**, 231 (1998) [arXiv:hep-th/9711200]

[33] S. S. Gubser, I. R. Klebanov and A. M. Polyakov, Phys. Lett. B **428**, 105 (1998) [hep-th/9802109]

[34] E. Witten, Adv. Theor. Math. Phys. **2**, 253 (1998) [hep-th/9802150]

[35] G. 't Hooft, Nucl. Phys. B **72**, 461 (1974)

[36] E. Witten, Nucl. Phys. B **160**, 57-115 (1979)

[37] T. Sakai and S. Sugimoto, Prog. Theor. Phys. **113**, 843 (2005) [hep-th/0412141]

[38] J. Erlich, E. Katz, D. T. Son and M. A. Stephanov, Phys. Rev. Lett. **95**, 261602 (2005) [arXiv:hep-ph/0501128]

[39] L. Da Rold and A. Pomarol, Nucl.Phys. B**721** (2005) 79 [hep-ph/0501218]

[40] L. Da Rold and A. Pomarol, JHEP **01**, 157 (2006) [arXiv:hep-ph/0510268 [hep-ph]]

[41] J. Hirn and V. Sanz, JHEP **0512**, 030 (2005) [arXiv:hep-ph/0507049]

[42] O. Domenech, G. Panico and A. Wulzer, Nucl. Phys. A **853**, 97-123 (2011) [arXiv:1009.0711 [hep-ph]]

[43] D. K. Hong and D. Kim, Phys. Lett. B **680**, 480-484 (2009) [arXiv:0904.4042 [hep-ph]]

[44] L. Cappiello, O. Catà and G. D'Ambrosio, Phys. Rev. D **83**, 093006 (2011) [arXiv:1009.1161 [hep-ph]]

[45] J. Leutgeb, J. Mager and A. Rebhan, Phys. Rev. D **100**, 094038 (2019) [arXiv:1906.11795 [hep-ph]]

[46] J. Leutgeb and A. Rebhan, Phys. Rev. D **101**, no.11, 114015 (2020) [arXiv:1912.01596 [hep-ph]]

[47] L. Cappiello, O. Catà, G. D'Ambrosio, D. Greynat and A. Iyer, Phys. Rev. D **102**, no.1, 016009 (2020) [arXiv:1912.02779 [hep-ph]]

[48] G. Colangelo, F. Hagelstein, M. Hoferichter, L. Laub and P. Stoffer, Eur. Phys. J. C **81**, no.8, 702 (2021) [arXiv:2106.13222 [hep-ph]]

[49] J. Leutgeb and A. Rebhan, Phys. Rev. D **104**, no.9, 094017 (2021) [arXiv:2108.12345 [hep-ph]]

[50] J. Leutgeb, A. Rebhan and M. Stadlbauer, Phys. Rev. D **105** (2022) no.9, 094032 doi:10.1103/PhysRevD.105.094032 [arXiv:2203.16508 [hep-ph]]

[51] M. Hoferichter, D. R. Phillips and C. Schat, Eur. Phys. J. C **71**, 1743 (2011) [arXiv:1106.4147 [hep-ph]]

[52] B. Moussallam, Eur. Phys. J. C **71**, 1814 (2011) [arXiv:1110.6074 [hep-ph]]

[53] M. Hoferichter and P. Stoffer, JHEP **07**, 073 (2019) [arXiv:1905.13198 [hep-ph]]

[54] I. Danilkin, O. Deineka and M. Vanderhaeghen, Phys. Rev. D **101**, no.5, 054008 (2020) [arXiv:1909.04158 [hep-ph]]

[55] I. Danilkin, O. Deineka and M. Vanderhaeghen, Phys. Rev. D **103**, no.11, 114023 (2021) [arXiv:2012.11636 [hep-ph]]

[56] L. V. Fil'kov and V. L. Kashevarov, Eur. Phys. J. A **5**, 285-292 (1999) [arXiv:nucl-th/9810074 [nucl-th]]

[57] L. Cappiello, O. Catà and G. D'Ambrosio, Phys. Rev. D **105** (2022) no.5, 056020 doi:10.1103/PhysRevD.105.056020 [arXiv:2110.05962 [hep-ph]]
[58] P.A. Zyla *et al.* [Particle Data Group], PTEP **2020**, no.8, 083C01 (2020)
[59] L. Y. Dai and M. R. Pennington, Phys. Rev. D **90**, no.3, 036004 (2014) [arXiv:1404.7524 [hep-ph]]
[60] A. Cherman, T. D. Cohen and E. S. Werbos, Phys. Rev. C **79**, 045203 (2009) [arXiv:0804.1096 [hep-ph]]
[61] P. Colangelo, F. De Fazio, F. Giannuzzi, F. Jugeau and S. Nicotri, Phys. Rev. D **78**, 055009 (2008) [arXiv:0807.1054 [hep-ph]]
[62] I. Danilkin, M. Hoferichter and P. Stoffer, Phys. Lett. B **820**, 136502 (2021) [arXiv:2105.01666 [hep-ph]]
[63] M. Knecht, S. Narison, A. Rabemananjara and D. Rabetiarivony, Phys. Lett. B **787**, 111-123 (2018) [arXiv:1808.03848 [hep-ph]]
[64] V. Pauk and M. Vanderhaeghen, Eur. Phys. J. C **74**, 3008 (2014) [arXiv:1401.0832 [hep-ph]]
[65] I. Danilkin and M. Vanderhaeghen, Phys. Rev. D **95**, no.1, 014019 (2017) [arXiv:1611.04646 [hep-ph]]
[66] J. Bijnens, E. Pallante and J. Prades, Phys. Rev. Lett. **75**, 1447 (1995) [Phys. Rev. Lett. **75**, 3781 (1995)] [hep-ph/9505251]
[67] J. Bijnens, E. Pallante and J. Prades, Nucl. Phys. B **474**, 379 (1996) [hep-ph/9511388]
[68] F. Jegerlehner and A. Nyffeler, Phys. Rep. **477**, 1 (2009) [arXiv:0902.3360 [hep-ph]]
[69] J. Prades, E. de Rafael and A. Vainshtein, Adv. Ser. Direct. High Energy Phys. **20**, 303 (2009) [arXiv:0901.0306 [hep-ph]]
[70] F. Jegerlehner, Springer Tracts Mod. Phys. **274**, 1 (2017)

EPJ Web of Conferences 270, 00030 (2022) https://doi.org/10.1051/epjconf/202227000030
QCD@Work 2022

Pseudoscalars and the η' in the holographic soft-wall model

Floriana Giannuzzi[1,*] and *Stefano* Nicotri[1,**]

[1]INFN – Istituto Nazionale di Fisica Nucleare – Sezione di Bari, via Orabona 4, 70125, Bari, Italy

Abstract. Pseudoscalar mesons and glueballs are studied in a model based on the AdS/QCD correspondence. Glueball masses are obtained in the pure-gauge case, while the mixing between pseudoscalar glueballs and singlet pseudoscalar mesons produces the expected mass of the η' meson. The topological susceptibility is computed as a function of the quark mass. The model reproduces partial conservation of axial current, the anomaly equation, and the Witten-Veneziano relation.

1 Introduction

The η' meson has a mass of 958 MeV, too high for it to be a pseudo-Goldstone boson. Indeed, while $SU(2)_A$ symmetry is spontaneously and explicitly broken generating an octet of pseudo-Goldstone bosons, $U(1)_A$ is anomalous. We describe the $U(1)_A$ anomaly and reproduce the η' mass in a bottom-up holographic model by considering coupled singlet pseudoscalar mesons and glueballs.

2 Model and results

In this work, the η' mass and the topological susceptibility are computed in the soft-wall model [1], one of the bottom-up approaches that can be followed when applying the AdS/CFT correspondence to QCD. This is motivated by the fact that in AdS/CFT the nonperturbative regime of the gauge theory corresponds to the supergravity limit of the dual string theory, characterized by small coupling and a large radius of the spacetime. In a bottom-up approach, a $5d$ effective field theory is properly constructed in a anti-de Sitter (AdS) space with metric

$$ds^2 = \frac{R^2}{z^2}(dt^2 - d\bar{x}^2 - dz^2)\,, \tag{1}$$

with R the AdS curvature radius and $z \geqslant 0$ the additional bulk coordinate; $z = 0$ corresponds to the AdS boundary. The AdS/CFT correspondence links a gravity theory to a conformal gauge theory, but QCD is only asymptotically conformal, so in the soft-wall model conformal invariance is broken by inserting a background dilaton field $e^{-\phi}$ in the action, behaving as $e^{-c^2z^2}$, where c sets the mass scale of the model. The $5d$ theory contains the fields dual to the QCD operators under study, whose masses are fixed by the conformal dimension of their dual

*e-mail: floriana.giannuzzi@ba.infn.it
**e-mail: nicotri@infn.it

operator according to the relation $m_5^2R^2 = (\Delta - p)(\Delta + p - 4)$; their boundary values (i.e. their values at $z = 0$) are the sources of the corresponding QCD operators. A crucial feature of AdS/CFT is that the partition function of the $5d$ model is equal to the generating functional of the correlators in the gauge theory. The former has a simple form in the supergravity limit, given by the exponential of the on-shell action. This means that the correspondence provides a straightforward way to compute correlation functions of a strongly-coupled gauge theory, simply deriving the on-shell action of the bulk theory with respect to the boundary values of the fields. To this aim, it is useful to write a field in Fourier space as $\phi(q,z) = K(q,z)\phi_0(q)$, where $\phi_0(q)$ is the source of the operator and $K(q,z)$ is a the so-called bulk-to-boundary propagator.

The soft-wall model was introduced in [1] to study chiral symmetry breaking, so the involved fields and the Lagrangian are well known. The fields dual to the left and right currents are the vector fields $A^M_{L/R}(x,z) = A^{M,A}_{L/R}(x,z)T^A$, where T^A ($A = 0,...,n_f^2 - 1$) are the generators of $U(n_f)$, with $T^0 = \frac{1}{2n_f}I_{n_f}$ and $\mathrm{Tr}(T^aT^b) = \delta^{ab}/2$ for $a,b = 1,...,n_f^2 - 1$. Their masses vanish, so they are $5d$ gauge fields. Indeed, the global chiral symmetry of QCD is promoted to a local symmentry in the dual $5d$ theory. The axial field is then defined as $A = (A_L - A_R)/2$. We choose the gauge $A_5 = 0$, while the other components of the axial field can be written as a sum $A^A_\mu = A^A_{\perp\mu} + \partial_\mu\varphi^A$ of a transverse part, describing axial-vector mesons, and a longitudinal part, contributing to the descritpion of pseudoscalar mesons. The $5d$ field dual to the QCD $\bar{q}_Rq_L$ operator is a scalar field that has a vacuum expectation value (vev) $X_0(z)$, and a phase η^A describing pseudoscalar mesons: $X(x,z) = e^{i\eta^A(x,z)T^A}X_0(z)e^{i\eta^A(x,z)T^A}$. We consider a model with $n_f = 3$ identical quarks, so X_0 is proportional to the identity matrix: $X_0(z) = \sqrt{2}\,v_q(z)\,I_{n_f}$. We also assume [2]:

$$v_q(z) = \frac{m_q}{R}z + \frac{\sigma}{R}z^3\,, \tag{2}$$

where m_q is the quark mass, and σ is an independent coefficient related to the chiral condensate: $\langle\bar{q}q\rangle = -\frac{N_c}{2\pi^2}\sigma$. Finally, the field $Y(x,z) = Y_0(z)e^{2ia(x,z)}$ is a scalar dual to the square of the gluon field strength. The phase $a(x,z)$ is dual to the $\frac{\alpha_s}{8\pi}G_{\mu\nu}\tilde{G}^{\mu\nu}$ operator describing the pseudoscalar glueballs, and it mixes with the singlet pseudoscalar meson fields. The vev is given by:

$$Y_0(z) = \frac{y_0}{R} + \frac{2\,y_1}{R\,c^4}(e^{c^2z^2}(-1 + c^2z^2) + 1)\,, \tag{3}$$

a form obtained solving the equation of motion assuming no potential term. It contains two additional parameters, y_0 and y_1. We assume $a(x,z) = a_P(x,z) + V_a(z)a_f(x,z)$, where a_P is invariant under the $U(1)_A$ symmetry, while a_f transforms under $U(1)_A$ by a shift; $V_a(z) = e^{-v_q(z)^2}$ is a potential term which vanishes in the pure-gauge limit [3].

The Lagrangian describing these fields is:

$$\mathcal{L} = \frac{1}{k}\sqrt{g}\,e^{-\phi}\left[\mathrm{Tr}\left\{-\frac{1}{4g_5^2}(F_L^2 + F_R^2) + |DX|^2 - m_X^2|X|^2\right\} + \frac{1}{2}\mathcal{K}_a\right], \tag{4}$$

$$\mathcal{K}_a = |\partial_M Y_0(z) + 2iY_0(\partial_M a(x,z) - \eta^0(x,z)\partial_M V_a(z) - A^0_M(x,z)V_a(z))|^2\,. \tag{5}$$

The first term describes the octet of pseudoscalar mesons [1], while the second deals with the Y field and is invariant under $U(1)_A$ transformations:

$$\eta^0 \to \eta^0 - \alpha \tag{6}$$

$$\varphi^0 \to \varphi^0 - \alpha \tag{7}$$

$$a \to a - V_a\alpha\,. \tag{8}$$

The term in Eq. (5) has also been obtained from the Wess-Zumino action in another bottom-up model [4]. After expanding the Lagrangian, we obtain the following expression:

$$\begin{aligned}\mathcal{L} &= \frac{R}{k}e^{-\phi}\left[\frac{1}{4n_f g_5^2 z}(\partial_z\partial_\nu\varphi^0)^2 + \frac{1}{2g_5^2 z}(\partial_z\partial^\nu\varphi^8)^2 - \frac{2R^2 v_q^2}{n_f z^3}(\partial_z\eta^0)^2 + \right.\\ &+\frac{2R^2 v_q^2}{n_f z^3}(\partial_\nu\eta^0 - \partial_\nu\varphi^0)^2 - \frac{4R^2 v_q^2}{z^3}(\partial_z\eta^8)^2 + \frac{4R^2 v_q^2}{z^3}(\partial_\nu\eta^8 - \partial_\nu\varphi^8)^2 + \\ &\left. -\frac{2R^2}{z^3}Y_0^2(\partial_z a - \eta^0\partial_z V_a)^2 + \frac{2R^2}{z^3}Y_0^2(\partial_\nu a - V_a\partial_\nu\varphi^0)^2\right], \end{aligned} \tag{9}$$

containing the known terms for η_8 and ϕ_8, which describe the octet of pseudoscalar mesons, plus new terms involving a, η_0 and ϕ_0, describing the singlet pseudoscalar mesons. Notice that the a field is only coupled to singlet pseudoscalar mesons, since we are considering identical quarks.

Some of the parameters appearing in the Lagrangian are already known from previous studies in the soft-wall model: $R/k = N_c/16\pi^2$ and $g_5^2 = 3/4$ are fixed from scalar and vector meson two-point functions [5]; $c = 388$ MeV is obtained by fitting the ρ meson mass [1]. There are four unknown parameters, m_q, $\bar{q}q$, y_0, y_1, that we shall fix from phenomenology.

2.1 Pseudoscalar glueballs in pure gauge

Let us first consider the pure-gauge case, where the action only contains the a field describing pseudoscalar glueballs:

$$S_{PG} = \frac{R}{k}\int d^5x\, e^{-\phi}\left[-\frac{2R^2}{z^3}Y_0^2(\partial_z a_P)^2 + \frac{2R^2}{z^3}Y_0^2(\partial_\mu a_P)^2\right]. \tag{10}$$

The bulk-to-boundary propagator and the two-point function of the $G\tilde{G}$ operator can be computed analytically [2]. The two-point function has an infinite number of poles, the positions of which correspond to the masses of the pseudoscalar glueballs $m_n^2 = 4c^2(n+2)$, while the residues R_n are related to the decay constants: $R_n = m_n^4 f_n^2$. The lightest state has mass $m_{G\tilde{G},0} = 1.1$ GeV while the first radial excitation has mass $m_{G\tilde{G},1} = 1.34$ GeV, suggesting it could correspond to the $\eta(1405)$.
The first term in the high $Q^2 = -q^2$ expansion of the two-point function can be matched to QCD [6], thus fixing $y_0 = \frac{1}{\sqrt{N_c}}\frac{\alpha_s}{\pi}$ [2]. This shows that the Y field appears in the Lagrangian at a lower order in the large N_c expansion with respect to the other fields. Indeed, in the limit $N_c \to \infty$ the anomaly turns off, and the η' is degenerate with pions, while the degeneracy is lifted by $q\bar{q}$ annihilation diagrams, which are suppressed by $1/N_c$. Assuming $y_0 = \frac{1}{\sqrt{N_c}\pi}$, the decay constant of the ground state is 9.8 MeV.
The $q^2 \to 0$ limit of the two-point function gives the topological susceptibility [7]:

$$\chi_t = -\lim_{q^2\to 0}\Pi_{aa}(q^2). \tag{11}$$

In pure gauge we find $\chi_{PG} = \alpha_s y_1/\pi^3$. Assuming $\chi_{PG} \sim (191\text{ MeV})^4$, we can fix $y_1 = 0.041\text{ GeV}^4$.

2.2 Nonsinglet pseudoscalar mesons

Away from pure gauge, a tower of singlet and nonsinglet pseudoscalar mesons appears. The equations of motion for the fields describing each nonsinglet meson in the octet are:

$$\partial_z\left(\frac{v_q^2 e^{-\phi}}{z^3}\partial_z\eta^8\right) + q^2\frac{v_q^2 e^{-\phi}}{z^3}(\eta^8 - \varphi^8) = 0 \tag{12}$$

$$\partial_z \left(\frac{e^{-\phi}}{g_5^2 z} \partial_z \varphi^8 \right) + \frac{8 v_q^2 e^{-\phi}}{z^3} (\eta^8 - \varphi^8) = 0 \,. \tag{13}$$

A combination of the two equations gives a constraint equation:

$$\frac{q^2}{g_5^2 z} \partial_z \varphi^8 - \frac{8 v_q^2}{z^3} \partial_z \eta^8 = 0 \,. \tag{14}$$

To study correlation functions it is convenient to use the matrix formalism [8], since the two fields are given by a linear combination of the sources in Fourier space:

$$\Phi = \begin{pmatrix} \varphi^8 \\ \eta^8 \end{pmatrix} = F\,\Phi_0 \,, \tag{15}$$

where $\Phi_0(q^2) = (\varphi_0^8(q^2), -\eta_0^8(q^2))^T$ is the vector containing the sources of the two operators. Then, the two one-point functions are contained in a vector:

$$J^8 = \left. \frac{\partial S_{os}}{\partial \Phi_0} \right|_{z \to \varepsilon} = \begin{pmatrix} \langle J_\varphi^8 \rangle \\ \langle J_\eta^8 \rangle \end{pmatrix} , \tag{16}$$

where $\langle J_\varphi^8 \rangle$ and $\langle J_\eta^8 \rangle$ are the one-point functions of the longitudinal axial current $(\partial_\mu \bar{\psi} \gamma_5 \gamma^\mu T^8 \psi)$ and the pseudoscalar current $(2 m_q \bar{\psi} \gamma_5 T^8 \psi)$, respectively, in the nonsinglet sector. Using Eq. (14), we find

$$\langle J_\varphi^8 \rangle = \langle J_\eta^8 \rangle \,, \tag{17}$$

which establishes the partial conservation of axial current.
The two-point functions are collected in the matrix

$$\Pi^{88} = \left. \frac{\partial^2 S_{os}}{\partial \Phi_0 \partial \Phi_0} \right|_{z \to \varepsilon} = \begin{pmatrix} \Pi_{\varphi\varphi}^{88} & \Pi_{\varphi\eta}^{88} \\ \Pi_{\eta\varphi}^{88} & \Pi_{\eta\eta}^{88} \end{pmatrix} . \tag{18}$$

We have checked that in the large $Q^2 = -q^2$ limit they behave as in QCD [2]. The pion decay constant in the chiral limit is:

$$f_\pi^2 = -\frac{R}{k} \frac{e^{-\phi}}{g_5^2 z} \partial_z F_{00} \Big|_{\substack{q^2=0 \\ z \to \varepsilon}} = (91.6 \text{ MeV})^2 \,, \tag{19}$$

a value obtained after setting $|\langle q\bar{q} \rangle| = (0.281 \text{ GeV})^3$. The strange quark mass can be set requiring that the first pole of the two-point function is located at the mass of the η meson, obtaining $m_{\eta_8} = 548$ MeV if $m_q = m_s = 59.5$ MeV.

2.3 Singlet pseudoscalar mesons

Next, we have solved the three coupled equations of motion for the singlet states η_0, ϕ_0 and a:

$$\partial_z \left(\frac{v_q^2 e^{-\phi}}{n_f z^3} \partial_z \eta^0 \right) + q^2 \frac{v_q^2 e^{-\phi}}{n_f z^3} (\eta^0 - \varphi^0) + \frac{Y_0^2 e^{-\phi}}{z^3} (\partial_z V_a)(\partial_z a - \eta^0 \partial_z V_a) = 0 \tag{20}$$

$$\partial_z \left(\frac{e^{-\phi}}{2 n_f g_5^2 z} \partial_z \varphi^0 \right) + \frac{4 v_q^2 R^2 e^{-\phi}}{n_f z^3} (\eta^0 - \varphi^0) + \frac{4 Y_0^2 R^2 e^{-\phi}}{z^3} V_a (a - \varphi^0 V_a) = 0 \tag{21}$$

https://doi.org/10.1051/epjconf/202227000030

$$\partial_z\left(\frac{Y_0^2 e^{-\phi}}{z^3}(\partial_z a-\eta^0\,\partial_z V_a)\right)+q^2\frac{Y_0^2 e^{-\phi}}{z^3}(a-\varphi^0 V_a)=0\,. \tag{22}$$

These equations can be properly combined and integrated to get a constraint equation:

$$\frac{4v_q^2R^2}{n_f z^3}\partial_z\eta^0-\frac{q^2}{2n_f g_5^2 z}\partial_z\varphi^0+\frac{4Y_0^2R^2}{z^3}V_a(\partial_z a-\eta^0\partial_z V_a)=0\,. \tag{23}$$

In the matrix formalism, the three fields are aggregated in a vector

$$\Psi=\begin{pmatrix}\varphi^0\\ \eta^0\\ a\end{pmatrix}=H\,\Psi_0\,, \tag{24}$$

where $\Psi_0(q^2)=(\varphi_0^0(q^2),-\eta_0^0(q^2),a_0(q^2))^T$ is the vector containing the sources of the three operators.
The masses of the singlet states can be obtained from the poles of the two-point functions:

$$\Pi^{00}=\left.\frac{\partial^2 S_{os}}{\partial\Psi_0\partial\Psi_0}\right|_{z\to\varepsilon}=\begin{pmatrix}\Pi^{00}_{\varphi\varphi} & \Pi^{00}_{\varphi\eta} & \Pi_{\varphi a}\\ \Pi^{00}_{\eta\varphi} & \Pi^{00}_{\eta\eta} & \Pi_{\eta a}\\ \Pi_{a\varphi} & \Pi_{a\eta} & \Pi_{aa}\end{pmatrix}. \tag{25}$$

The lightest state, which we indentify with the η' meson, has mass 958 MeV, while in the chiral limit its mass is 903 MeV. In Fig. 1 we show the values of the squared mass of the lightest state and of the first radial excitation as a function of the quark mass, together with the result for the lighest state in the octet.

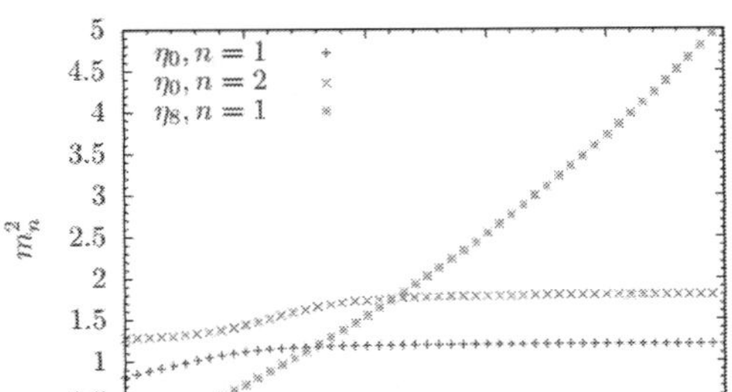

Figure 1. Squared mass (GeV^2) of the indicated states versus the quark mass (GeV).

The one-point functions are:

$$J^0=\left.\frac{\partial S_{os}}{\partial\Psi_0}\right|_{z\to\varepsilon}=\begin{pmatrix}\langle J^0_\varphi\rangle\\ \langle J^0_\eta\rangle\\ \langle J^0_a\rangle\end{pmatrix}=\begin{pmatrix}-\frac{R}{k}\frac{e^{-\phi}q^2}{4n_f g_5^2 z}\varphi_0'\Big|_{z\to\varepsilon}\\ -\frac{R}{k}\frac{e^{-\phi}2v_q^2R^2}{n_f z^3}\eta_0'\Big|_{z\to\varepsilon}\\ \frac{R}{k}\frac{e^{-\phi}2Y_0^2R^2}{z^3}a'\Big|_{z\to\varepsilon}\end{pmatrix}, \tag{26}$$

where $\langle J^0_\varphi \rangle$ and $\langle J^0_\eta \rangle$ are the one-point functions of the longitudinal axial current $(\partial_\mu \bar{\psi}\gamma_5\gamma^\mu T^0\psi)$ and of the pseudoscalar current $(2m_q\bar{\psi}\gamma_5 T^0\psi)$ in the singlet sector, respectively, while $\langle J^0_a \rangle$ is the one-point function of the topological charge density $(\langle \frac{\alpha_s}{8\pi} G_{\mu\nu}\tilde{G}^{\mu\nu} \rangle)$. The constraint equation (23) establishes a relation between the three one-point functions:

$$-\langle J^0_\eta \rangle + \langle J^0_\varphi \rangle + \langle J^0_a \rangle = 0\,, \tag{27}$$

a Ward identity representing the anomaly equation in this formalism.
From Π_{aa}, the two-point function of the $G\tilde{G}$ operator, we can compute the topological susceptibility as in Eq. (11). Expanding the equations of motion in the chiral limit for large N_c [2], at lowest order the constraint equation reproduces the Witten-Veneziano relation $\chi_{PG} = \frac{m^2_{\eta'} f^2_\pi}{2n_f}$. At $q^2 = 0$ the equations of motion can be integrated and the following relation holds [2]:

$$\frac{1}{\chi_t} = \frac{1}{\chi_{PG}} + \frac{1}{\chi_f} \tag{28}$$

with

$$\frac{1}{\chi_f} = \frac{k}{R} n_f \int_0^\infty dz\, \frac{e^{\phi(z)} z^3}{4} \left(\frac{V_a(z)}{v_q(z) R} \right)^2 , \tag{29}$$

so the full (inverse) topological susceptibility gets a contribution from the pure-gauge one and another one accounting for finite quark masses. The numerical result is represented by the black curve in Fig. 2. As $m_q \to \infty$ it approaches the pure-gauge value, while for small m_q it depends linearly on the quark mass, in agreement with chiral perturbation theory, which predicts $\chi_t \xrightarrow[m_q \to 0]{} \chi_f \sim \frac{\langle \bar{q}q \rangle}{n_f} m_q$ [9], shown by the blue dashed line in Fig. 2. χ_t vanishes at $m_q = 0$, as expected. The relation

$$\frac{1}{\chi_t} = \frac{1}{\chi_{PG}} + \frac{n_f}{m_q |\langle \bar{q}q \rangle|} \tag{30}$$

represented by the orange curve in Fig. 2, was proposed in [10][11]. Finally, in Fig. 3 we

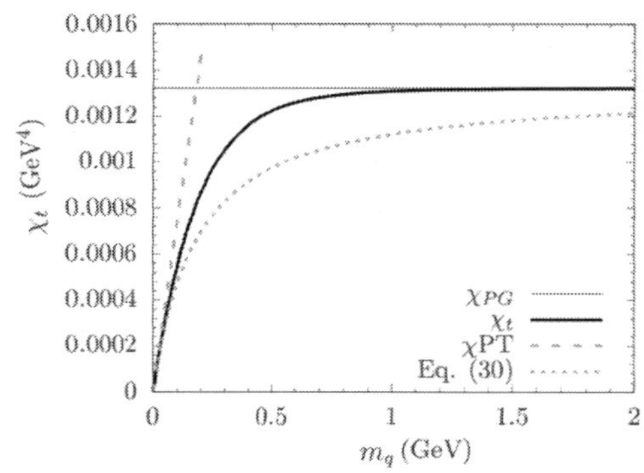

Figure 2. Topological susceptibility (GeV^4) vs quark mass (GeV) in this model (black curve), in pure gauge (horizontal grey line), from Eq. (30) (dotted orange line); expected topological susceptibility at small quark mass (dashed cyan line).

have compared our results for $n_f = 2$ with some lattice data [12][13][14]. It is not possible to get a quantitative comparison since the various points have been obtained with different values of the chiral condensate, producing different initial slopes.

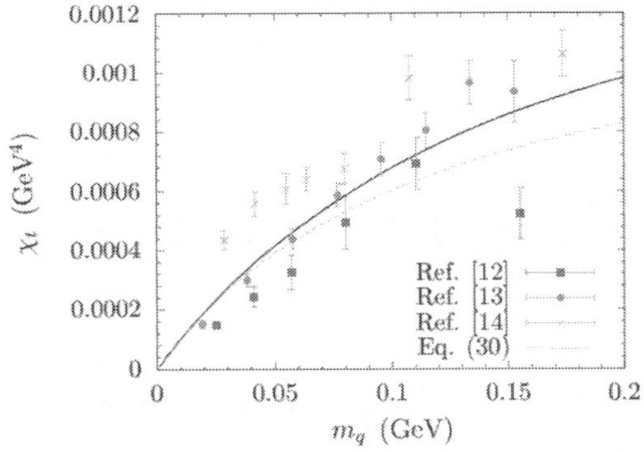

Figure 3. Topological susceptibility computed in this model (solid black line), and from Eq. (30) (dashed green line) at $n_f = 2$ and $|\langle\bar{q}q\rangle| = (0.281\ \text{GeV})^3$, compared with lattice data.

3 Conclusions

The masses of singlet and nonsinglet pseudoscalar mesons have been computed in the soft-wall holographic model of QCD. The mixing among the fields dual to the axial current, pseudoscalar current and the $G\tilde{G}$ operator can explain the large mass of the η'. Partial conservation of axial current, anomaly equation and Witten-Veneziano relation have been derived from the constraint equations (14) and (23), obtained by a combination of the equations of motion of the involved fields. The topological susceptibility χ_t has been computed for any value of the quark mass. The correction χ_f to the pure-gauge value depends linearly on the quark mass for small values of m_q, as in chiral perturbation theory, while it diverges as m_q^4 at high m_q.

References

[1] A. Karch, E. Katz, D.T. Son, M.A. Stephanov, Phys. Rev. **D74**, 015005 (2006), hep-ph/0602229
[2] F. Giannuzzi, S. Nicotri, Phys. Rev. D **104**, 014021 (2021), 2105.00923
[3] D. Arean, I. Iatrakis, M. Jarvinen, E. Kiritsis, Phys. Rev. D **96**, 026001 (2017), 1609.08922
[4] R. Casero, E. Kiritsis, A. Paredes, Nucl. Phys. B **787**, 98 (2007), hep-th/0702155
[5] P. Colangelo, F. De Fazio, F. Giannuzzi, F. Jugeau, S. Nicotri, Phys. Rev. D **78**, 055009 (2008), 0807.1054
[6] G. Gabadadze, Phys. Rev. D **58**, 055003 (1998), hep-ph/9711380
[7] E. Katz, M.D. Schwartz, JHEP **08**, 077 (2007), 0705.0534
[8] M. Kaminski, K. Landsteiner, J. Mas, J.P. Shock, J. Tarrio, JHEP **02**, 021 (2010), 0911.3610
[9] E. Vicari, H. Panagopoulos, Phys. Rept. **470**, 93 (2009), 0803.1593
[10] P. Di Vecchia, G. Veneziano, Nucl. Phys. B **171**, 253 (1980)
[11] H. Leutwyler, A.V. Smilga, Phys. Rev. D **46**, 5607 (1992)
[12] S. Aoki et al. (JLQCD, TWQCD), Phys. Lett. B **665**, 294 (2008), 0710.1130
[13] T.W. Chiu, T.H. Hsieh, Y.Y. Mao (TWQCD), Phys. Lett. B **702**, 131 (2011), 1105.4414
[14] T. DeGrand, Phys. Rev. D **101**, 114509 (2020), 2004.09649

From PT quantum mechanics to PT holography

Karl Landsteiner[1,*]

[1]Instituto de Física Teórica UAM-CSIC, C/ Nicolás Cabrera 13-15, 28049 Campus Cantoblanco, Spain

Abstract. $\mathcal{PT}$ quantum mechanics studies the physics of non-Hermitian Hamiltonians having an anti-linear involution denoted by $\mathcal{PT}$ as a symmetry. We extend the concept of $\mathcal{PT}$ non-Hermitian quantum theory to gauge-gravity duality. Non-Hermiticity is introduced via boundary conditions in asymptotically AdS spacetimes. We identify how key properites such as the $\mathcal{PT}$ phase transition are implemented in the gauge-gravity duality. Then we move on to study time dependent systems (holographic quantum quenches) and discuss how Hermitian and non-Hermitian time evolution arise in $\mathcal{PT}$ quantum mechanics and how it is implemented in the gauge-gravity duality. A key feature is that the non-Hermitian time evolution violates the null energy condition (NEC) and speculate about possible applications.

1 Introduction

Quantum mechanics is based on the axiom that observable quantities are represented by Hermitian operators. It comes then as a surprise that some seemingly non-Hermitian Hamiltonians can have real energy spectra: [1–4] found that the underlying reason for this is that such Hamiltonians have an anti-linear Z_2 symmetry. This is commonly denoted as $\mathcal{PT}$. The reason is that in many cases it can be thought of as the product of parity $\mathcal{P}$ and time-reversal $\mathcal{T}$. The study of such Hamiltonian has raised much interest and is reviewed in the recent book [5]. A basic property of a $\mathcal{PT}$ symmetric Hamiltonian is the existence of two regimes separated by a $\mathcal{PT}$ critical point. In the so-called $\mathcal{PT}$ symmetric regime the eigenvectors of the Hamiltonian are simultaneous eigenvectors of $\mathcal{PT}$ which implies that the energies are real. In the $\mathcal{PT}$ broken regime the eigenvectors of the Hamiltonian come in doublets under the $\mathcal{PT}$ symmetry and the energy eigenvalues are complex conjugate to each other.

In the $\mathcal{PT}$ symmetric regime the Hamiltonian is actually pseudo-Hermitian, i.e. there exists a Hermitian similarity transformation η such that $\eta H \eta^{-1} = h$ with $h^\dagger = h$ being Hermitian [6–9].

A key observation for the implementation of $\mathcal{PT}$ in holography (the gauge-gravity) duality is that the similarity transformation η can be thought of switching on certain non-Hermitian operators by analytically continuing in the couplings of these operators. The simplest possible example is a complex operator O. One can add this to a given (Hermitian) Hamiltonian via a coupling $\lambda O + \bar{\lambda} O^\dagger$ with the coupling λ being just a complex number. If we write the coupling as $|\lambda| e^{i\alpha}$ we can think of the non-Hermitian theory to arise from analytically continuing in the angle $\alpha \to i\beta$. Now we have two real couplings coupling λ, $\bar{\lambda}$ and the

*e-mail: karl.landsteiner@csic.es

resulting theory is not-Hermitian! But as long as these couplings obey the constraint $\lambda\bar{\lambda} > 0$ we can think of it as arising from the analytic continuation in the angle α of the Hermitian theory. This indeed describes the $\mathcal{PT}$ symmetric regime in which eigenvalues are real. The $\mathcal{PT}$ critical point is the point where $\lambda\bar{\lambda} = 0$. If however $\lambda\bar{\lambda} < 0$ this regime can not be reached by analytic continuation in the angle α and eigenvalues will come in complex pairs.

It has to be emphasized that despite the the fact that the $\mathcal{PT}$ symmetric regime is mathematically equivalent to quantum mechanics with a Hermitian Hamiltonian the physical interpretation is very different. A standard Hermitian Hamiltonian describes a closed isolated quantum system whereas a $\mathcal{PT}$ symmetric Hamiltonian describes an open quantum system in which the non-Hermitian couplings describe a dynamical equilibrium with an exact balance between inflow and outflow of the quantity measured by the operator O.

Let us suppose now that we can manipulate the in- and outflow in a time dependent manner while maintaining exact balance at all times. In this situation the non-Hermitian Hamiltonian at any time t would be pseudo-Hermitian at all times

$$H(t) = \eta(t)^{-1} h(t) \eta(t)\,. \tag{1}$$

The question is if such a time dependent system is still equivalent to a Hermitian system and answer is "it depends".

We can insist that the time evolution is determined by a Schrödinger equation based on the Hamiltonian (1)

$$i\partial_t = H|\Psi\rangle\,. \tag{2}$$

This time evolution does not respect unitarity. It is easy to see that the product of two vectors in the Hilbert space is not conserved using (2)

$$\partial_t(\langle\Psi|\Phi\rangle_\eta) = 2\langle\Psi|(\eta^{-1}\partial_t\eta)|\Phi\rangle_\eta\,. \tag{3}$$

A cure to this problem is as follows. One defines the modified time dependent Schrödinger equation

$$i(\partial_t + \eta^{-1}\dot{\eta})|\Psi\rangle = H|\Psi\rangle\,. \tag{4}$$

It is easy to check that indeed now time evolution is unitary.

A natural interpretation is that formally we have introduced a time component of a gauge field for the gauged Schrödinger equation

$$iD_t|\Psi\rangle = i(\partial_t - iA_t)|\Psi\rangle = H|\Psi\rangle\,. \tag{5}$$

Formally treating the time dependent similarity transformation as a gauge transformation one sees that the Hermitian system described by h, $|\psi\rangle$, $A_t = 0$ is gauge equivalent to $H = \eta^{-1}h\eta$, $|\Psi\rangle = \eta^{-1}|\psi\rangle$ and $A_t = i\eta^{-1}\partial_t\eta$. A more detailed discussion of the time dependent pseudo-Hermitian systems can be found in the recent reviews [10] and [11] and a more general situation allowing also space dependence has recent been investigated in [12].

It is worth emphasizing that this is not a true gauge transformation. A gauge transformation does not change in any way the physical system under consideration. It merely describes it in a different language. In contrast here we do have to carefully engineer a quantum system with balanced in- and outflow at each point in time and in addition switch on the coupling corresponding to $A_t = i\eta^{-1}\partial_t\eta$ in order to achieve a time evolution that is equivalent to the one of a closed quantum system with Hermitian Hamiltonian!

2 PT Holography

Gravity with asymptotically anti-de Sitter boundary conditions allows a dual interpretation in terms of strongly coupled quantum systems. Useful introductions to applications fo this to strongly coupled quantum many body physics are [13, 14]. We will now review the holographic dual to a non-Hermitian quantum field theory [15, 16]. The key is that in the holographic duality the asymptotic values of the fields encode the couplings of the dual field theory. The model of holographic field theory is defined by the action

$$S = \frac{1}{2\kappa^2}\int_{\mathcal{M}} d^4x\,\sqrt{-g}\left[R + \frac{6}{L^2} - \overline{D_\mu\phi}D^\mu\phi - m^2\phi\bar{\phi} - \frac{V}{2}\phi^2\bar{\phi}^2 - \frac{1}{4}F_{\mu\nu}F^{\mu\nu}\right], \tag{6}$$

The important point is that it contains a complex scalar field ϕ dual to a complex scalar operator O. According to the holographic dictionary the boundary values of the scalar field encode the couplings to the operator O. The non-Hermitian holographic field theory will then be defined by choosing appropriate boundary conditions.

We use coordinates such that the boundary is located at $u = 0$, and the boundary coordinates (v, x_1, x_2) are collectively denoted as x

$$ds^2 = -f dv^2 - \frac{2}{u^2}e^g du dv + S^2(dx_1^2 + dx_2^2) \tag{7}$$

where (f, g, S) are functions of (u, v) whose near boundary expansion recover the AdS geometry

$$\lim_{u\to 0}(u^2 f) = 1\,, \quad \lim_{u\to 0} g = 0\,, \quad \lim_{u\to 0} S = 1\,. \tag{8}$$

With the choice $m^2L^2 = -2$ for the mass of the scalar field its asymptotic expansion is

$$\phi(u, x) = u\phi_1(x) + O(u^2)\,, \tag{9}$$

$$\bar{\phi}(u, x) = u\bar{\phi}_1(x) + O(u^2)\,. \tag{10}$$

The leading terms ϕ_1, $\bar{\phi}_1$ acts as the couplings of the scalar operator O, $\bar{O}$ of conformal dimension $\Delta = 2$. The non-Hermitian theory is obtained with the boundary conditions

$$\phi_1(x) = (1 - \xi(x))\varphi_1(x)\,, \tag{11}$$

$$\overline{\phi}_1(x) = (1 + \xi(x))\varphi_1(x)\,. \tag{12}$$

We can also define the analytically continued (hyperbolic) angle

$$e^{\beta(x)} = \sqrt{\frac{1 + \xi(x)}{1 - \xi(x)}}\,. \tag{13}$$

For $\xi(x) = 0$ we recover the standard Hermitian boundary conditions.

So far we have only considered the boundary conditions on the scalar fields. As we emphasized in the introduction, such a procedure will result in a non-unitary time evolution as soon as the parameter ξ or equivalently β becomes time dependent. A unitary time evolution can be obtained by also introducing a gauge field. In our case and generalizing to full spacetime dependence this gauge field would be given by

$$A_\mu = i\partial_\mu\beta = \frac{i\partial_\mu\xi}{1 - \xi^2}\,. \tag{14}$$

We will use such a gauge field to show explicitly that the time evolution in that case is equivalent to a usual time evolution with Hermitian boundary conditions obeying $\phi_1^* = \bar{\phi}_1$ where the star denotes complex conjugation.

2.1 Time independent case ξ = const.

We will consider first solutions at zero temperature. In the dual gravitational theory these are solutions which are regular in the interior and do not contain a black hole. These solutions correspond to domain wall geometries interpolating between two AdS regions of different cosmological constant. This is so as long as $|\xi| < 1$. At the critical points $|\xi| = 1$ the scalar field equation actually decouples from the metric and one finds a simple AdS−4 geometry as solution. For $|\xi| > 1$ on the other hand one find two solutions with complex conjugate metrics for the same boundary conditions. This is expected since this $|\xi| = 1$ are precisely the critical points separating the $\mathcal{PT}$ symmetric from the broken regime. Gravity backgrounds containing a black hole are dual to the field theory at finite temperature (the Hawking temperature of the black hole). In this case real solutions exist even for $|\xi| > 1$. It turns out however that they are all unstable since a quasinormal mode has moved into the upper half-plane. Despite much numerical effort no stable ground state could be found in this regime [15]. This is a qualitatively new behavior that has not appeared before in non-Hermtian quantum systems or quantum field theories. It would be interesting to see if it persists in a weakly coupled $\mathcal{PT}$ theory at finite temperature.

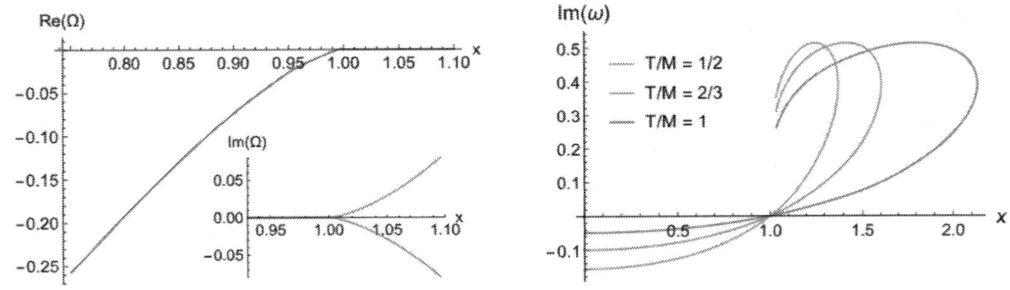

Figure 1. *Left panel:* The free energy (Ω) of the dual field theory as function of the non-Hermiticity parameter ξ (denoted by "x" in the plot). For $|\xi| > 1$ there are two complex conjugate solutions as expected for the $\mathcal{PT}$ broken regime. *Right panel:* Lowest quasinormal mode for solutions containing a black hole (finite temperature solutions). While real solution can be found even for $|\xi| > 1$ the are all unstable since there is a quasi-normal mode in the upper half plane.

2.2 Time independent case $\xi(v)$.

Now we focus on the time evolution of the system for some interesting examples when the non-Hermiticiy parameter is a function of time [16]. Initial and final states correspond to equilibrium finite temperature solutions. To study the response of the system when ξ interpolates between two different values we use the profile function

$$\xi(v) = \xi_i + \frac{\xi_f - \xi_i}{2}\left[1 + \tanh\left(\frac{v - v_m}{\tau}\right)\right] . \tag{15}$$

Here τ is the timescale over which the non-Hermitian coupling varies. Following standard usage call this a holographic quantum quench. We start at the Hermitian point $\xi = 0$ and end either in the $\mathcal{PT}$-symmetric phase, i.e. $|\xi| < 1$, or at the exceptional point. In these type of quenches no additional gauge field is switched on. Therefore it is expected to be dual to a non-unitary time evolution. Two signature of this in the gravity theory are as follows. First the expectation value of the charge operator J^0 dual to the gauge field is purely imaginary

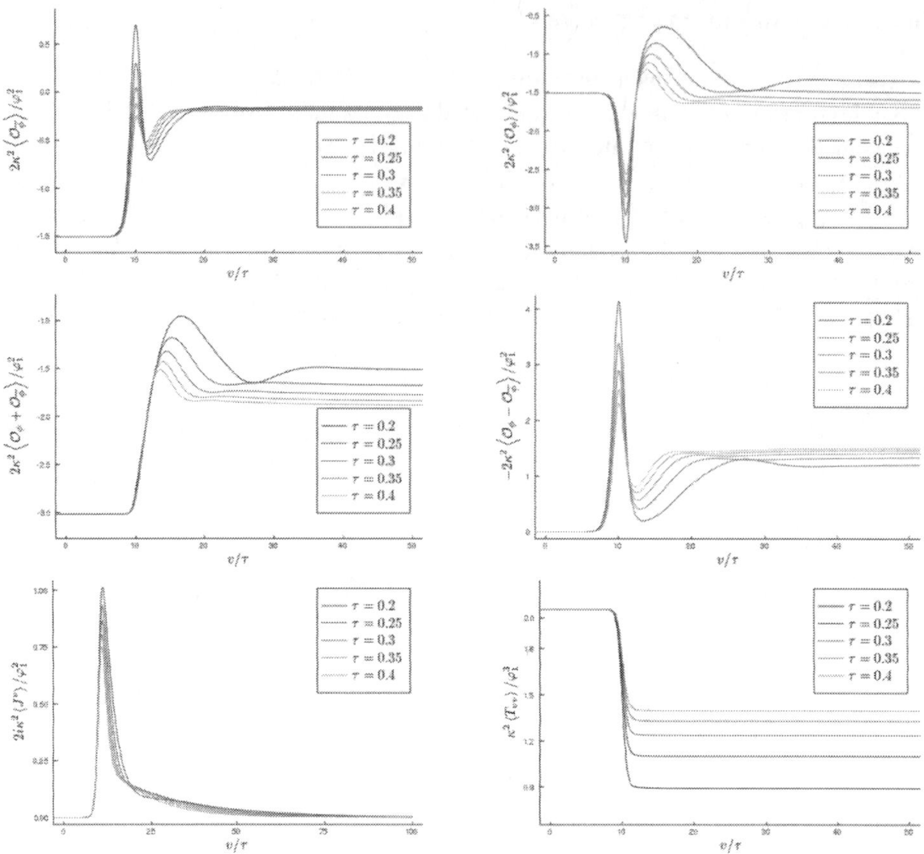

Figure 2. Expectation values of the holographic operators for a quench with profile as in eq. (15) for several values of τ which interpolates between the Hermitian point $\xi_i = 0$ and a final value $\xi_f = 0.8$. We set $v_m = 10\tau$.

during the time evolution. The second one is that the energy (the expectation value of the T_{vv} component of the energy-momentum tensor) diminishes. This also means that the horizon of the dual geometry shrinks, a situation that does not arise is sufficiently "reasonable" types of matter fall into a black hole. Indeed the so called null energy condition is violated. This condition states that $T_{\mu\nu}d^\mu d^\nu > 0$ for any null vector d^μ. As long as this condition holds for matter falling into a black hole the horizon expands. In the non-Hermitian quantum quench however this is violated as can be clearly seen in figure 3. Another interesting type of holographic quantum quench is defined by the profile function

$$\xi(v) = \xi_i + \frac{\xi_m - \xi_i}{2}\left[1 + \tanh\left(\frac{v - v_m}{\tau}\right)\right] + \frac{\xi_f - \xi_m}{2}\left[1 + \tanh\left(\frac{v - \tilde{v}_m}{\tilde{\tau}}\right)\right]. \quad (16)$$

where ξ_m is some intermediate value. This is interesting because it is possible to chose this intermediate value to lie in the unstable regime $|\xi| > 1$. From the static analysis it is known that there is an instability. Indeed in figure 4 this is clearly seen as the onset of exponential growth. One can stay in this unstable regime for sime finite time and upon entering into the $\mathcal{PT}$ symmetric regime the system settles down again to a stationary state at late times after the quench has ended.

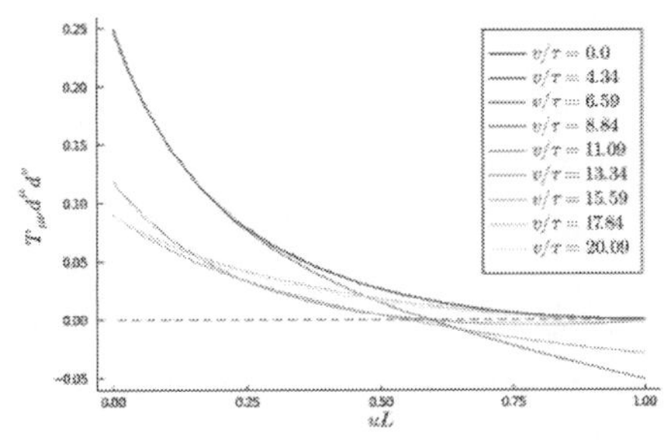

Figure 3. The NEC $T_{\mu\nu}d^{\mu}d^{\nu} \geq 0$ for the particular simulation of figure 2 with $\tau = 0.4$ at different stages v/τ of the evolution. It is clearly violated, especially at the horizon $uL = 1$.

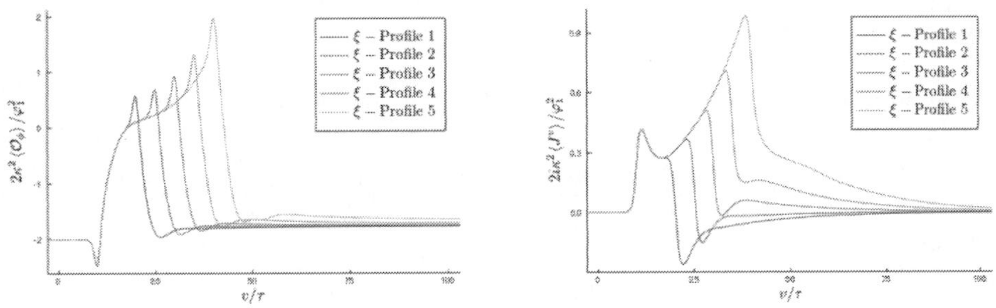

Figure 4. Expectation values of the scalar operator and the charge operator for a quench with profile 16 for $\tilde{v}_m/\tilde{\tau} = \{20, 25, 30, 35, 40\}$. The initial and final states are located at $\xi = 0.8$, besides we set $\xi_m = 1.1$ and $\tau = \tilde{\tau} = 0.25$. An instability starts to develop, but it is suppressed as we re-enter the $\mathcal{PT}$-symmetric region.

2.3 Time independent case $\xi(v)$: Hermitian evolution

We have also studied the time evolution corresponding in which one also introduces a gauge field. More precisely the gauge field in AdS is sourced by

$$A_0|_{u=0} = \frac{i\partial_v \xi}{1-\xi}, \tag{17}$$

corresponding to the time evolution according to (4). In this case the time evolution should be exactly equivalent to a Hermitian time evolution. This is indeed the case (see figure 5).

3 Conclusions

We have shown that the concept of $\mathcal{PT}$ symmetry can be implemented in strongly coupled holographic field theories. Several interesting observations can be made that give hints about what to look for in weakly coupled $\mathcal{PT}$ symmetry field theory. Finally we also would like to remark that during the conference we have been made aware of some problem arising in the renormaliation of weakly coupled $\mathcal{PT}$ quantum field theory [17, 18]. In contrast we found not obvious problems in carrying out the holographic renormalization. This is certainly a point that merits further investigation.

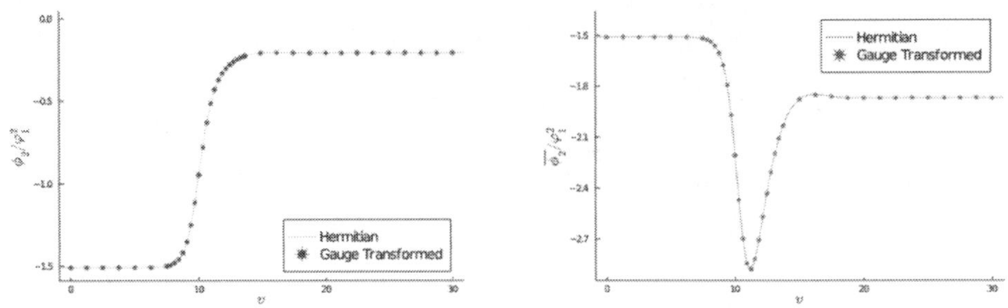

Figure 5. Vevs of the scalar operators for a purely Hermitian quench and the corresponding non-Hermitian one with the gauge field (17). Both give exactly the same results.

Acknowledgements

I would like to thanks D. Arean, I. Salazar-Landea, S. Morales-Tejera for an enjoyable and fruitful collaboration on this subject and V. Branchina for interesting discussion on $\mathcal{PT}$ field theories during the workshop and bringing the references [17, 18] to my attention. This work was supported by grants PID2021-123017NB-I00, PID2021-127726NB-I00 and Severo Ochoa excellence award CEX2020-001007-S.

References

[1] C.M. Bender, S. Boettcher, Phys. Rev. Lett. **80**, 5243 (1998), physics/9712001
[2] C.M. Bender, S. Boettcher, P. Meisinger, J. Math. Phys. **40**, 2201 (1999), quant-ph/9809072
[3] C.M. Bender, Contemp. Phys. **46**, 277 (2005), quant-ph/0501052
[4] C.M. Bender, Rept. Prog. Phys. **70**, 947 (2007), hep-th/0703096
[5] C.M. Bender, P.E. Dorey, C. Dunning, A. Fring, D.W. Hook, H.F. Jones, S. Kuzhel, G. Lévai, R. Tateo, *PT Symmetry* (WSP, 2019)
[6] A. Mostafazadeh, J. Math. Phys. **43**, 205 (2002), math-ph/0107001
[7] A. Mostafazadeh, J. Math. Phys. **43**, 2814 (2002), math-ph/0110016
[8] A. Mostafazadeh, J. Math. Phys. **43**, 3944 (2002), math-ph/0203005
[9] A. Mostafazadeh, J. Phys. A **36**, 7081 (2003), quant-ph/0304080
[10] A. Mostafazadeh, Entropy **22**, 471 (2020), 2004.05254
[11] A. Fring, *An introduction to PT-symmetric quantum mechanics – time-dependent systems*, in *9th Quantum Fest: InterInternational Conference on Quantum Phenomena, Quantum Control and Quantum Optics* (2022), 2201.05140
[12] M.N. Chernodub, P. Millington (2021), 2110.05289
[13] M. Ammon, J. Erdmenger, *Gauge/gravity duality: Foundations and applications* (Cambridge University Press, Cambridge, 2015), ISBN 978-1-107-01034-5, 978-1-316-23594-2
[14] J. Zaanen, Y.W. Sun, Y. Liu, K. Schalm, *Holographic Duality in Condensed Matter Physics* (Cambridge Univ. Press, 2015), ISBN 978-1-107-08008-9
[15] D. Areán, K. Landsteiner, I. Salazar Landea, SciPost Phys. **9**, 032 (2020), 1912.06647
[16] S. Morales-Tejera, K. Landsteiner (2022), 2203.02524
[17] V. Branchina, A. Chiavetta, F. Contino, Phys. Rev. D **104**, 085010 (2021), 2104.12702
[18] V. Branchina, A. Chiavetta, F. Contino, Eur. Phys. J. C **81**, 535 (2021)

Gravitational coupling of QED and QCD: 3- and 4- point functions in momentum space

Matteo Maria Maglio[1,*] and *Riccardo* Tommasi[2,**]

[1]Institute for Theoretical Physics, University of Heidelberg, Germany
[2]Dipartimento Matematica e Fisica "Ennio de Giorgi", Università del Salento, Lecce, Italy

Abstract. Conformal symmetry has important consequences for strong interactions at short distances and provides powerful tools for practical calculations. Even if the Lagrangians of Quantum Chromodynamics (QCD) and Electrodynamics (QED) are invariant under conformal transformations, this symmetry is broken by quantum corrections. The signature of the symmetry breaking is encoded in the presence of massless poles in correlators involving stress-energy tensors. We present a general study of the correlation functions $\langle TJJ \rangle$ and $\langle TTJJ \rangle$ of conformal field theory (CFT) in the flat background limit in momentum space, following a reconstruction method for tensor correlators. Furthermore, our analysis also focuses on studying the dimensional degeneracies of the tensor structures related to these correlators.

1 Introduction

Correlation functions play a significant role in conformal field theories (CFT), and their functional form can easily be obtained by methods in coordinate space rather than in momentum space. Although these methods are very powerful, they apply when the correlation functions are at separate points. On the other hand, anomalies occur in coordinate space at short distances and thus with configurations of a correlator in which some points coalesce. Then the coordinate space approach provides limited information on the origin of the anomaly, except for telling us that its origin is a short-distance effect. One of the main reasons for studying CFT correlation functions directly in momentum space is to see the effects of anomalies more directly. When a theory is classically conformal invariant, then as a consequence, its energy momentum tensor will have a null trace. If the quantum theory has an anomaly, then the trace of the expectation value of the energy momentum tensor in a metric ($g_{\mu\nu}$) background develops a non-zero value. This phenomenon is essentially related to divergences in the quantum theory that break the conformal invariance once the theory is renormalized. In particular, two counterterms are needed in $d = 4$ in the renormalization of correlators with multiple stress-energy tensors, responsible for the generation of the anomaly: E and C^2, the Euler-Poincarè density and the Weyl tensor squared respectively, which find application in the $\langle TTT \ldots \rangle$ (n-graviton) vertex [1–3]. For other correlators, such as the $\langle TJJ \rangle$, the renormalization of the vector 2-point function $\langle JJ \rangle$ is sufficient to generate a finite correlator, which is reflected in the F^2 term of the anomaly functional, with F the field strength of the gauge field (see [4, 5] for related studies).

*e-mail: maglio@thphys.uni-heidelberg.de
**e-mail: riccardo.tommasi@unisalento.it

2 Anomalies as light cone processes

The perturbative realization of CFT correlators allows us to handle far more simplified expressions proceeding with analyzing an ordinary Feynman expansion. The analysis in momentum space of such correlators provides additional information on the emergence of the conformal anomaly. Indeed, the appearance of the anomaly can be described by the emergence of massless effective scalar degrees of freedom in the 3-point functions containing insertions of stress energy tensors, which can be interpreted as light-cone interactions. As discussed in [6–10], this phenomenon points towards an interpretation of the origin of the conformal anomaly as mediated by correlated pairs of fermions/scalars, as emerging from the spectral representation of a given perturbative correlator. Such interactions also play a role in the context of Weyl semimetals, with the paired electrons (representing the massless pole) interacting with the lattice of such materials [11, 12]. It is worth mentioning that both conformal and chiral anomalies play a key role in this phenomenon [13]. These interactions are associated with renormalization and are not related to specific parametrization of the tensor correlators. The proof can be illustrated more quickly in the case of the TJJ correlator in QED. A similar analysis can be performed for QCD, even if it is more involved.

3 Decomposition of correlators

We briefly review the method to decompose any n-point functions involving tensorial operators, first presented for the case of the three-point functions [14]. This method is based on the reconstruction of the full n-point function involving stress-energy tensors, currents, and scalar operators starting from the expressions of transverse and traceless part only. We will present the decomposition of TJJ and $TTJJ$ correlation functions directly in momentum space. Defining the projectors

$$\pi^{\mu}_{\alpha} = \delta^{\mu}_{\alpha} - \frac{p^{\mu} p_{\alpha}}{p^2}, \qquad \Pi^{\mu\nu}_{\alpha\beta} = \pi^{(\mu}_{\alpha}\pi^{\nu)}_{\beta} - \frac{1}{d-1}\pi^{\mu\nu}\pi_{\alpha\beta}, \tag{1}$$

with the properties

$$p_{i\mu_i}\,\pi^{\mu_i\nu_i}(p_i) = 0, \qquad p_{i\mu_i}\,\Pi^{\mu_i\nu_i}_{\alpha_i\beta_i}(p_i) = 0, \qquad \delta_{\mu_i\nu_i}\,\Pi^{\mu_i\nu_i}_{\alpha_i\beta_i}(p_i) = 0, \tag{2}$$

we consider the decomposition of the energy-momentum tensor $T^{\mu\nu}$ and the current J^{μ} as

$$\begin{aligned}
&j^{\mu}(p) = \pi^{\mu}_{\alpha}(p)\, J^{\alpha}(p), && j^{\mu}_{loc}(p) = \frac{p^{\mu}p_{\alpha}}{p^2} J^{\alpha}(p),\\
&t^{\mu\nu}(p) = \Pi^{\mu\nu}_{\alpha\beta}(p)\, T^{\alpha\beta}(p), && t^{\mu\nu}_{loc}(p) = \left(\mathcal{I}^{\mu\nu}_{\alpha\beta} + \frac{1}{d-1}\pi^{\mu\nu}\delta_{\alpha\beta}\right) T^{\alpha\beta}(p),\\
&\mathcal{I}^{\mu\nu}_{\alpha\beta} = \frac{p_{\beta}}{p^2}\left[2p^{(\mu}\delta^{\nu)}_{\alpha} - \frac{p_{\alpha}}{d-1}\left(\delta^{\mu\nu} + (d-2)\frac{p^{\mu}p^{\nu}}{p^2}\right)\right].
\end{aligned} \tag{3}$$

The decomposition of the operators $T^{\mu\nu} = t^{\mu\nu}_{loc} + t^{\mu\nu}$ and $J^{\mu} = j^{\mu} + j^{\mu}_{loc}$, allows to split any correlation function into a sum of correlators containing j^{μ}, j^{μ}_{loc}, $t^{\mu\nu}$ and $t^{\mu\nu}_{loc}$. However, as shown in [9, 14, 15], by using the conservation Ward Identities, that relate n-point functions to lower point, it is possible to completely fix the longitudinal parts, i.e. those terms containing at least one t_{loc} or j_{loc}. Therefore, the only term to be studied in order to reconstruct the entire correlator is the transverse traceless part consisting only of operators $t^{\mu\nu}$ and j^{μ}.
The transverse traceless part, as we will show, can be expressed in a number of minimal tensor structures and form factors. Furthermore, due to the presence of dimensional tensor degeneracies, the number of independent tensor structures contributing to the decomposition can be properly reduced.

4 *TJJ* reconstruction

In this section we give the decomposition of the correlator TJJ, and in particular, by using (3), we obtain

$$\begin{aligned}\langle T^{\mu_1\nu_1}\, J^{\mu_2}\, J^{\mu_3}\rangle &= \langle t^{\mu_1\nu_1}\, j^{\mu_2}\, j^{\mu_3}\rangle + \langle T^{\mu_1\nu_1}\, J^{\mu_2}\, j^{\mu_3}_{loc}\rangle + \langle T^{\mu_1\nu_1}\, j^{\mu_2}_{loc}\, J^{\mu_3}\rangle + \langle t^{\mu_1\nu_1}_{loc}\, J^{\mu_2}\, J^{\mu_3}\rangle \\ &- \langle T^{\mu_1\nu_1}\, j^{\mu_2}_{loc}\, j^{\mu_3}_{loc}\rangle - \langle t^{\mu_1\nu_1}_{loc}\, j^{\mu_2}_{loc}\, J^{\mu_3}\rangle - \langle t^{\mu_1\nu_1}_{loc}\, J^{\mu_2}\, j^{\mu_3}_{loc}\rangle + \langle t^{\mu_1\nu_1}_{loc}\, j^{\mu_2}_{loc}\, j^{\mu_3}_{loc}\rangle\,.\end{aligned}$$

It is important to emphasise that all the terms except the first one can be rewritten as two-point functions via Ward identities. The explicit form of the transverse traceless part $\langle t^{\mu_1\nu_1}\, j^{\mu_2}\, j^{\mu_3}\rangle$ is

$$\langle t^{\mu_1\nu_1}(p_1) j^{\mu_2}(p_2) j^{\mu_3}(p_3)\rangle = \Pi^{\mu_1\nu_1}_{\alpha_1\beta_1}(p_1)\pi^{\mu_2}_{\alpha_2}(p_2)\pi^{\mu_3}_{\alpha_3}(p_3) X^{\alpha_1\beta_1\alpha_2\alpha_3}, \tag{4}$$

where $X^{\alpha_1\ldots\alpha_3}$ is a general rank four tensor built by products of metric tensors and momenta with the appropriate choice of indices. As a consequence of the projectors in (4), $X^{\alpha_1\ldots\alpha_3}$ can not be constructed by using $\delta^{\alpha_1\beta_1}$, nor by $p_i^{\alpha_i}$, $i = 1,\ldots,3$. In addition, the conservation of the total momentum

$$p_1^{\alpha_i} + p_2^{\alpha_i} + p_3^{\alpha_i} = 0, \tag{5}$$

allows selecting for each index α_i a pair of momenta to be used in the general construction of X. The choice of the independent momenta of the expansion can be different for each set of contracted tensor indices. One can choose

$$\begin{aligned}\{\alpha_1,\beta_1\} &\leftrightarrow p_1, p_2,\\ \{\alpha_2\} &\leftrightarrow p_2, p_3\,,\\ \{\alpha_3\} &\leftrightarrow p_3, p_1\,,\end{aligned} \tag{6}$$

as basis of the expansion for each pair of indices shown above. The linear dependence of one momentum, for instance p_3, which we will impose at a later stage, is not in contradiction with this choice, which allows to reduce the number of form factors, due to the presence of a single t projector for each external momentum. For which concerns the metric delta the only non vanishing terms appearing in $X^{\alpha_1\ldots\alpha_3}$ are

$$\delta^{\alpha_1\alpha_2},\ \delta^{\alpha_1\alpha_3},\ \delta^{\alpha_2\alpha_3}, \tag{7}$$

together with the terms obtained by the exchange $\alpha_1 \leftrightarrow \beta_1$. To construct the transverse traceless part, we must use these tensors to build all possible four rank tensors. Still, we must keep in mind that, due to symmetries of the correlator, form factors associated with structures linked by a $2 \leftrightarrow 3$ transformation are dependent. Then the transverse traceless part is written as

$$\begin{aligned}&\langle t^{\mu_1\nu_1}(p_1) j^{\mu_2}(p_2) j^{\mu_3}(p_3)\rangle = \\ &= \Pi^{\mu_1\nu_1}_{\alpha_1\beta_1}(p_1)\,\pi^{\mu_2}_{\alpha_2}(p_2)\,\pi^{\mu_3}_{\alpha_3}(p_3)\,(A_1\; p_2^{\alpha_1} p_2^{\beta_1} p_3^{\alpha_2} p_1^{\alpha_3} + A_2\;\delta^{\alpha_2\alpha_3} p_2^{\alpha_1} p_2^{\beta_1} \\ &\quad + A_3\;\delta^{\alpha_1\alpha_2} p_2^{\beta_1} p_1^{\alpha_3} + A_3(p_2 \leftrightarrow p_3)\delta^{\alpha_1\alpha_3} p_2^{\beta_1} p_3^{\alpha_2} + A_4\;\delta^{\alpha_1\alpha_3}\delta^{\alpha_2\beta_1}),\end{aligned} \tag{8}$$

It is worth mentioning that the form factors A_i are fixed by the conformal invariance and in particular by the CWIs

$$K^\kappa \langle T^{\mu_1\nu_1}\, J^{\mu_2}\, J^{\mu_3}\rangle = 0, \qquad D\,\langle T^{\mu_1\nu_1}\, J^{\mu_2}\, J^{\mu_3}\rangle = 0,$$

where D and K^κ are the dilatation and the special conformal operators respectively [14, 16–18]

5 *TJJ* perturbative realization

A perturbative calculation of the TJJ correlator has been performed in two different realization of QED, namely with complex scalars and fermions [4]. The actions considered are given by

$$S_s = \int d^dx\,\sqrt{-g}\,\left(\partial^\mu\phi^\dagger\,\partial_\mu\phi + ie\,A^\mu\,(\partial_\mu\phi^\dagger\,\phi - \phi^\dagger\,\partial_\mu\phi)\;+\;e^2A^\mu A_\mu\,\phi^\dagger\phi\;+\;\chi R\,\phi^\dagger\phi\right), \quad (9)$$

$$S_f = \int d^dx\,V\left(\frac{i}{2}(\bar\psi\gamma^\lambda\,\partial_\lambda\psi - \partial_\gamma\bar\psi\gamma^\lambda\,\psi)\;-\;e\,\bar\psi\,\gamma^\lambda A_\lambda\,\psi\;-\;\frac{i}{4}\,\omega_{\mu ab}\,V^\mu_c\,\bar\psi\gamma^{abc}\psi\right) \quad (10)$$

where V^μ_a is the vielbein and $\omega_{\mu ab}$ is the spin connection. The correlator TJJ is obtained from the sum of the Feynmann diagrams that can be constructed, and as an example we show some of them in Fig. 1.

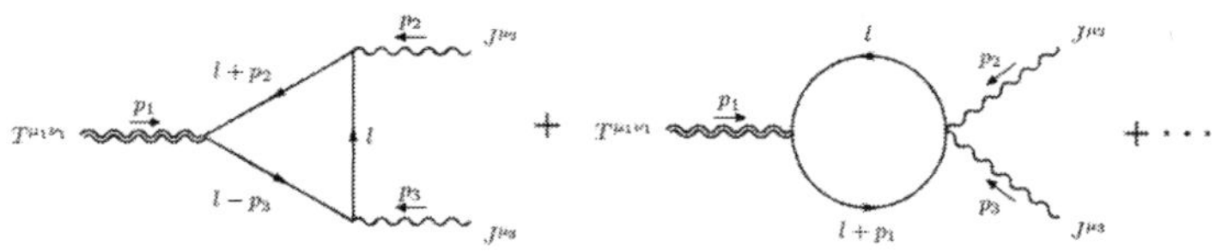

Figure 1. One-loop diagrams for the TJJ in QED.

From the calculation of the Feynman diagrams, the form factors in the decomposition can be written explicitly in terms of master integrals B_0 and C_0. In $d = 4$, this correlator manifests UV divergences that can be renormalized by adding the gauge and Weyl invariant counterterm

$$S_{ct} = -\frac{c}{\epsilon}\int d^dx\,\sqrt{-g}\,F_{\mu\nu}F^{\mu\nu}\,. \quad (11)$$

The renormalization procedure of the correlator induces a breaking of the conformal invariance, manifested in the presence of an anomaly pole. The correlator, which was classically conformal invariant, acquires, after renormalization, a trace anomaly contribution. The effective action related to the anomaly contribution at the first order in the perturbation h is

$$S_{pole} = -\frac{e^2}{36\pi^2}\int d^4x\,d^4y\,\left(\Box h(x) - \partial_\mu\partial_\nu h^{\mu\nu}(x)\right)\Box^{-1}_{xy}\,F_{\alpha\beta}(x)F^{\alpha\beta}(y). \quad (12)$$

6 Dimensional dependent degeneracies

The number of independent tensorial structures corresponds to the number of independent form factors that characterize the transverse traceless part. At first sight, it seems evident that different tensorial sectors (i.e. terms involving only metric δ and terms involving δ and momenta p_i) are not connected under symmetry transformations, and we can consider each sector individually when counting the number of independent form factors.
However, this is the case only when we work in general dimension d. If we are interested in a specific dimension, there may be relations that relate to terms among different sectors. One way to analyze such identities is by using Lovelock's double antisymmetrization method [19, 20].
The main point of this analysis is to consider a (k, l)-rank tensor S in d dimensions, with $k, l < d$. Then the relation

$$S_{[\alpha_1\ldots\alpha_k}{}^{\beta_1\ldots\beta_l}\,\delta^{\beta_{l+1}}_{\alpha_{k+1}}\cdots\delta^{\beta_{l+m}}_{\alpha_{k+m}]} = 0 \qquad \text{for}\;\; k+m > d\,, \quad (13)$$

is trivially satisfied, and further contractions of these identities may lead to non trivial relations.

This result can be reformulated in a different way [19]. Let T be an anti-symmetric traceless (k,l)-rank tensor, then the Lovelock's theorem states that

$$T_{[\alpha_1\ldots\alpha_k}{}^{[\beta_1\ldots\beta_l}\,\delta^{\beta_{l+1}}_{\alpha_{k+1}}\cdots\delta^{\beta_{l+m}]}_{\alpha_{k+m}]}=0,\qquad \text{for } m\geq d+1-(k+l)\,. \tag{14}$$

This relation may seem less intuitive, but it is the same principle of (13). Since the number of indices exceeds the dimensionality, two indices will be either equal on the same line or contracted. In the latter case, being the tensor traceless, the identity will be satisfied. Lovelock's theorem ensures that any contraction of (14) results only in trivial identities.

This theorem provides evidence for tensor identities based on the dimension considered. Indeed, given the dimension d, we build an antisymmetric traceless (k,l) rank tensor. We obtain a family of tensor identities using a double antisymmetrization in the form of (14). For example, we could construct a Riemann-like tensor using p_1 and p_2

$$R_{\alpha_1\alpha_2}{}^{\beta_1\beta_2}=p_{1[\alpha_1}p_{2\alpha_2]}\,p_1{}^{[\beta_1}p_2{}^{\beta_2]}\,. \tag{15}$$

The antisymmetrization of this tensor with two metric tensors defines the traceless tensor W

$$W_{\alpha_1\alpha_2}{}^{\beta_1\beta_2}=R_{\alpha_3\alpha_4}{}^{[\alpha_3\alpha_4}\delta^{\beta_1}_{\alpha_1}\delta^{\beta_2]}_{\alpha_2} \tag{16}$$

that has the same properties as the Weyl tensor. In $d\leq 3$ this tensor vanishes identically, i.e.

$$W_{\alpha_1\alpha_2}{}^{\beta_1\beta_2}=0,\quad d\leq 3, \tag{17}$$

as specific case of (14) with $k=l=2$, that in $d=3$ imply $m=0$ as pointed out in [15].

Then there are possible values of (k,l) depending on the specific dimensions that produce different identities on tensor structures, and they may reduce the number of independent form factors.

7 *TTJJ* reconstruction and perturbative realization

The general decomposition presented can be extended in studying higher point functions. In the case of $\langle TTJJ\rangle$, the conservation of the total momentum is not as strict as the three-point case, and the symmetry of the correlator is now twofold, having both $(1\leftrightarrow 2)$ and $(3\leftrightarrow 4)$ permutation symmetry to be considered. The choice of the independent momenta in the transverse traceless part, as we did in (6), is similar for the four point function but with the association of two independent moments for each index, as pointed out in [2, 21]. More details to obtain the decomposition into minimal independent form factors will be presented in [22].

The perturbative realization of the $\langle TTJJ\rangle$ can be obtained analogously to the case of the three-point function by calculating the corresponding Feynman diagrams. In both cases, when $d=4$, the renormalization procedure, due to the presence of divergences, breaks the conformal invariance, which is reflected in anomaly massless poles. We will present the form of the anomaly effective action for the $\langle TTJJ\rangle$ correlator studying its implications [22]. In addition, a detailed study on tensor degeneracies for the $\langle TTJJ\rangle$ decomposition will be investigated.

8 Conclusions

We have briefly presented the general method to construct 3- correlation functions using the decomposition into a longitudinal part and a transverse traceless one. This method can be extended straightforwardly to the case of 4-point functions. The transverse traceless part of the correlator can be written in terms of independent form factors and tensor structures. The number of independent form factors is related to the number of independent tensorial structures that depends on the specific dimension chosen. Indeed, we have illustrated the case where tensor identities have to be considered in specific dimensions. These constraints change the number of independent form factors in the specific dimensions. Finally, in $d = 4$, we have mentioned how the renormalization procedure of correlators involving energy momentum tensors leads to the presence of anomalous massless poles and the identification of the anomaly effective action for the TJJ. The tensor identities presented in this paper can play a crucial role in identifying the anomalous part of the $TTJJ$ in $d = 4$.

Acknowledgement

M. M. M. is supported by the European Research Council (ERC) under the European Union as Horizon 2020 research and innovation program (grant agreement No818066) and by Deutsche Forschungsgemeinschaft (DFG, German Research Foundation) under Germany's Excellence Strategy EXC-2181/1 - 390900948 (the Heidelberg STRUCTURES Cluster of Excellence). The work of R.T. is supported by INFN iniziativa specifica QFT-HEP.

References

[1] C. Coriano, M.M. Maglio, E. Mottola, Nucl. Phys. B **942**, 303 (2019), 1703.08860
[2] C. Corianò, M.M. Maglio, D. Theofilopoulos, Eur. Phys. J. C **81**, 740 (2021), 2103.13957
[3] C. Corianò, M. Cretì, S. Lionetti, M.M. Maglio, R. Tommasi (2022), 2205.03535
[4] C. Corianò, M.M. Maglio, Nucl. Phys. B **938**, 440 (2019), 1802.07675
[5] C. Corianò, M.M. Maglio, Phys. Lett. B **781**, 283 (2018), 1802.01501
[6] M. Giannotti, E. Mottola, Phys. Rev. D **79**, 045014 (2009), 0812.0351
[7] R. Armillis, C. Coriano, L. Delle Rose, Phys. Rev. D **82**, 064023 (2010), 1005.4173
[8] R. Armillis, C. Coriano, L. Delle Rose, AIP Conf. Proc. **1317**, 185 (2010), 1007.2141
[9] C. Corianò, M.M. Maglio, Phys. Rept. **952**, 1 (2022), 2005.06873
[10] C. Coriano, L. Delle Rose, E. Mottola, M. Serino, JHEP **08**, 147 (2012), 1203.1339
[11] M.N. Chernodub, C. Corianò, M.M. Maglio, Phys. Lett. B **802**, 135236 (2020), 1910.13727
[12] M.N. Chernodub, A. Cortijo, M.A.H. Vozmediano, Phys. Rev. Lett. **120**, 206601 (2018), 1712.05386
[13] M.N. Chernodub, Y. Ferreiros, A.G. Grushin, K. Landsteiner, M.A.H. Vozmediano, Phys. Rept. **977**, 1 (2022), 2110.05471
[14] A. Bzowski, P. McFadden, K. Skenderis, Journal of High Energy Physics **2014** (2014)
[15] A. Bzowski, P. McFadden, K. Skenderis, JHEP **11**, 153 (2018), 1711.09105
[16] C. Coriano, L. Delle Rose, E. Mottola, M. Serino, JHEP **07**, 011 (2013), 1304.6944
[17] A. Bzowski, P. McFadden, K. Skenderis, JHEP **03**, 066 (2016), 1510.08442
[18] C. Corianò, M.M. Maglio, JHEP **09**, 107 (2019), 1903.05047
[19] D. Lovelock, Mathematical Proceedings of the Cambridge Philosophical Society **68**, 345 (1970)

[20] S.B. Edgar, A. Hoglund, J. Math. Phys. **43**, 659 (2002), gr-qc/0105066

[21] C. Corianò, M.M. Maglio, D. Theofilopoulos, Eur. Phys. J. C **80**, 540 (2020), 1912.01907

[22] C. Corianò, M.M. Maglio, R. Tommasi, Work in preparation (2022)

Tsallis statistics and QCD thermodynamics*

Airton Deppman[1,**] and *Eugenio* Megías[2,***]

[1]Instituto de Física, Universidade de São Paulo, Rua do Matão 1371-Butantã, São Paulo-SP, CEP 05580-090, Brazil

[2]Departamento de Física Atómica, Molecular y Nuclear and Instituto Carlos I de Física Teórica y Computacional, Universidad de Granada, Avenida de Fuente Nueva s/n, 18071 Granada, Spain

Abstract. We summarize recent progress on the applications of Tsallis statistics to high energy and heavy ion physics. We also address the possible connections of this statistics with a fractal structure of hadrons.

1 Introduction

Before the quark structure of the hadrons was widely accepted, another approach resulted in an interesting description of many observed phenomena. Supporting this approach is the self-consistent model (Bootstrap Model) proposed by Frautschi and others [1], which gave an exponential mass spectrum for the hadrons that was able to describe the high mass sector of the observed hadrons.

Hagedorn proposed a Self-Consistent Thermodynamics for the fireball formed at high energy collisions, where fireball was defined as a thermodynamical system formed by fireballs [2]. With this definition, it was possible to describe the thermodynamical aspects of the fireballs, as well as the exponential mass spectrum. Another fundamental result of the theory was the limiting temperature, called Hagedorn's temperature, representing the maximum temperature at which the fireball could be heated. According to the Hagedorn's theory, increasing the energy given to the system would result in the production of a larger number of particles, keeping the temperature below its asymptotic value.

The success of the Self-Consistent Thermodynamics motivated the development of the Hadron Resonance Gas models, where different exponential mass spectra were used to investigate the thermodynamical characteristics of the fireballs [3–6]. With the advance of the theory based on the quark structure, the Hagedorn temperature was reinterpreted as a phase-transition temperature, corresponding to the interface between the confined and deconfined regimes of the hadronic matter. The deconfined regime is known as Quark-Gluon Plasma, and has being identified and studied experimentally since 2005.

The Self-Consistent Thermodynamics predicts exponential distributions for the energy and momentum of the particles produced in the collisions between high energy hadrons. However, with the increasing of the collision energy promoted by the advance of the colliders, it was observed that this was not observed experimentally, that is, $(\frac{d^2N}{dp_\perp dy} \not\approx e^{-p_\perp/T})$.

*Talk given by E. Megías at the QCD@Work: International Workshop on QCD, 27-30 June 2022, Lecce, Italy.
**e-mail: deppman@usp.br
***e-mail: emegias@ugr.es

After 2000, it was clear that Hagedorn theory could not describe the full range of the multiparticle production, and the distributions should be described by the Tsallis distributions. When using Tsallis statistics, then power-law distributions are obtained and the agreement between theory and experiment is perfect in many orders of magnitude [7, 8].

2 Tsallis statistics

The Tsallis statistics is a generalization of Boltzmann-Gibbs (BG) statistics for systems with correlations that can emerge from non Markovian evolution, long-range forces or complex internal structure. It is based on a new definition for the Entropy, given by

$$S_q \equiv -k_B \sum_i p_i^q \ln_q p_i , \qquad \ln_q(p) \equiv \frac{p^{1-q}-1}{1-q} \quad \left[\underset{q\to 1}{\longrightarrow} \log(p) \right] . \tag{1}$$

A consequence is that the entropy of the system is non-additive, and for two independent systems A and B [9]

$$S_{A+B} = S_A + S_B + k_B^{-1}(1-q) S_A S_B , \tag{2}$$

where the entropic index q measures the degree of non-extensivity. Tsallis generalized statistics leads to q-exponential and q-logarithm functions,

$$e_q^{(\pm)}(x) = [1 \pm (q-1)x]^{\pm 1/(q-1)} . \tag{3}$$

It was proved that the thermodynamical relations are kept intact if one maximize the Tsallis entropy, and the whole thermodynamics can be derived from the new entropic formula, as for instance, the occupation number, given by

$$n_q^{(\pm)}(x) = \frac{1}{\left(e_q^{(\pm)}(\beta(\varepsilon-\mu)) - \xi \right)^{1 \pm (q-1)}} \qquad \left[\underset{q\to 1}{\longrightarrow} \frac{1}{e^{\beta(\varepsilon-\mu)} - \xi} \right] , \tag{4}$$

where $\xi = \pm 1$ for bosons and fermions, respectively, and the density of hadron species in an ideal gas of massive particles, given by

$$\rho_i \equiv \frac{\langle N_i \rangle}{V} = \frac{g_i}{2\pi^2} \left[\int_0^{p_{i\star}} dp\, p^2 n_q^{(-)}(x_i) + \int_{p_{i\star}}^{\infty} dp\, p^2 n_q^{(+)}(x_i) \right] , \tag{5}$$

where

$$x_i \equiv \beta \left(\varepsilon_i - \sum_a \mu_a q_{ai} \right) , \quad \varepsilon_i \equiv \sqrt{p^2 + m_i^2} , \quad \sum_q \mu_a q_{ai} \equiv \mu_u u_i + \mu_d d_i + \mu_s s_i , \tag{6}$$

and $p_{i\star} > 0$ if $\sum_a \mu_a q_{ai} > m_i$.

3 Non-Extensive Self-Consistent Thermodynamics

The Self-Consistent Thermodynamic Theory was generalized by the inclusion of Tsallis statistics. It was shown that the critical temperature is obtained in this generalization, but with a new spectrum mass formula that increases as a power-law instead of the exponential increase from the Hagedorn theory, that is,

$$\rho(m) = \rho_o \cdot [1 + (q-1) m / T_H]^{1/(q-1)} .$$

This new spectrum mass formula describes the observed data better than the Hagedorn's spectrum mass formula, describing the full range of observed masses from the pion mass up to approximately 2.5 GeV. Moreover, it predicts a constant value for the parameter q after the limiting temperature is reached.

The energy and momentum distributions also change, resulting in q-exponential distribution instead of the exponential ones, that is,

$$\frac{d^2N}{dp_\perp dy} = gV\frac{p_\perp m_\perp}{(2\pi)^2}e_q^{(+)}\left(-\frac{m_\perp}{T}\right). \tag{7}$$

4 Thermofractals

Emergence of the non-extensive behavior has been attributed to different causes: 1) long-range interactions, correlations and memory effects; 2) temperature fluctuations; 3) finite size of the system [10]. We will study a natural derivation of non-extensive statistics in terms of thermofractals.

Thermofractals are systems in thermodynamical equilibrium presenting the following properties [11]:

1. Total energy is given by
$$U = F + E, \tag{8}$$
where F is the kinetic energy, and E is the internal energy of N constituent subsystems, so that $E = \sum_{i=1}^{N}\varepsilon_i^{(1)}$.

2. Constituent subsystems are thermofractals, so that the distribution $P_{\mathrm{TF}}(E)$ is self-similar or self-affine, i.e. at level n of the subsystem hierarchy, $P_{\mathrm{TF}(n)}(E)$ is equal to the distribution in the other levels:
$$P_{\mathrm{TF}\,(n)}(\varepsilon) \propto P_{\mathrm{TF}\,(n+m)}(\varepsilon). \tag{9}$$

3. At level n the phase space is so narrow that one can consider $P_{\mathrm{TF}}(E_n)dE_n = \rho dE_n$.

In thermofractals the phase space must include momentum degrees of freedom of free particles as well as internal degrees of freedom. According to property 2 of self-similar thermofractals

$$P_{\mathrm{TF}(0)}(U)dU = A'F^{\frac{3N}{2}-1}\exp\left(-\frac{\alpha F}{kT}\right)dF\left[P_{\mathrm{TF}(1)}(\varepsilon)\right]^{\nu}d\varepsilon, \tag{10}$$

with $\alpha = 1 + \frac{\varepsilon}{kT}$ and $\frac{\varepsilon}{kT} = \frac{E}{F}$, and ν is an exponent to be determined. By imposing self-similarity

$$P_{\mathrm{TF}(0)}(U) \propto P_{\mathrm{TF}(1)}(\varepsilon) \tag{11}$$

one finds

$$P_{\mathrm{TF}}(\varepsilon) = A\left[1 + \frac{\varepsilon}{kT}\right]^{-\frac{3N}{2}\frac{1}{1-\nu}} \quad\Longrightarrow\quad P_{\mathrm{TF}(n)}(\varepsilon) = A_{(n)}e_q\left[-\frac{\varepsilon}{k\tau}\right]. \tag{12}$$

Therefore, the distribution of thermofractals then obeys Tsallis statistics with $q-1 = \frac{2}{3N}(1-\nu)$.

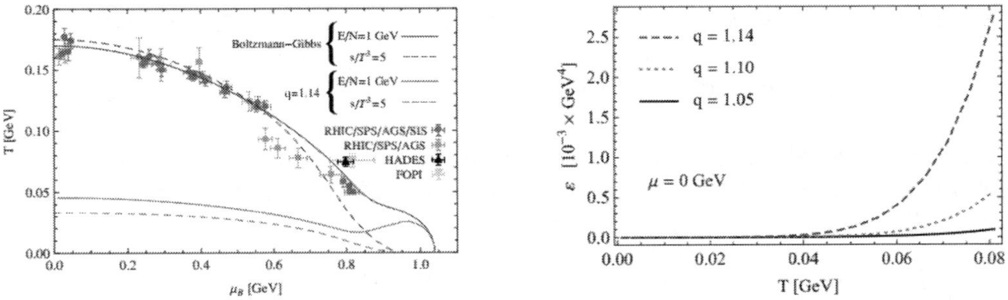

Figure 1. Chemical freeze-out line $T = T(\mu_B)$ (left panel) and energy density as a function of temperature (right panel). We plot the result by using BG statistics, and Tsallis statistics.

From the generalization of Hagedorn's thermodynamics, it is possible to derive the thermodynamical functions that will show a non-extensive character [12]. Indeed,

$$\log Z_q(V,T,\mu) = -\xi V \int \frac{d^3p}{(2\pi)^3} \sum_{r=\pm} \Theta(rx) \log_q^{(-r)} \left(\frac{e_q^{(r)}(x) - \xi}{e_q^{(r)}(x)} \right), \tag{13}$$

which results in the Boltzmann partition function in the limit $q \to 1$. The non-extensive thermodynamics, then, can describe all the observables of the thermodynamically equilibrated system formed at high energy collisions. On this respect, the thermodynamics of QCD in the confined phase can be studied within the Hadron Resonance Gas approach, for which physical observables can be expressed in terms of hadronic states [3]. In particular, the partition function is given by

$$\log Z_q(V,T,\{\mu\}) = \sum_{i \in \text{hadrons}} \log Z_q(V,T,\mu_i), \tag{14}$$

and from that one can compute all the QCD thermodynamics quantities:

- Energy density: $\longrightarrow \quad \varepsilon \equiv \frac{\langle E \rangle}{V} = -\frac{1}{V}\frac{\partial}{\partial \beta} \log Z_q \Big|_{\mu_B} + \frac{1}{V}\frac{\mu_B}{\beta}\frac{\partial}{\partial \mu_B} \log Z_q \Big|_{\beta}.$
- Pressure: $\longrightarrow \quad P = \frac{1}{\beta}\frac{\partial}{\partial V} \log Z_q.$
- Entropy: $\longrightarrow \quad S = -\beta^2 \frac{\partial}{\partial \beta} \left(\frac{\log Z_q}{\beta} \right) \Big|_{\mu_B}.$
- Baryon density: $\longrightarrow \quad \rho_B \equiv \frac{\langle B \rangle}{V} = \frac{1}{3V} \left(\langle N_{\text{quarks}} \rangle - \langle N_{\text{antiquarks}} \rangle \right).$

The non-extensive hadronic matter presents interesting aspects. The equation of states becomes harder as the parameter q increases, therefore the hadronic matter with $q = 1.14$, as it results from the thermofractal QCD, leads to a higher pressure. Applying this reasoning to a possible deconfined phase in Neutron Stars, one obtains large masses for the star for the same radius, giving a possible explanation for the massive objects observed. Some results on the QCD thermodynamics are displayed in Fig. 1.

5 Scales in YM theory

At this point, it is interesting to investigate the relations between the thermofractal structure and the QCD. The two main features of fractals are the scaling invariance and the complex internal structure. These two aspects are also present in the description of the effective partonic structure of hadronic matter.

The effective parton has a complex structure arising from the self-energy interactions, which becomes apparent when one describes the high-order contributions. The inclusion of the self-energy interaction leads, indeed, to the infinities in the QCD theory, or in general, in any Yang-Mills theory.

It is possible to cut-off the infinity contributions because of another property of the quantum fields, that is represented mathematically by the renormalization group equation given by [13, 14]

$$\left[M\frac{\partial}{\partial M}+\beta_{\bar{g}}\frac{\partial}{\partial \bar{g}}+\gamma\right]\Gamma=0.$$

The equation above establishes the scaling invariance of the field theory, and fixing a maximum value for the momentum is equivalent to renormalising the parameters, like mass and charge, of the partons described by the theory.

With the two characteristics above, a fractal theory of the quantum fields was proposed in Ref. [15]. The fractal structure leads to a recursive equation that allows to obtain the running-coupling as

$$\bar{g}(\varepsilon)=\prod_{i=1}^{\tilde{N}} G\left[1-(q-1)\frac{\varepsilon_i}{\lambda}\right]^{\frac{1}{q-1}}. \tag{15}$$

With this coupling, and comparing with the predictions of the QCD, one can obtain the value for the parameter q in terms of the number of colours and the number of flavours, given by

$$\frac{1}{q-1}=\left[\frac{11}{3}N_f-\frac{4}{3}N_c\right].$$

For $N_f=3$ and $N_c=6$, it results $q=8/7=1.14$, in excellent agreement with the experimental analysis.

6 Conclusions

We have reviewed the non-extensive statistics in the form of Tsallis statistics of a quantum gas at finite T and μ. It is shown that the Tsallis statistics allows for a generalization of the self-consistent thermodynamics first proposed by Hagedorn. The phase-transition temperature is still obtained, and a new spectrum mass formula for hadrons can describe the observed states better than the original Hagedorn's formula.

It is discussed that the emergence of the generalized statistics in HEP can be due to the formation of fractal structures of the thermodynamics distributions, which can be associated with a fractal structure in the momentum space. These structures, called thermofractals, would lead to the correlations needed for the emergence of Tsallis statistics.

Finally, recent work is discussed where thermofractal structures can appear in any field described on the basis of a Yang-Mills theory. The structure would be due to the scaling properties that allow the renormalization of the theory. This result allows the calculation of the entropic index, q, in terms of the fundamental parameters of the field theory, as the numbers of colours and flavours in QCD. In this case, the value contained, $q=8/7$, is in very good agreement with the experimental value.

Finally, let us point out that Tsallis statistics has many applications, including high energy collisions [8, 16, 17], hadron models [18, 19], hadron mass spectrum [7], hadron structure [20], neutron stars [21], lattice QCD [22], Bose-Einstein condensation [23, 24], and non-extensive statistical mechanics [12, 25, 26]. The study of all these features deserves further investigation.

Acknowledgement

E.M. would like to thank the Instituto de Física and the Instituto de Astronomia e Geofísica of the Universidade de São Paulo, Brazil, for hospitality during the process of writing this work. A.D. is supported by the Project INCT-FNA (Instituto Nacional de Ciência e Tecnologia - Física Nuclear Aplicada) Proc. No. 464898/2014-5, by the Conselho Nacional de Desenvolvimento Científico e Tecnológico (CNPq-Brazil), grant 304244/ 2018-0, by Project INCT- FNA Proc. No. 464 898/2014-5, and by FAPESP, Brazil grant 2016/17612-7. The work of E.M. is supported by the project PID2020-114767GB-I00 financed by MCIN/AEI/10.13039/501100011033, by the FEDER/Junta de Andalucía-Consejería de Economía y Conocimiento 2014-2020 Operational Programme under Grant A-FQM-178-UGR18, by Junta de Andalucía under Grant FQM-225, and by the Consejería de Conocimiento, Investigación y Universidad of the Junta de Andalucía and European Regional Development Fund (ERDF) under Grant SOMM17/6105/UGR. The research of E.M. is also supported by the Ramón y Cajal Program of the Spanish MICIN under Grant RYC-2016-20678.

References

[1] S.C. Frautschi, Phys. Rev. D **3**, 2821 (1971)
[2] R. Hagedorn, Nuovo Cim. Suppl. **3**, 147 (1965)
[3] R. Hagedorn, Lect. Notes Phys. **221**, 53 (1985)
[4] J. Cleymans, K. Redlich, Phys. Rev. **C60**, 054908 (1999), nucl-th/9903063
[5] E. Megías, E. Ruiz Arriola, L.L. Salcedo, Phys. Rev. D **80**, 056005 (2009), 0903.1060
[6] J. Noronha-Hostler, J. Noronha, C. Greiner, Phys. Rev. C **86**, 024913 (2012), 1206.5138
[7] L. Marques, E. Andrade-II, A. Deppman, Phys. Rev. **D87**, 114022 (2013), 1210.1725
[8] L. Marques, J. Cleymans, A. Deppman, Phys. Rev. **D91**, 054025 (2015), 1501.00953
[9] C. Tsallis, J. Statist. Phys. **52**, 479 (1988)
[10] L. Borland, Phys. Lett. **A**, 67 (1998)
[11] A. Deppman, Phys. Rev. **D93**, 054001 (2016), 1601.02400
[12] E. Megías, D.P. Menezes, A. Deppman, Physica **A421**, 15 (2015)
[13] F.J. Dyson, Phys. Rev. **75**, 1736 (1949)
[14] M. Gell-Mann, F.E. Low, Phys. Rev. **95**, 1300 (1954)
[15] A. Deppman, E. Megías, D.P. Menezes, Phys. Rev. **D101**, 034019 (2020), 1908.08799
[16] J. Cleymans, D. Worku, J. Phys. **G39**, 025006 (2012), 1110.5526
[17] C.Y. Wong, G. Wilk, L.J.L. Cirto, C. Tsallis, Phys. Rev. **D91**, 114027 (2015), 1505.02022
[18] P.H.G. Cardoso, T. Nunes da Silva, A. Deppman, D.P. Menezes, Eur. Phys. J. **A53**, 191 (2017), 1706.02183
[19] E. Andrade, A. Deppman, E. Megías, D.P. Menezes, T. Nunes da Silva, Phys. Rev. D **101**, 054022 (2020), 1906.08301
[20] E. Megías, M.J. Teixeira, V.S. Timóteo, A. Deppman (2022), 2203.11080
[21] D.P. Menezes, A. Deppman, E. Megías, L.B. Castro, Eur. Phys. J. **A51**, 155 (2015)
[22] A. Deppman, J. Phys. **G41**, 055108 (2014), 1212.0379
[23] J. Chen, Z. Zhang, G. Su, L. Chen, Y. Shu, Phys. Lett. A **300**, 65 (2002)
[24] E. Megías, V.S. Timóteo, A. Gammal, A. Deppman, Physica A: Statistical Mechanics and its Applications **585**, 126440 (2022), 2105.07548
[25] A. Deppman, Physica **A391**, 6380 (2012), 1205.0455
[26] A. Deppman, T. Frederico, E. Megías, D.P. Menezes, Entropy **20**, 633 (2018)

Asymptotic symmetries and soft theorems in higher-dimensional gravity

Stefano Lionetti[1,*]

[1]Dipartimento di Matematica e Fisica "Ennio De Giorgi", Università del Salento and INFN-Lecce, Via Arnesano, 73100 Lecce, Italy

Abstract. Soft theorems can be recast as Ward identities of asymptotic symmetries. We review such relation for the leading and subleading soft graviton theorems in arbitrary even dimensions. While soft theorems are trivially generalized to dimensions higher than four, the charges of asymptotic symmetries are plagued by divergences requiring a renormalization. We argue that the renormalized charges of these symmetries can be determined by rewriting soft theorems as Ward identities. In order to show that the charges of such identities generate asymptotic symmetries, we propose a suitable commutation relation among certain components of the metric fields

1 Introduction

Over the last few years a triangular equivalence relation was discovered connecting three apparently different topics: asymptotic symmetries, soft theorems and memory effects [1–3]. This equivalence relation can be drawn potentially in every theory with a massless particle, for example in QED, QCD, SUSY and gravity (for a review of the topic, see [4]). In this paper we focus on the relationship between asymptotic symmetries and soft theorems at leading and subleading order in gravity in arbitrary even dimensions.

In gravity the asymptotic symmetries are diffeomorphisms that preserve the asymptotic flatness of spacetime [5–7]. Such transformations map an asymptotically flat metric into an other asymptotically flat metric. The asymptotic symmetry group is an infinite-dimensional extension of the Poincaré group. The Ward identitity of asymptotic symmetries turn out to be nothing but soft theorems.

The soft graviton theorem [8, 9] is a universal formula relating scattering amplitudes that differ only by the addition of a graviton whose energy ω is taken to zero

$$\mathcal{M}_{n+n'+1}\left(q=\omega\hat{q};p_1,\dots,p_{n+n'}\right)=\left[S^{(1)}+S^{(2)}\right]\mathcal{M}_{n+n'}\left(p_1,\dots,p_{n+n'}\right)+O(\omega) \tag{1}$$

where $\{p_1,\dots p_n\}$ are the momenta of the incoming particles, while $\{p_{n+1},\dots p_{n+n'}\}$ are the momenta of the outgoing particles. $S^{(1)}$ and $S^{(2)}$ are given by

$$\begin{aligned} S^{(1)} &\equiv \frac{\kappa}{2}\left(\sum_{k=n+1}^{n+n'}\frac{\varepsilon_{\mu\nu}p_k^\mu p_k^\nu}{p_k\cdot q}-\sum_{k=1}^{n}\frac{\varepsilon_{\mu\nu}p_k^\mu p_k^\nu}{p_k\cdot q}\right)\\ S^{(2)} &\equiv -\frac{i\kappa}{2}\left(\sum_{k=n+1}^{n+n'}\frac{\varepsilon_{\mu\nu}p_k^\mu q_\rho J_k^{\rho\nu}}{p_k\cdot q}-\sum_{k=1}^{n}\frac{\varepsilon_{\mu\nu}p_k^\mu q_\rho J_k^{\rho\nu}}{p_k\cdot q}\right) \end{aligned} \tag{2}$$

*e-mail: stefano.lionetti@unisalento.it

where $\kappa^2 = 32\pi G$ and $J_k^{\rho\nu}$ is the total angular momentum (orbital+spin) of the k-th particle and $\varepsilon_{\mu\nu}$ is the polarization tensor of the graviton. In the soft limit $\omega \to 0$, the leading term given by $S^{(1)}$ is of order $1/\omega$, while the subleading term given by $S^{(2)}$ is constant in ω. The formula indeed comprises the leading and subleading soft graviton theorem. We are not concerned with higher order terms in our analysis.

In four dimensions the leading soft theorem is equivalent to the supertranslation Ward identity [1, 2], while the subleading soft theorem is equivalent to the Diff(S^2) Ward identity [10, 11]. When computing the Diff(S^2) charges, we encounter divergences and we need to take care of them by adding appropriate counterterms to the action as shown in [12–14]. As we will see, in $d > 4$ such divergences are even more present (supertranslations charges will require a renormalization too). Together, supertranslations and Diff(S^2) make up the asymptotic symmetry group as defined in [10, 11]

$$\mathcal{AG} = \text{Diff}\left(S^2\right) \ltimes \text{ Supertranslations} \tag{3}$$

The equivalence relation between asymptotic symmetries and soft theorems seems to be somewhat more obscure in dimensions higher than four. Indeed soft gravitons theorems hold in any dimension $d = 2m + 2$ while both supertranslations and Diff(S^{2m}) charges seem to diverge in $d > 4$. Many researchers avoided such divergences' problem by imposing strong falloff conditions, leaving only the Poincaré group as part of the asymptotic symmetry group [15–17]. In $d = 4$, such argument would not hold. Falloff conditions disallowing supertranslations and Diff(S^2) automatically exclude all the generic radiative solutions in $d = 4$. In any case, we will not consider such strong falloff conditions in $d > 4$ either, since our aim is to preserve the equivalence relation between asymptotic symmetries and soft theorems. However, a renormalization seems to be mandatory in order to solve the divergence problems.[1]

In this paper we present the proper asymptotic symmetry group that can be linked to soft graviton theorems. We then rewrite the leading and subleading soft graviton theorems as Ward identities. We argue that such identities are associated to asymptotic symmetries by proposing a suitable commutation relation among certain components of the metric fields. We expect the need for a renormalization in order to prove such commutation relation. We work in the Bondi gauge and linearized gravity coupled to massless matter throughout the paper. Our discussion is restricted to tree-level. Asymptotic symmetries have also been studied in higher dimensions in the full nonlinear theory, for any spin, in odd-dimensional cases and in asymptotically (A)dS spacetimes [19–24].

2 Asymptotic symmetries in higher dimensions

In gravity the asymptotic symmetries are diffeomorphisms that preserve the asymptotic flatness of spacetime. In order to define precisely asymptotic flatness, we need to specify the rate at which the metric approaches a Minkowski metric at asymptotically large distances. Unfortunately there is no unambiguous method of determining such asymptotic falloff conditions which are often only a posteriori justified. There are however guidelines that must be followed: the falloff conditions should be weak enough so that all interesting solutions are allowed, but strong enough to rule out unphysical solutions, such as those with infinite energy. Various options are discussed in the literature leading to different results [15, 25–27]. In this paper, we advocate for falloff conditions leading to non-trivial relations among $\mathcal{S}$–matrix

[1] In [18] the author proposed boundary conditions that eliminate divergences in the supertranslations charges. However these boundary conditions are only consistent in linearized gravity and a renormalization is still needed in the full theory.

elements. Indeed Ward identities, associated to proper asymptotic symmetries, turn out to be nothing but soft theorems.

We are interested in asymptotically flat spacetimes at both future and past null infinity $\mathcal{I}^{\pm}$. For concreteness, let us focus on future null infinity. We choose the coordinate system (u, r, z^a), where $u = t - r$ is the retarded time, r is the radial coordinate and z^a $(a = 1, \ldots, 2m)$ are the coordinates on the sphere S^{2m}. We work in the Bondi gauge imposing the following $2m + 2$ conditions

$$g_{rr} = 0, \qquad g_{ra} = 0, \qquad \det g_{ab} = r^{4m} \det \gamma_{ab} \tag{4}$$

where γ_{ab} is the unit S^{2m} sphere metric[2] with covariant derivative D_a. All angular indices $(a, b, c \ldots)$ are raised and lowered with γ_{ab}.

We can now define an asymptotically flat spacetime by imposing the following falloff conditions on the metric

$$g_{uu} = O(1), \qquad g_{ur} = -1 + O\left(\frac{1}{r}\right), \qquad g_{ua} = O(1), \qquad g_{ab} = O(r^2) \tag{5}$$

The asymptotic symmetry group is the group of all non-trivial diffeomorphisms preserving the gauge-fixing conditions (4) and the asymptotic falloffs (5). Such diffeomorphisms are generated by the vector

$$\begin{aligned}
\zeta^u &= f(z) + \frac{u}{2m} D_a Y^a(z) + \ldots \\
\zeta^a &= -\frac{1}{r} D^a f(z) + Y^a(z) - \frac{u}{2mr} D^a D_b Y^b(z) + \ldots \\
\zeta^r &= \frac{1}{2m} D^2 f(z) - \frac{r}{2m} D_a Y^a(z) + \frac{u}{4m^2} D^2 D_a Y^a(z) + \ldots
\end{aligned} \tag{6}$$

Transformations with $Y^a = 0$ and an arbitrary function f on the sphere are known as supertranslations and they are a generalization of the global translations. Transformations with $f = 0$ and an arbitrary smooth vector Y^a on the sphere are known as Diff(S^{2m}) and they are a generalization of the global Lorentz transformations. Note that not all ζ are asymptotic killing vector fields

$$\lim_{r \to \infty} \nabla_{(a} \zeta_{b)} \neq 0 \tag{7}$$

but they all are asymptotically divergence-free

$$\lim_{r \to \infty} \nabla_a \zeta^a = 0 \tag{8}$$

In $d = 4$ an alternative and more restrictive description of asymptotic flatness has been previously proposed, leading to the requirement that Y is a conformal killing vector (CKV) of the sphere [28–30]. In $d = 4$, there are infinitely many local solutions Y to the CKV equation, corresponding to infinitely many transformations generated by ζ. Such transformations are called superrotations and they generalize the Lorentz transformations given by global solutions to the CKV equation. However the number of independent constraints, imposed by the CKV equation, grows with the dimensions of the space. In $d = 4$ the CKV equations are the Cauchy-Riemann equations and so there are infinite independent local solutions. For $d > 4$, the system of equations is over-determined and there are only a finite number of independent solutions: the global Lorentz transformations. In dimensions higher than four, since one cannot extend Lorentz transformations to superrotations, the Diff(S^{2m}) seem to be the proper asymptotic symmetries to link to the subleading soft graviton theorem. Therefore, here we

[2] We are not assuming that γ_{ab} is the standard round sphere metric.

will consider arbitrary smooth vectors Y_a of the sphere. The relationship between superrotations and Diff(S^2) in $d = 4$ has been studied in [31] and it involves a shadow transform. For $d > 4$ such shadow transform is not well defined and this is consistent with the fact that superrotations don't exist in higher dimensions.

3 From soft theorems to asymptotic symmetries

Let us start by considering the leading soft graviton theorem

$$\lim_{\omega\to 0} \mathcal{M}_{n+n'+1}(q = \omega\hat{q}; p_1, \dots, p_{n+n'}) = S^{(1)}\mathcal{M}_{n+n'}(p_1, \dots, p_{n+n'}) \tag{9}$$

where

$$S^{(1)} \equiv \frac{\kappa}{2}\left(\sum_{k=n+1}^{n+n'} \frac{\varepsilon_{\mu\nu}p_k^\mu p_k^\nu}{p_k \cdot q} - \sum_{k=1}^{n} \frac{\varepsilon_{\mu\nu}p_k^\mu p_k^\nu}{p_k \cdot q}\right) \tag{10}$$

Such theorem can be recast into a Ward identity (for the derivation see [25])

$$\langle out|Q^+ \mathcal{S} - \mathcal{S}Q^-|in\rangle = 0 \tag{11}$$

where the charges can be decomposed into a soft and a hard part

$$Q^\pm = Q_H^\pm + Q_S^\pm \tag{12}$$

The hard part comes from the soft factor $S^{(1)}$ on right-hand side of eq (9) after some manipulation, while the soft part comes from the left-hand side of the equation. Therefore, Q_H generates asymptotic symmetries on the external hard states, while Q_S is responsible for the creation of a soft graviton. Focusing for now on Q^+, the hard charge takes the form

$$\langle z_{n+1}, \dots| \, Q_H^+ = \langle z_{n+1}, \dots| \sum_{k=n+1}^{n+n'} E_k f(z_k) \tag{13}$$

which represent the action of supertranslations on each outgoing particle. The soft charge is given by

$$Q_S^+ \propto \int d^{2m}z\, \sqrt{\gamma} f(z) \prod_{l=m+1}^{2m-1} \left(D^2 - (2m-l)(l-1)\right) I^{(m-2)} D^a D^b g_{ab}^{(m-2)} \tag{14}$$

where $g_{\mu\nu}^{(n)}$ is the n-component of the metric in the $1/r$ expansion and we defined the integral operator $I^{(m-2)} = \left(\int du\right)^{(m-2)}$. [3] In [25, 26] the following commutation relation has been proposed

$$\left[g_{uu}^{(2m-1)}(z)\big|_{\mathcal{I}^+_-}, C(u', z')\right] = \frac{8\pi i G}{m} \frac{\delta^{2m}(z - z')}{\sqrt{\gamma}} \tag{15}$$

where $g_{ab}^{(-1)} = \left(\frac{1}{m}\gamma_{ab}D^2 - 2D_a D_b\right)C$. Thanks to this commutation relation, it is possible to show that the charges $Q = Q_S + Q_H$ generate supertanslations. The eq. (15) has not been proven yet but it is consistent with the four-dimensional case [2]. In order to derive the formula in higher dimensions, a renormalization will be required.

[3] In the four-dimensional case ($m = 1$), we use the following prescriptions: $\prod_{l=m+1}^{2m-1}\left(D^2 - (l-1)(2m-l)\right) = 1$ and $I^{(m-2)} = \partial_u$.

Let us now consider the subleading soft graviton theorem

$$\lim_{\omega\to 0}(1+\omega\partial_\omega)\mathcal{M}_{n+n'+1}(q=\omega\hat{q};p_1,\ldots,p_{n+n'})=S^{(2)}\mathcal{M}_{n+n'}(p_1,\ldots,p_{n+n'}) \tag{16}$$

where

$$S^{(2)}=-\frac{i\kappa}{2}\left[\sum_{k=n+1}^{n+n'}\frac{p_{k\mu}\varepsilon^{\mu\nu}q^{\lambda}J_{k\lambda\nu}}{p_k\cdot q}-\sum_{k=1}^{n}\frac{p_{k\mu}\varepsilon^{\mu\nu}q^{\lambda}J_{k\lambda\nu}}{p_k\cdot q}\right] \tag{17}$$

The operator $(1+\omega\partial_\omega)$ projects out the pole $\frac{1}{\omega}$ at the leading order in $S^{(1)}$.

The subleading soft theorem can also be recast as a Ward identity (for the derivation see [26]). The hard charge of such identity is given by

$$\langle out|\,Q_H^{+}=i\sum_{k=n+1}^{n+n'}\left(Y^a(z_k)\,\partial_{z_k^a}-\frac{E_k}{2m}D_aY^a(z_k)\,\partial_{E_k}\right)\langle out| \tag{18}$$

which represent the action of a Diff(S^{2m}) on each outgoing particle. The soft charge is given by

$$Q_S^{+}\propto\int du\int d^{2m}z\,\sqrt{\gamma}\,uD.Y\prod_{l=m+1}^{2m-1}\left(D^2-(l-1)(2m-l)\right)I^{(m-2)}\left(D^aD^bg_{ab}^{(m-2)}\right) \tag{19}$$

Using again the commutation relation (15), we are able to show that such charges generate Diff(S^{2m}).

4 Conclusions and Outlook

We examined the relation between asymptotic symmetries and soft theorems at leading and subleading order in higher-dimensional gravity. In order to preserve the equivalence relation in $d>4$, the asymptotic charges need to be renormalized. Such renormalization procedure has already been proven to be effective in $d=4$ in [12–14]. The search for the right counterterms in $d>4$ is currently an open problem that needs further investigation. However we already know the final form of the renormalized charges since we can derive them from soft theorems. Indeed, we can recast the soft graviton theorems as Ward Identities. We then argued that such identities are associated to asymptotic symmetries, provided a suitable commutation relation between metric fields holds. Recent analyses also support the form of asymptotic charges we derived [19]. As a result, supertranslations and Diff(S^{2m}) are symmetries of gravitational scattering.

There are still many other open topics that need to be tackled in the future. Soft theorems and asymptotic symmetries are part of a triangular equivalence relation called IR Triangle. The last corner of the triangle is a memory effect and it represents an exciting experimental prospect for the coming future of the gravitational wave physics. The spin memory effect has been studied in relation to the subleading soft graviton in $d=4$ in [32]. It would be interesting to examine such effect in higher dimensions.

Moreover asymptotic symmetries and soft theorems play a central role in the celestial holography program. Celestial holography proposes a duality between gravitational scattering in asymptotically flat spacetimes and a conformal field theory (CFT) living in two dimensions lower on the celestial sphere (for reviews of the topic see [33–35]). This field of research has been extensively studied in $d=4$ but it can also be generalized to higher dimensions [36, 37] where further investigations are needed.

Lastly, the sub-subleading soft graviton theorem has been recently studied in relation to asymptotic symmetries in $d=4$ [38–40]. It would be worthwhile to also extend such analyses to arbitrary dimensions.

References

[1] A. Strominger, JHEP **07** (2014), 152
[2] T. He, V. Lysov, P. Mitra and A. Strominger, JHEP **05** (2015), 151
[3] A. Strominger and A. Zhiboedov, JHEP **01** (2016), 086
[4] A. Strominger, "Lectures on the Infrared Structure of Gravity and Gauge Theory,"
[5] H. Bondi, M. G. J. van der Burg and A. W. K. Metzner, Proc. Roy. Soc. Lond. A **269** (1962), 21-52
[6] R. K. Sachs, Proc. Roy. Soc. Lond. A **270** (1962), 103-126
[7] R. Sachs, Phys. Rev. **128** (1962), 2851-2864
[8] S. Weinberg, Phys. Rev. **140** (1965), B516-B524
[9] F. Cachazo and A. Strominger, [arXiv:1404.4091 [hep-th]]
[10] M. Campiglia and A. Laddha, Phys. Rev. D **90** (2014) no.12, 124028
[11] M. Campiglia and A. Laddha, JHEP **04** (2015), 076
[12] G. Compère, A. Fiorucci and R. Ruzziconi, JHEP **11** (2018), 200
[13] G. Compère, A. Fiorucci and R. Ruzziconi, JHEP **10** (2020), 205
[14] V. Chandrasekaran, E. E. Flanagan, I. Shehzad and A. J. Speranza, Int. J. Mod. Phys. A **37** (2022) no.17, 2250105
[15] S. Hollands, A. Ishibashi and R. M. Wald, Class. Quant. Grav. **34** (2017) no.15, 155005
[16] S. Hollands and A. Ishibashi, J. Math. Phys. **46** (2005), 022503
[17] K. Tanabe, S. Kinoshita and T. Shiromizu, Phys. Rev. D **84** (2011), 044055
[18] A. Aggarwal, Phys. Rev. D **99** (2019) no.2, 026015
[19] A. Campoleoni, D. Francia and C. Heissenberg, JHEP **12** (2020), 129
[20] X. Bekaert and B. Oblak, [arXiv:2209.02253 [hep-th]]
[21] F. Capone, JHEP **10** (2021), 158
[22] O. Fuentealba, M. Henneaux, J. Matulich and C. Troessaert, JHEP **07** (2022), 149
[23] A. Fiorucci and R. Ruzziconi, JHEP **05** (2021), 210
[24] C. Chowdhury, R. Mishra and S. G. Prabhu, [arXiv:2201.07813 [hep-th]]
[25] D. Kapec, V. Lysov, S. Pasterski and A. Strominger, Ann. Math. Sci. Appl. **02** (2017), 69-94
[26] D. Colferai and S. Lionetti, Phys. Rev. D **104** (2021) no.6, 064010
[27] M. Pate, A. M. Raclariu and A. Strominger, JHEP **06** (2018), 138
[28] G. Barnich and C. Troessaert, PoS **CNCFG2010** (2010), 010
[29] G. Barnich and C. Troessaert, JHEP **12** (2011), 105
[30] D. Kapec, V. Lysov, S. Pasterski and A. Strominger, JHEP **08** (2014), 058
[31] L. Donnay, S. Pasterski and A. Puhm, JHEP **09** (2020), 176
[32] S. Pasterski, A. Strominger and A. Zhiboedov, JHEP **12** (2016), 053
[33] S. Pasterski, Eur. Phys. J. C **81** (2021) no.12, 1062
[34] A. M. Raclariu, [arXiv:2107.02075 [hep-th]]
[35] T. McLoughlin, A. Puhm and A. M. Raclariu, [arXiv:2203.13022 [hep-th]]
[36] D. Kapec and P. Mitra, JHEP **05** (2018), 186
[37] S. Pasterski and S. H. Shao, Phys. Rev. D **96** (2017) no.6, 065022
[38] M. Campiglia and A. Laddha, JHEP **01** (2017), 036
[39] M. Campiglia and A. Laddha, Phys. Lett. B **764** (2017), 218-221
[40] L. Freidel, D. Pranzetti and A. M. Raclariu, JHEP **05** (2022), 186

Dynamics of bubble walls at the electroweak phase transition

Stefania De Curtis[1,*], *Luigi* Delle Rose[2,**], *Andrea* Guiggiani[1,***], *Ángel* Gil Muyor[3,****], and *Giuliano* Panico[1,†]

[1]INFN Sezione di Firenze and Dipartimento di Fisica e Astronomia, Università di Firenze, Via G. Sansone 1, I-50019 Sesto Fiorentino, Italy
[2]Dipartimento di Fisica, Universit'a della Calabria, I-87036 Arcavacata di Rende, Cosenza, Italy
[3]IFAE and BIST, Universitat Autònoma de Barcelona, 08193 Bellaterra, Barcelona, Spain

Abstract. First order phase transitions in the early universe naturally lead to the production of a stochastic background of gravitational waves and to the generation of a matter-antimatter asymmetry. The dynamics of the phase transition is affected by the density perturbations in the hot plasma. We address this topic by providing, for the first time, a full numerical solution to the linearized Boltzmann equation for the top quark species coupled to the Higgs field during a first order phase transition at the electroweak scale. Differently from the traditional approaches, our results do not depend on any ansatz and can fully describe the non-equilibrium distribution functions of the particle species in the plasma.

1 Introduction

The recent observation of gravitational waves has renewed a vivid interest in the study of the dynamics of first order phase transitions. Indeed, future gravitational wave experiments, such as the European proposal LISA [1, 2], the Japanese interferometer DECIGO [3, 4] and the Chinese projects Taiji [5, 6] and TianQin [7] will probe a range of the expected peak frequencies of stochastic background of gravitational waves sourced by PhTs at the electroweak scale. These interferometers will provide us with a new handle that, supporting already existing collider experiments, will help us in the quest for the physics beyond the Standard Model (BSM), and in particular for all the models affecting the dynamics of the electroweak (EW) symmetry breaking.

It is worth mentioning that the stochastic gravitational wave background is not the only cosmological relic left after the completion of a PhT. Indeed, the same mechanism can also produce dark matter remnants, an asymmetry in the matter-antimatter abundances, primordial black holes, magnetic fields and other topological defects. Obviously, a reliable characterization of these processes requires an accurate modelling of the PhT dynamics, which is controlled, among other parameters, by the propagation speed of the PhT front, namely, the bubble wall.

*e-mail: stefania.decurtis@fi.infn.it
**e-mail: luigi.dellerose@unical.it
***e-mail: andrea.guiggiani@unifi.it
****e-mail: agil@ifae.es
†e-mail: giuliano.panico@unifi.it

In the steady state regime, the velocity of the bubble wall is a result of a balance of the external friction exerted by the particles in the plasma impinging on the wall and of the internal pressure, caused by the potential difference between the two phases. In fact, the plasma is driven out of equilibrium by the presence of the moving bubble and induces a backreaction that slows down its propagation. This is an extremely relevant topic, as shown by the huge amount of literature, for an incomplete list see for instance [8–40], nevertheless, the speed of the wall is one of the parameters characterizing the PhT dynamics on which we have less theoretical control.

The first computation of the bubble speed in the SM has been carried out by Moore and Prokopec in their seminal work [8, 9]. The friction on the bubble wall exerted by the plasma was calculated by evaluating all the relevant interactions between the plasma particles and the bubble and required the solution of the Boltzmann equation to determine the out of equilibrium distribution functions of the different species in the plasma through. Besides this microphysical calculation, other phenomenological approaches have also been explored. These relied, instead, on a modelling of the friction and on its parameterization in terms of the viscosity parameter [32–40].

At the core of the formalism developed in [8, 9], that in the following we will denote as the "old formalism", there is the exploitation of an ansatz for the distribution functions. This is necessarily needed in order to parametrize the momentum dependence of the out of equilibrium distributions and to compute the collision integral appearing in the Boltzmann equation. In particular, in ref. [9] the fluid approximation was chosen as an ansatz: it assumes that the out of equilibrium distribution function of each species in the plasma is entirely described by only three perturbations: the chemical potential, the temperature fluctuation and the velocity perturbation. These corrections are extracted by taking suitable moments of the Boltzmann equation with specific weights. With such a method, the integro-differential Boltzmann equation is recast into a much more manageable system of ordinary differential equations.

Two the peculiar features of the fluid approximation explored in [9] are: a) the Liouville operator of the Boltzmann equation develops a zero eigenvalue at the speed of sound c_s and, b) for velocities larger than c_s all the perturbations trail the source term. This has a large impact on models of non-local EW baryogenesis because any non-equilibrium dynamics in front of the bubble wall would be suppressed for bubble walls faster than c_s [12], resulting to a very inefficient production of the matter-antimatter asymmetry. Recently in refs. [15, 16] it has been suggested that the singular behavior at the speed of sound is only an artifact of the truncation in momenta introduced by the fluid approximation and of the corresponding weights used to compute the out-of-equilibrium perturbations. Indeed, changing the set of weights can modify the location of the singularity, which clearly suggests that the speed of sound cannot be a critical value that characterizes particle diffusion in the fluid equations. In ref. [16] the authors overcame this problem by developing a "new formalism" that relies on a different parameterization of the non-equilibrium distributions, different weights as well as on a factorization ansatz [41]. The new formalism erases the discontinuity at the speed of sound and still provides results quantitatively similar to those of the fluid approximation for small enough velocities.

The same topic has been recently explored also in ref. [30], specifically for the calculation of the matter-antimatter asymmetry, and in ref. [31] for the calculation of the friction on the bubble wall. In these works the fluid approximation of ref. [9] has been generalized by including higher orders in the small momenta expansion and the absence of the singularity for the out-of-equilibrium distribution functions of the heavy species has been confirmed (notice that the fluid approximation can be seen as a first order expansion in the momenta). Besides that, the speed of sound in the plasma still played a peculiar role in the old formalism (even if higher orders in the momenta expansion are included, assuming that the thickness of the wall is not too large and the interaction strength among the plasma particles is not too strong) as it provides peaks in the integrated friction, one for each vanishing eigenvalue of the Liouville operator. For the massless background species, instead, as already pointed out in ref. [9] and confirmed in ref. [31, 33, 42–44], a discontinuity at $v_w \simeq c_s$ of the out-of-equilibrium perturbations in the temperature and the velocity profiles of the fluid could be present. This discontinuity manifests itself in a singularity at c_s if the Boltzmann equations are linearized. Besides the singularity, large corrections in the friction, with respect to the old formalism, have also been pointed out in [30, 31] when higher orders in momenta are taken into account. These results confirms that the fluid approximation cannot provide reliable predictions, neither quantitatively nor qualitatively.

The methodologies discussed above are affected by two main drawbacks: first of all, they must rely on an ansatz describing the dependence in the momenta of the non-equilibrium distribution functions. This is unavoidable to allow for the computation the collision integrals; secondly, the choice of the momentum basis used in the ansatz and the choice of the weights used to construct the system of ordinary differential equations are not unique: different ansatzes have qualitative and quantitative impacts on the resulting distribution functions. This clearly indicates that a full solution to the Boltzmann equation that does not impose any specific momentum dependence is necessary to provide reliable quantitative predictions for both the non-equilibrium distribution functions and the friction exerted on the bubble wall. As a byproduct this will also allow to clarify the issue of the presence of a singularity. A first step in this direction has been carried out in [45], where we have presented, for the first time, a fully quantitative numerical solution to the Boltzmann equation. For the purpose of presenting the methodology and to quantitatively asses the differences among the aforementioned formalisms, we have considered the EWPhT and we have focused on the study of top quark species, the one that provides the largest contribution to the friction. By feeding the non-equilibrium distribution functions into the equation of motion of the Higgs field, one will be able to carry out a precise determination of the bubble profile and of the the wall velocity. As we have already pointed at the beginning of the introduction, this is necessary in order to clearly identify, at the quantitative level, the potential of a given BSM scenario to yield interesting predictions for the cosmological relics listed above.

2 The Boltzmann equation

The aim is to find the solution to the Boltzmann equation for the distribution function of the plasma species in the presence of a bubble wall expanding in the true vacuum. For large enough bubbles we can adopt the planar limit and we can assume a steady state regime. In this case the solutions to the Boltzmann equation are stationary in the wall frame. Orienting the z axis along the velocity of the bubble wall, the equation for the distribution function f of a particle species is

$$\mathcal{L}[f] \equiv \left(\frac{p_z}{E}\partial_z - \frac{(m^2(z))'}{2E}\partial_{p_z}\right) f = -C[f] \,, \tag{1}$$

where $m(z)$ is the position-dependent mass of the particle. The term C is the collision integral, describing local microscopic interactions among the plasma particles, while $\mathcal{L}$ is the Liouville operator.

The collision term ensures that far from the wall each particle species approaches local thermal equilibrium described by the standard Fermi or Bose–Einstein distributions for a fluid moving with velocity v along the z axis, namely

$$f_v = \frac{1}{e^{\beta\gamma(E - v p_z)} \pm 1} \,, \tag{2}$$

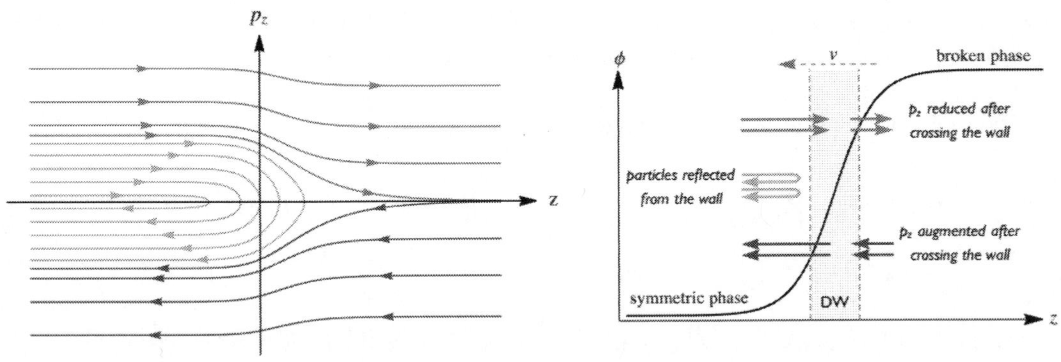

Figure 1. Left panel: Paths with fixed energy and transverse momentum in the $z - p_z$ phase space for the choice $m(z) \propto 1 + \tanh(z/L)$. The red, green and purple colors denote sets of contours with different behavior. The arrows show the flow of a particle within the phase space. Right panel: Schematic representation of the behavior of the particles across the bubble wall.

with $\beta = 1/T$ and $\gamma = 1/\sqrt{1 - v^2}$.

Deviations with respect to the local equilibrium distribution are present close to the wall and vanish at $z \to \pm\infty$. For small enough perturbations, the distribution function can be written as $f = f_v + \delta f$ and the Boltzmann equation can be linearized in δf:

$$\left(\frac{p_z}{E}\partial_z - \frac{(m(z)^2)'}{2E}\partial_{p_z}\right)\delta f + \overline{C}[\delta f] = \frac{(m(z)^2)'}{2E}\partial_{p_z} f_v = \beta\gamma v \frac{(m(z)^2)'}{2E} f_v' , \qquad (3)$$

where we defined

$$f_v' \equiv -\frac{e^{\beta\gamma(E - v p_z)}}{(e^{\beta\gamma(E - v p_z)} \pm 1)^2} , \qquad (4)$$

and $\overline{C}[\delta f]$ denotes the collision integral linearized in δf. Sizable values for the source term in the equation above are present only close to the wall, where the non-trivial Higgs profile generates a non-negligible z dependence. Away from the wall, where the Higgs profile is almost constant, the source term is suppressed, in agreement with the expectation that deviations from thermal equilibrium are only present close to the wall and should decrease to zero away from it.

2.1 The Liouville operator and the collision integral

Along the paths on which both the transverse momentum $p_\perp$ and the quantity $p_z^2 + m^2(z)$ are constant, the Liouville differential operator reduces to a total derivative with respect to z:

$$\mathcal{L} = \left(\frac{p_z}{E}\partial_z - \frac{(m^2(z))'}{2E}\partial_{p_z}\right) \quad \to \quad \frac{p_z}{E}\frac{d}{dz} . \qquad (5)$$

The paths represent the trajectories of the particles in the $(p_\perp, p_z, z)$ phase space in the collisionless limit in which the energy and the momentum parallel to the wall are conserved. As such the trajectories of the particles are given by the condition $p_z^2 + m^2(z) = const$ where $m(z) \to 0$ for $z \to -\infty$, and $m(z) \to m_0 > 0$ for $z \to +\infty$. The three classes of curves are shown schematically in fig. 1 for the choice $m(z) \propto 1 + \tanh(z/L)$, with L denoting the wall thickness. The first type of paths describes particles that travel in the positive z direction, and eventually enter into the bubble. The second type of paths describes particles that initially travel in the positive z direction, hit the wall and are reflected. The third type of paths describes particles that travel in the negative z direction, and eventually exit from the bubble.

The collision integral for the $2 \to 2$ processes is given by

$$C[f] = \sum_i \frac{1}{4N_p E_p} \int \frac{d^3\mathbf{k}\, d^3\mathbf{p}'\, d^3\mathbf{k}'}{(2\pi)^5 2E_k 2E_{p'} 2E_{k'}} |\mathcal{M}_i|^2 \delta^4(p + k - p' - k')\mathcal{P}[f] \,, \tag{6}$$

with the population factor

$$\mathcal{P}[f] = f(p)f(k)(1 \pm f(p'))(1 \pm f(k')) - f(p')f(k')(1 \pm f(p))(1 \pm f(k)) \,, \tag{7}$$

where the sum is over all the scattering processes described by a squared scattering amplitude $|\mathcal{M}_i|^2$. Moreover, N_p is the number of degrees of freedom of the incoming particle with momentum p, k is the momentum of the second incoming particle, while p' and k' are the momenta of the outgoing particles. The $\pm$ signs are $+$ for bosons and $-$ for fermions.

3 Numerical analysis

To solve the Boltzmann equation we have developed a numerical iterative method described in [45]. For the sake of simplicity, we considered a single species in the plasma, the top quark, which is expected to provide the most relevant effects to the bubble dynamics, being the state with largest coupling to the Higgs. The main contribution to the collision integral comes from the annihilation process $t\bar{t} \to gg$, whereas the scattering of tops on gluons and light quarks gives smaller contributions.
The bubble wall is modelled assuming that the Higgs profile can be approximated by

$$\phi(z) = \frac{\phi_0}{2}[1 + \tanh(z/L)] \,, \tag{8}$$

where $L = 5/T$ is the thickness of the bubble wall and $\phi_0 = 150$ GeV is the Higgs VEV in the broken phase. We fixed the phase transition temperature to $T = 100$ GeV.
An important quantity that can be obtained from the numerical solution is the friction acting on the bubble wall [9]

$$F(z) = \frac{dm^2}{dz} N \int \frac{d^3\mathbf{p}}{(2\pi)^3 2E} \delta f(p) \,, \tag{9}$$

where N denotes the number of degrees of freedom.

In fig. 2 we show the friction integrated over z as a function of the wall velocity (solid black line). We also compare our result (full solution, FS) with the ones obtained with the weighted methods. The green lines correspond to the total friction computed in the old formalism (OF) of ref. [9], taking also into account higher-order corrections (1, 2 and 3) to the fluid approximation [30]. The red line represents, instead, the result obtained in new formalism (NF) of ref. [16].
For small and intermediate velocities, $v \lesssim 0.3$, the old formalism is in a fair numerical agreement with the full solution. At higher velocities, instead, the old formalism develops peaks corresponding to the zero eigenvalues of the Liouville operator and highlighting an instability in the methodology. Our results for the full solution to the Boltzmann equation confirm that the peaks are an artifact and that no strong effect is present close to the speed of sound if the top contributions are taken into account.
Concerning the new formalism, it correctly predicts a smooth behavior for the total friction for all the velocities but the quantitative agreement with the full solution is good only for very low ones.

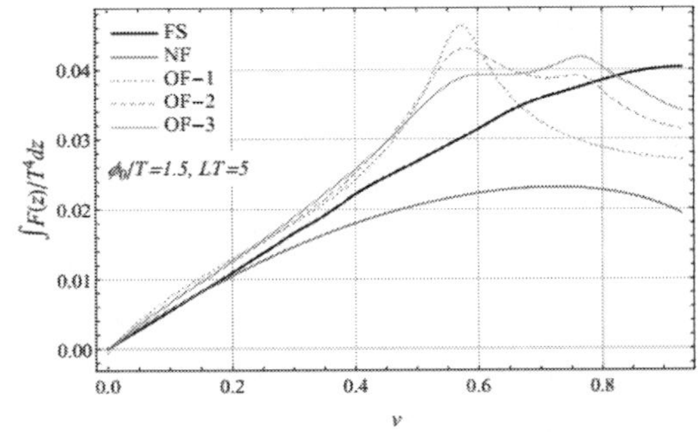

Figure 2. Friction acting on the bubble wall as a function of the velocity. The black solid line corresponds to the solution of the full Boltzmann equation (FS, our result), the dotted, dashed and solid green lines are obtained with the old formalism (OF) at order 1, 2 and 3 respectively [9, 30], while the red line corresponds to the new formalism (NF) [16].

4 Conclusions and Outlook

In [45] we presented for the first time the fully quantitative solution of the Boltzmann equation that describes the distribution functions of plasma particles in the presence of an expanding bubble. Differently to the existing approaches, we did not exploited any ansatz nor we imposed any momentum dependence on the distribution functions away from equilibrium. We have then critically compared our results with the ones obtained using the formalisms developed so far in the literature, namely the fluid approximation [9], its extended version [30, 31] and the "new formalism" [16].

This clearly represents a necessary step towards a reliable understanding of the bubble wall dynamics. Using the friction obtained with the numerical method developed in our work [45], one can solve the equation of motion of the Higgs profile and extract the velocity of the bubble wall as well as the features of its shape, such as the wall thickness. These parameters crucially impact on the prospects of any BSM scenario to predict cosmological signals, such as a stochastic gravitational wave background and the amount of matter-antimatter asymmetry.

References

[1] C. Caprini et al., JCAP **04**, 001 (2016), 1512.06239
[2] C. Caprini et al., JCAP **03**, 024 (2020), 1910.13125
[3] S. Kawamura et al., Class. Quant. Grav. **23**, S125 (2006)
[4] S. Kawamura et al., Class. Quant. Grav. **28**, 094011 (2011)
[5] W.R. Hu, Y.L. Wu, Natl. Sci. Rev. **4**, 685 (2017)
[6] W.H. Ruan, Z.K. Guo, R.G. Cai, Y.Z. Zhang, Int. J. Mod. Phys. A **35**, 2050075 (2020), 1807.09495
[7] J. Luo et al. (TianQin), Class. Quant. Grav. **33**, 035010 (2016), 1512.02076
[8] G.D. Moore, T. Prokopec, Phys. Rev. Lett. **75**, 777 (1995), hep-ph/9503296
[9] G.D. Moore, T. Prokopec, Phys. Rev. D **52**, 7182 (1995), hep-ph/9506475
[10] P. John, M.G. Schmidt, Nucl. Phys. B **598**, 291 (2001), [Erratum: Nucl.Phys.B 648, 449–452 (2003)], hep-ph/0002050
[11] G.D. Moore, JHEP **03**, 006 (2000), hep-ph/0001274
[12] T. Konstandin, G. Nardini, I. Rues, JCAP **09**, 028 (2014), 1407.3132
[13] J. Kozaczuk, JHEP **10**, 135 (2015), 1506.04741

[14] D. Bodeker, G.D. Moore, JCAP **05**, 025 (2017), 1703.08215
[15] J.M. Cline, K. Kainulainen, Phys. Rev. D **101**, 063525 (2020), 2001.00568
[16] B. Laurent, J.M. Cline, Phys. Rev. D **102**, 063516 (2020), 2007.10935
[17] M. Barroso Mancha, T. Prokopec, B. Swiezewska, JHEP **01**, 070 (2021), 2005.10875
[18] S. Höche, J. Kozaczuk, A.J. Long, J. Turner, Y. Wang, JCAP **03**, 009 (2021), 2007.10343
[19] A. Azatov, M. Vanvlasselaer, JCAP **01**, 058 (2021), 2010.02590
[20] S. Balaji, M. Spannowsky, C. Tamarit, JCAP **03**, 051 (2021), 2010.08013
[21] R.G. Cai, S.J. Wang, JCAP **03**, 096 (2021), 2011.11451
[22] X. Wang, F.P. Huang, X. Zhang (2020), 2011.12903
[23] A. Friedlander, I. Banta, J.M. Cline, D. Tucker-Smith, Phys. Rev. D **103**, 055020 (2021), 2009.14295
[24] J.M. Cline, A. Friedlander, D.M. He, K. Kainulainen, B. Laurent, D. Tucker-Smith, Phys. Rev. D **103**, 123529 (2021), 2102.12490
[25] J.M. Cline, B. Laurent, Phys. Rev. D **104**, 083507 (2021), 2108.04249
[26] F. Bigazzi, A. Caddeo, T. Canneti, A.L. Cotrone, JHEP **08**, 090 (2021), 2104.12817
[27] W.Y. Ai, B. Garbrecht, C. Tamarit (2021), 2109.13710
[28] M. Lewicki, M. Merchand, M. Zych (2021), 2111.02393
[29] Y. Gouttenoire, R. Jinno, F. Sala (2021), 2112.07686
[30] G.C. Dorsch, S.J. Huber, T. Konstandin, JCAP **08**, 020 (2021), 2106.06547
[31] G.C. Dorsch, S.J. Huber, T. Konstandin (2021), 2112.12548
[32] A. Megevand, A.D. Sanchez, Nucl. Phys. B **825**, 151 (2010), 0908.3663
[33] J.R. Espinosa, T. Konstandin, J.M. No, G. Servant, JCAP **06**, 028 (2010), 1004.4187
[34] L. Leitao, A. Megevand, Nucl. Phys. B **844**, 450 (2011), 1010.2134
[35] A. Mégevand, JCAP **07**, 045 (2013), 1303.4233
[36] S.J. Huber, M. Sopena (2013), 1302.1044
[37] A. Megevand, F.A. Membiela, Phys. Rev. D **89**, 103507 (2014), 1311.2453
[38] L. Leitao, A. Megevand, Nucl. Phys. B **891**, 159 (2015), 1410.3875
[39] A. Megevand, F.A. Membiela, Phys. Rev. D **89**, 103503 (2014), 1402.5791
[40] A. Megevand, F.A. Membiela, A.D. Sanchez, JCAP **03**, 051 (2015), 1412.8064
[41] J.M. Cline, M. Joyce, K. Kainulainen, JHEP **07**, 018 (2000), hep-ph/0006119
[42] M. Laine, Phys. Rev. D **49**, 3847 (1994), hep-ph/9309242
[43] J. Ignatius, K. Kajantie, H. Kurki-Suonio, M. Laine, Phys. Rev. D **49**, 3854 (1994), astro-ph/9309059
[44] H. Kurki-Suonio, M. Laine, Phys. Rev. D **51**, 5431 (1995), hep-ph/9501216
[45] S. De Curtis, L.D. Rose, A. Guiggiani, A.G. Muyor, G. Panico, JHEP **03**, 163 (2022), 2201.08220

AUTHOR INDEX

EDP Sciences – Web of Conferences
17, Avenue du Hoggar
Parc d'Activité de Courtabouf
BP 112
F-91944 Les Ulis Cedex A
France

ISBN 978-1-7138-6618-3

9 781713 866183

CURRAN ASSOCIATES INC.
proceedings.com